The Spatial Dimensions of Water Management – Redistribution of Benefits and Risks

7th International Water Resources Management Conference of ICWRS

Proceedings of the International Association of Hydrological Sciences (PIAHS) Volume 373, 2016

Bochum, Germany
18 – 20 May 2016

Editors:

H. Schumann
G. Blöschl
A. Castellarin
D. Rosbjerg
A. Viglione
J. Dietrich
S. Grimaldi
U. Haberlandt
A. Montanari
S. Vorogushyn

ISBN: 978-1-5108-3909-0

Printed from e-media with permission by:

Curran Associates, Inc.
57 Morehouse Lane
Red Hook, NY 12571

Some format issues inherent in the e-media version may also appear in this print version.

Printed by Curran Associates, Inc. (2017)

International Association of Hydrological Sciences
Centre for Ecology and Hydrology
Wallingford, Oxfordshire OX10 8BB
United Kingdom

Phone: 441 491 692 442
Fax: 441 491 692 448

http://iahs.info/

Additional copies of this publication are available from:

Curran Associates, Inc.
57 Morehouse Lane
Red Hook, NY 12571 USA
Phone: 845-758-0400
Fax: 845-758-2633
Email: curran@proceedings.com
Web: www.proceedings.com

TABLE OF CONTENTS

Proc. IAHS, 373, 221–222, 2016
proc-iahs.net/373/221/2016/
doi:10.5194/piahs-373-221-2016

Open Access
PIAHS

Preface: The spatial dimensions of water management – Redistribution of benefits and risks

Andreas H. Schumann

Institute of Hydrology, Water Management and Environmental Engineering, Ruhr-University Bochum, Bochum, 44780, Germany

Correspondence to: Andreas H. Schumann (andreas.schumann@rub.de)

Published: 12 May 2016

Many spatial aspects of hydrology and water management are discussed in these proceedings of the 7th International Water Resources Management Conference of IAHS-ICWRS.

The main topics of the conference, specified in the call for papers, were:

1. Spatial aspects of water supply and the redistribution of benefits from water
2. Floods and spatial aspects of flood risks
3. Spatial aspects of water scarcity
4. Scale problems of water management – when the summation of individual measures becomes a problem
5. Hydrological regionalization issues
6. The spatial dimension in socio-hydrology

The response of the scientific community was very unevenly distributed. The most submissions were dedicated to the topics 2 and 4. Floods are hotspots of applied hydrology and so many different research questions were raised in these contributions, e.g. methods for design flood estimations, impact assessments and operational flood forecasting.

The second main component of the conference were hydrological regionalization issues. It became evident that after the completion of the PUB-Initiative of IAHS, which was summarized by Günter Blöschl at the conference, many research activities are still going on to transfer hydrologic information from gauged to ungauged basins by new methodological approaches.

Surprisingly, the "classical" fields of water management, water supply and water scarcity, were less represented. This could be caused by two reasons: there are some tools available which are widely used (e.g. stochastic optimization) but not further improved, or (and this seems to be more probable) there is a low interest to apply new scientific tools in practical water management to optimize existing systems or to improve the planning of new ones. Nevertheless very interesting and (in some cases) sophisticated methodological approaches were presented in this field at the conference under the topics 1, 3 and 4.

With regard to future challenges of water research, the dynamic interactions and feedbacks between water and people becomes more and more important. One part of the conference was dedicated to these questions. Unfortunately the interesting presentations and discussions cannot be reflected completely in the proceedings. Nevertheless the participants got a comprehensive overview about the problems of socio-hydrology and the need for interdisciplinary research.

This conference was organized the second time at the Ruhr-University Bochum. In September 2006 scientists from all over the world met in Bochum for the 3th IWRM Symposium. In that time the conference accounted for the heterogeneity of world water problems by addressing the basic questions of Integrated Water Resources Management. Now, ten years later, the coherence between surface and groundwater, water quality and quantity or water and ecosystems are widely accepted aspects of modern water management. However the differences between hydrological conditions and the human expectations are still in the focus of research. More and more we become aware that water management is not able to serve societies in the solution of their water problems in a sustainable way if interactions and feedbacks between hydrological conditions and societies are not considered.

The stationarity assumption has long been compromised by human disturbances in river basins. The dynamics of changing societies will face water management with new and unexpected problems. To give an example: If a water poor country like Jordan with 6 Million inhabitants host 1.5 Million refugees, the long-term planning of water management issues will be overtaken by operational constraints. I hope that we will be able to find ways to direct our research activities towards these new problems of a fast changing world. Conferences like this one can make a contribution in this direction.

Proc. IAHS, 373, 1–6, 2016
proc-iahs.net/373/1/2016/
doi:10.5194/piahs-373-1-2016

Open Access
PIAHS

A flexible and efficient multi-model framework in support of water management

Vincent Wolfs[1], **Quan Tran Quoc**[1], **and Patrick Willems**[1,2]

[1]KU Leuven, Hydraulics Division, Kasteelpark Arenberg 40 box 2448, 3001 Leuven, Belgium
[2]Vrije Universiteit Brussel, Department of Hydrology and Hydraulic Engineering, Brussels, Belgium

Correspondence to: Vincent Wolfs (vincent.wolfs@bwk.kuleuven.be)

Published: 12 May 2016

Abstract. Flexible, fast and accurate water quantity models are essential tools in support of water management. Adjustable levels of model detail and the ability to handle varying spatial and temporal resolutions are requisite model characteristics to ensure that such models can be employed efficiently in various applications. This paper uses a newly developed flexible modelling framework that aims to generate such models. The framework incorporates several approaches to model catchment hydrology, rivers and floodplains, and the urban drainage system by lumping processes on different levels. To illustrate this framework, a case study of integrated hydrological-hydraulic modelling is elaborated for the Grote Nete catchment in Belgium. Three conceptual rainfall-runoff models (NAM, PDM and VHM) were implemented in a generalized model structure, allowing flexibility in the spatial resolution by means of an innovative disaggregation/aggregation procedure. They were linked to conceptual hydraulic models of the rivers in the catchment, which were developed by means of an advanced model structure identification and calibration procedure. The conceptual models manage to emulate the simulation results of a detailed full hydrodynamic model accurately. The models configured using the approaches of this framework are well-suited for many applications in water management due to their very short calculation time, interfacing possibilities and adjustable level of detail.

1 Introduction

Water management is constantly evolving: there is a shift from more traditional end-of-pipe solutions to a holistic and source-oriented approach, new technologies emerge and multiple trends have a major impact on the water system, such as climate change and the increasing urbanization. Integrated analyses are needed that require broadening the investigated scale and scope to identify cost-effective and future-proof solutions.

To cope with these changes and to ensure that such strategies can be identified, flexible modelling frameworks are needed that allow the water manager to easily and efficiently combine available data bases, real-time measurements and the most suitable models for every component of the holistic water system. The water system is characterized by multiple processes, each acting on different spatial and temporal scales. For instance, CSOs can spill into rivers during a short time frame, but their pollution can disperse and deteriorate the river's water quality for a much longer period. Therefore, the efficient use of spatially distributed data sets, and models with different degrees of model detail and varying spatial and temporal resolutions play a pivotal role in water management analyses. Within this context, we developed a flexible and computationally efficient modelling methodology to simulate the catchment hydrology and the hydraulics of rivers and urban drainage systems. The model detail can easily be adjusted, enabling the modeller to focus on the dominating processes at various scales.

Section 2 discusses this hydrological-hydraulic modelling framework. The approach is demonstrated for the Grote Nete catchment in Belgium. A flexible hydrological model was set up to simulate the catchment rainfall runoff at different spatial resolutions, and a conceptual hydraulic model was created of all main rivers in the catchment. Section 3 describes the study area. Section 4 shows the main results. Finally, the potential of the developed approach is discussed and a view on future developments is given.

2 Methodology

Figure 1 schematizes the developed flexible modelling framework to parsimoniously model the water system at the catchment scale. The modelling framework includes tailored approaches to model the hydrological and hydraulic (both the river and floodplains, and the sewer system) components of the water system. The following paragraphs elaborate on the developed hydrological and hydraulic modelling approaches.

2.1 Hydrological modelling approach

For catchment rainfall-runoff modelling, numerous different hydrological model conceptualizations and parameterizations exist. In order to allow flexibility in the spatial resolution, while maintaining consistency in results when spatial resolution is changed, and to allow flexible changes in process equations, we developed a flexible hydrological modelling approach (Tran Quoc and Willems, 2016). This methodology reasons from a generalized lumped model structure that can be disaggregated at various scales to add spatial detail. The use of such an approach that requires only a few parameters to be calibrated is frequently suggested in literature to obviate equifinality issues encountered in hydrological modelling (e.g. Jakeman et al., 1990; Wheather et al., 1993). Also, their parsimonious configuration not only facilitates and speeds up model calibration, but their performance is on average higher compared to more complex distributed models (e.g. Ajami et al., 2004; Breuer et al., 2009). The entire modelling framework is implemented in PCRaster – Python. Currently, three conceptual rainfall-runoff models (PDM, Moore, 2007; NAM, Nielsen and Hansen, 1973; and VHM, Willems, 2014) are incorporated because of their widespread use in Belgium. The concept of a flexible model structure also allows the use of an ensemble approach involving multiple model types to simulate the rainfall-runoff, thereby obviating the election of one model type over others.

Our modelling framework allows convenient, semi-automated disaggregation of the lumped model components to higher spatial resolutions. This allows tailoring model detail to the characteristics of specific areas or to the needs for intended applications. This also broadens the range of possible applications, including spatial scenario simulations. The disaggregation is achieved by mapping the different model components on a grid with varying cell sizes. To ensure that consistent results are obtained and the parameters have smooth transitions with varying spatial scales, the lumped parameters are disaggregated based on spatial catchment characteristics, such as the topography, land use and soil parameters. Only few additional calibration parameters are required to scale the relative spatial differences in model parameters. In the opposite direction, aggregation allows to spatially lump processes and transfer a highly spatially detailed model to a more lumped model. For further details on this approach, the reader is referred to Tran Quoc and Willems (2016).

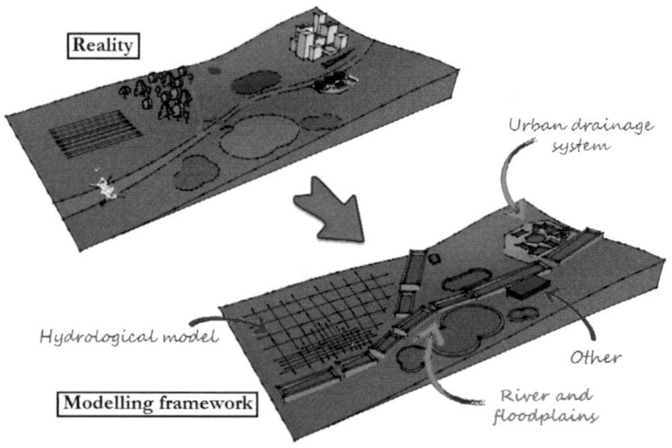

Figure 1. Schematization of the proposed modelling framework.

2.2 Hydraulic modelling approach

Hydrodynamic models, which are based on the de Saint-Venant equations, have become the standard tool for water practitioners to simulate rivers and sewer systems. Such models are generally very accurate, but the level of model detail cannot be adjusted, leading to overly complex models for many applications and analyses. Their complicatedness leads to prolonged calculation times. In addition, it is very often difficult or even impossible to interface different models. Linking such models would in turn lead to an even increased complicatedness.

Therefore, we developed a new flexible conceptual modelling approach for rivers and sewer systems. The modelling approach simplifies the network topology and momentum equations of the detailed hydrodynamic models according to a lower-fidelity approach (Ravazi et al., 2012) to strike the balance between model accuracy, and computational efficiency and model detail. The modelling approach is incorporated in the CMD software.

The modelling approach reasons from the storage cell concept, in which the entire system is divided into multiple interconnected reservoirs. In each such reservoir, the continuity equation is applied to close the water balance. The delineation of rivers and sewer systems into reservoirs is very flexible: additional reservoirs and model detail can be added in areas where enhanced spatial detail is necessary, while processes can be lumped on larger scales at other locations. A broad range of conventionally applied model structures, such as the linear reservoir theory and transfer functions, were combined with advanced machine learning techniques in a mechanistic and modular framework to fill in the momentum equations. This modularity allows the modeller to pick the most appropriate model structure based on the dynamics of the river or sewer system, the desired accuracy, available data and the intended applications. Due to variety of model structures, more complex dynamics can also be emulated accurately on various scales, such as backwater effects, and pressurized and reverse flows in pipes. Details of this modelling approach and software can be found in Wolfs et al. (2015).

This conceptual modelling approach has been used before for various applications and case studies, including flood probability mapping (Wolfs et al., 2012), real-time control of hydraulic structures to prevent flooding (Vermuyten et al., 2015), an impact analysis of CSOs on river water quality (Keupers et al., 2015) and the quantification of source control versus end-of-pipe solutions on flood probabilities in a coupled river-sewer system (De Vleeschauwer et al., 2014). In these studies, the conceptual models were calibrated to simulation results of detailed hydrodynamic models.

2.3 Model interfacing

Interfacing between hydrological and hydraulic models may be bidirectional, but unidirectional coupling was considered so far in this study, where results from the hydrological model are transferred to the hydraulic model at its spatial resolution. The hydrological model is coded in PCRaster-Python, while the hydraulic model is implemented in a C++ executable that can run on Windows and Unix systems. The results of the hydrological model are stored in ASCII format, those of the hydraulic model in netCDF. A simple routine handles the requisite data processing steps when transferring data from the hydrological to the hydraulic model. These include the transformation of data formats, and the adjustment of the save interval (spatial and temporal) of all time series, since the hydraulic model itself does not alter the save interval of its input time series. This routine also enables the hydrological and hydraulic models to run in different software environments, thereby obviating model code adaptations. The data transferring routine was coded in MATLAB, since this environment can easily and efficiently handle the used data standards of the hydrological and hydraulic models. Interfacing was done manually for this study, but this process is currently being formalized and automated by the authors. A web processing service interfacing is also being set up to enable interfacing with models and databases in a network.

3 Case study

The hydrological and hydraulic modelling approaches were tested on the Grote Nete catchment in Belgium (see Fig. 2). The catchment has a total area of 385 km^2. The most dominant land use types in the watershed are cropland (33 %), forest (23 %), grassland (19 %) and built-up area (19 %). The region has predominant sandy soil coverage with parabolic land dunes in the upstream parts, while the downstream

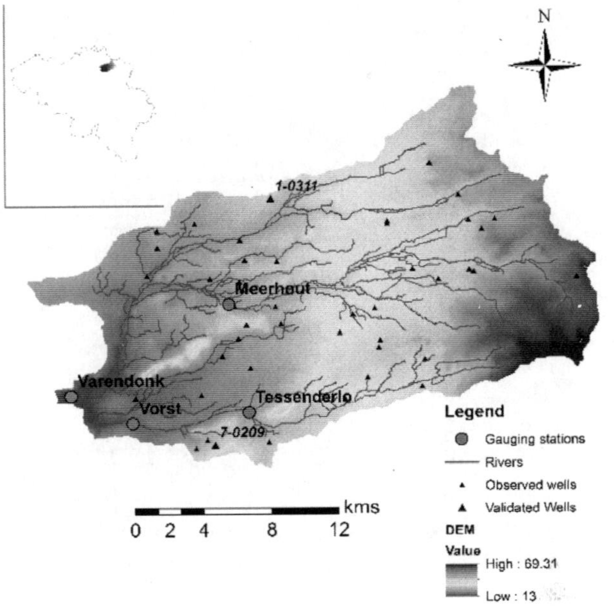

Figure 2. The Grote Nete catchment with the internal flow stations at Meerhout, Vorst and Tessenderlo, and the outlet station at Varendonk.

parts are characterized by wide alluvial planes with shallow groundwater. The catchment contains numerous river tributaries, and a dense network of ditches and subsurface drains that feed into the larger Grote Nete, Molse Nete and Grote Laak rivers.

Four flow gauging stations, six rain gauges and 30 observation wells which monitor groundwater heads are present in the catchment (see Fig. 2). In an earlier study (Vansteenkiste et al., 2014), NAM, PDM and VHM lumped conceptual models were calibrated for this region using a step-wise approach based on the rainfall and flow measurements. The reader is referred to that study for details on this calibration procedure. In addition, a detailed hydrodynamic MIKE11 model was configured that comprises these rivers and their main tributaries. This MIKE11 model uses the simulation results of the VHM model as rainfall-runoff input.

First, the generalized structure of the hydrological models was set-up by implementing the parameters of the earlier calibrated NAM, PDM and VHM models. Next, the parameters were spatially disaggregated using land use and soil texture maps with a 250 m resolution, leading to six hydrological models in total (NAM, PDM and VHM implementation; each lumped and spatially disaggregated). Two adjustment factors were calibrated for the spatial disaggregation to compensate for the changes in subflow distribution by the grid-to-grid routing compared with the lumped model. In addition, two parameters were calibrated to account for spatial variations in catchment characteristics. The hydrological models use hourly catchment averaged rainfall after applying the

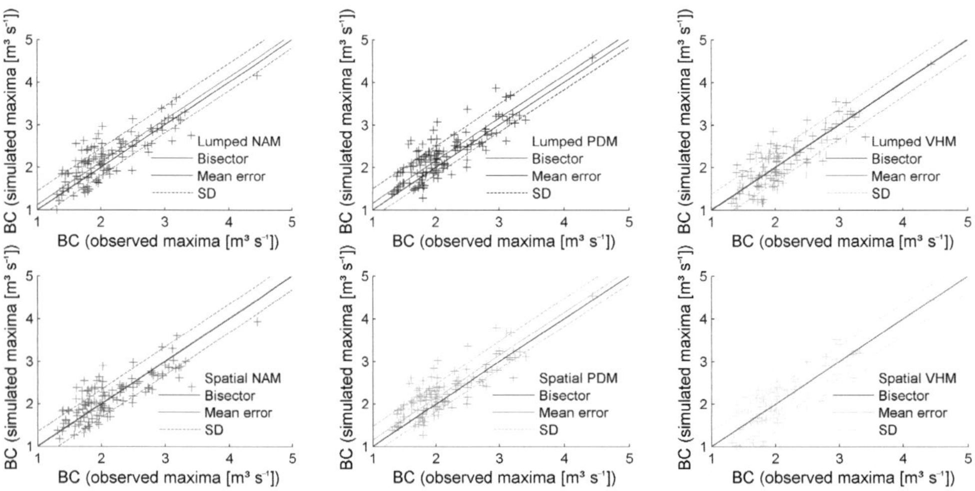

Figure 3. Simulated versus observed peak flows after Box-Cox transformation ($\lambda = 0.25$), for the spatial and lumped models at the Grote Nete downstream catchment station, for the full simulation period 13 August 2002–31 December 2008.

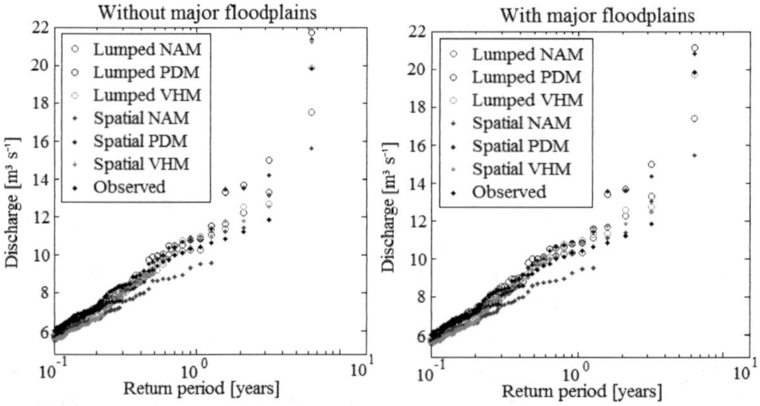

Figure 4. Simulated versus observed peak flow frequency distributions, for the spatial and lumped models at the Grote Nete downstream catchment station, for the full simulation period 13 August 2002–31 December 2008, with and without the effect of the three floodplains with highest capacity.

Thiessen polygon method to the six available rain gauges. The period of 13 May 2002–31 December 2005 was used for calibration, and 1 January 2006–31 December 2008 for validation.

The conceptual hydraulic model was configured for the main rivers in the Grote Nete catchment (see also Fig. 2). Given the availability of an existing and validated detailed hydrodynamic MIKE11 model, and due to the lack of enough measurement data, simulation results of the detailed model were used to calibrate and validate the conceptual model. Calibration was performed on the period 8 March 2001–11 April 2001, while validation was done for 12 April 2001–31 December 2008. The investigated river reaches were divided in 7 reservoirs. The boundaries of most reservoirs mostly coincide with bridges or culverts. Note that it is easily possible to adjust the number of calculation nodes, and thus to include additional model detail in specific areas. The main floodplains with a total capacity of 200 000 m^3 were also incorporated in the conceptual model. These floodplains are situated just upstream of the outlet station of the catchment. After the model topology was defined, the most appropriate model structures were identified and calibrated. The flow was predominantly modelled using transfer functions, while rating curves were applied to model the water levels. Besides visual comparison of the conceptual and detailed hydrodynamic model results, the NSE values were also calculated using the conceptual and MIKE11 simulation results. The performance of the hydraulic model is exceptionally good, with NSE values for all flows and water level results exceeding 0.985 for both the calibration and validation events.

Each of the six conceptual rainfall-runoff models (NAM, PDM, VHM; and for both the lumped and spatially distributed versions) were interfaced with the conceptual river model. In this way, six different integrated models were obtained.

4 Results

The results of the interfaced hydrological – hydraulic conceptual models were evaluated based on the flow observed at the catchment downstream gauging station. This is done using four methods: (i) visual comparison of the observed and simulated river flows at the gauging stations, (ii) evaluation of a set of five statistical measures, (iii) control of the water balance and (iv) analysis of evaluation plots focusing on extremes and cumulative values as proposed in Willems (2009).

For the peak flow extremes (Figs. 3, 4), the spatial and lumped models show similar performance and frequency distributions: whereas the peak flow error (PFE) after Box-Cox (BC) transformation (Box and Cox, 1964) ranges between -0.10 and $0.15\,\mathrm{BC(m^3\,s^{-1})}$ (BC parameter $\lambda = 0.25$) for the lumped models, its value range between 0.11 and $0.18\,\mathrm{BC(m^3\,s^{-1})}$ for the spatial models. For PDM, the spatial models are slightly more accurate or of equal accuracy for the peak flows as the lumped models, whereas for other models (NAM, VHM) the opposite is observed. The differences are, however, for all models minor.

Figure 4 shows the empirical peak flow frequency distribution for all configured models at the most downstream station. In order to investigate the influence of the river flooding on the river peak flow dynamics and statistics, the hydraulic model simulations were conducted with and without the floodplains with the highest capacity. It is seen in Fig. 4 that the floodplains only affect the highest flows. Closer inspection of the simulation results revealed that only two of the three floodplains were filled during the entire simulation period of more than six years, explaining the minor influence on the peak flows for the lower return periods. Note that all models overestimate the flow for empirical return periods between one and three years. Including also the numerous much smaller floodplains along the more upstream tributaries in the catchment will likely reduce the simulated flows in that range of return periods.

A similar conclusion holds for the low flow extremes. The low flow accuracy is fair for both the lumped and spatial models: LFE between 0.02 and $0.18\,\mathrm{BC(m^3\,s^{-1})}$ ($\lambda = 0.25$). One again does not see strong differences in the low flow performance between the lumped models (LFE between 0.04 and $0.18\,\mathrm{BC(m^3\,s^{-1})}$) and the spatial models (LFE between 0.02 and $0.18\,\mathrm{BC(m^3\,s^{-1})}$). The cumulative runoff volumes are for many models slightly underestimated but limited to 10.63 %, which is fair. Most models show lower differences: WBE between -10.63 and $+4.11$ % for the spatial models and between -10.11 and -0.80 % for the lumped models.

Given this good accuracy, the models hence can be considered useful for impact studies on peak flow extremes (flood studies), low flow extremes (water availability studies) and cumulative runoff volumes (water balance studies). The spatial models are applicable for scenario studies that do involve spatial land use changes, whereas the lumped models are useful for other types of scenarios studies and to reduce the computational times. For the lumped models, the calculation times were 10^3 to 10^6 shorter when compared to the detailed models. The lumped and spatial models moreover produce consistent results, as is shown by Tran Quoc and Willems (2016) for impact analysis of climate change scenarios.

5 Discussion and conclusions

The proposed approach successfully derived for the study catchment an accurate and fast combined conceptual rainfall-runoff and river hydraulic model. It was based on a modelling framework developed to incorporate several approaches to parsimoniously model the catchment hydrology and the hydraulics of rivers, floodplains and sewer systems. Albeit not shown in this paper, this flexible modelling framework can deliver hydrological and hydraulic models that can be configured with varying resolutions and provide consistent results. Further extensions are possible to the sewer systems in the catchment. The level of model detail can be tailored to the investigated scale, dynamics of the system and the intended application. Overly complicated models are both data- and time-demanding to set up. In addition, such complicatedness can lead to equifinality issues, prolonged calculation times, increased model uncertainties and even inaccurate simulations due to the accumulation of model uncertainties and discrepancies.

These models can easily be interfaced and employed for various water management studies. However, additional research remains necessary. Firstly, the incorporated modelling approaches require further developments. More model structures can be considered to identify and select the optimal one and/or to apply an ensemble approach to account for the model structure uncertainty. A sensitivity analysis may be performed to evaluate the parameters of the spatial aggregation and disaggregation steps. Also, the application of the models for scenario studies and water management decision making needs to be further studied. Both the hydraulic and hydrological modelling approach should be tested on different case studies. Secondly, future research can focus on the development of a decision support system for water management on catchment scale that incorporates the modelling framework. This requires further enhancement of the PCRaster-Python and CMD software tools to ensure that the hydrological and hydraulic conceptual models can be configured easily and quickly. Also, additional developments are necessary to (semi-automatically) interface the obtained models.

Acknowledgements. The authors would like to thank DHI for the MIKE11 license. The Province of Antwerp and the Flemish Environment Agency (VMM) are also gratefully acknowledged for providing measurement data of the cross-sections of the rivers in the Grote Nete catchment such that the hydrodynamic river model could be set up.

References

Ajami, N. K., Gupta, H., Wagener, T., and Sorooshian, S.: Calibration of a semi-distributed hydrologic model for streamflow estimation along a river system, J. Hydrol., 298, 112–135, 2004.

Box, G. E. P. and Cox, D. R.: An analysis of transformations, J. Roy. Stat. Soc., 26, 211–243, 1964.

Breuer, L., Huisman, J. A., Willems, P., Bormann, H., Bronstert, A., Croke, B. F. W., Frede, H.-G., Gräff, T., Hubrechts, L., Jakeman, A. J., Kite, G., Lanini, J., Leavesley, G., Lettenmaier, D. P., Lindström, G., Seibert, J., Sivapalan, M., and Viney, N. R.: Assessing the impact of land use change on hydrology by ensemble modeling (LUCHEM) I: Model intercomparison with current land use, Adv. Water Resour., 32, 129–146, 2009.

De Vleeschauwer, K., Weustenraad, J., Nolf, C., Wolfs, V., De Meulder, B., Shannon, K., and Willems, P.: Green-blue water in the city: quantification of impact of source control versus end-of-pipe solutions on sewer and river floods, Water Sci. Technol., 70, 1825–1837, 2014.

Jakeman, A. J., Littlewood, I. G., and Whitehead, P. G.: Computation of the instantaneous unit hydrograph and identifiable component flows with application to two small upland catchments, J. Hydrol., 117, 275–300, 1990.

Keupers, I., Wolfs, V., Kroll, S., and Willems, P.: Impact analysis of sewer overflows on the receiving river water quality using an integrated conceptual model, Proceedings of the 10th conference on Urban Drainage Modelling (10UDM), 20–23 September 2015, Québec, Canada, 2015.

Moore, R. J.: The PDM rainfall-runoff model, Hydrol. Earth Syst. Sci., 11, 483–499, doi:10.5194/hess-11-483-2007, 2007.

Nielsen, S.-A. and Hansen, E.: Numerical simulation of the rainfall-runoff process on a daily basis, Nord. Hydrol., 4, 171–190, 1973.

Ravazi, S., Tolson, B. A., and Burn, D. H.: Review of surrogate modeling in water resources, Water Resour. Res., 48, W07401, doi:10.1029/2011WR011527, 2012.

Tran Quoc, Q. and Willems, P.: Flexible conceptual hydrological modelling – Disaggregation from lumped catchment scale to higher spatial resolutions, submitted, 2016.

Vansteenkiste, T., Tavakoli, M., Ntegeka, V., De Smedt, F., Batelaan, O., Pereira, F., and Willems, P.: Intercomparison of hydrological model structures and calibration approaches in climate scenario impact projections, J. Hydrol., 519, 743–755, 2014.

Vermuyten, E., Meert, P., Wolfs, V., and Willems, P.: Using a fast conceptual river model for floodplain inundation forecasting and real-time flood control – a case study in Flanders, Belgium. 21st International Congress on Modelling and Simulation (MODSIM2015), 29 November–4 December 2015, Broadbeach, Queensland, Australia, 2015.

Wheather, H. S., Jakeman, A. J., and Beven, K. J.: Progress and directions in rainfall-runoff modeling, in: Modelling change in environmental systems, edited by: Jakeman, A. J., Beck, M. B., and McAleer, M. J., John Wiley & Sons, 102–132, 1993.

Willems, P.: A time series tool to support the multi-criteria performance evaluation of rainfall-runoff models, Environ. Model. Softw., 24, 311–321, 2009.

Willems, P.: Parsimonious rainfall-runoff model construction supported by time series processing and validation of hydrological extremes – Part 1: Step-wise model-structure identification and calibration approach, J. Hydrol., 510, 578–590, 2014.

Wolfs, V., Van Steenbergen, N., and Willems, P.: Flood probability mapping by means of conceptual modelling, in: River Flow 2012: Vol. 2. International Conference on Fluvial Hydraulics, edited by: Muñoz, R., Costa Rica, 5–7 September 2012, 1081–1085, London: CRC Press, Taylor & Francis Group, 2012.

Wolfs, V., Meert, P., and Willems, P.: Modular conceptual modelling approach and software for river hydraulic simulations, Environ. Model. Softw., 71, 60–77, 2015.

Proc. IAHS, 373, 7–12, 2016
proc-iahs.net/373/7/2016/
doi:10.5194/piahs-373-7-2016

Impact of urbanization on rainfall-runoff processes: case study in the Liangshui River Basin in Beijing, China

Zongxue Xu[1,2] **and Gang Zhao**[1,2]

[1]College of Water Sciences, Beijing Normal University, Beijing 100875, China
[2]Key Laboratory of Water and Sediment Sciences, Ministry of Education, Beijing 100875, China

Correspondence to: Zongxue Xu (zongxuexu@vip.sina.com)

Published: 12 May 2016

Abstract. China is undergoing rapid urbanization during the past decades. For example, the proportion of urban population in Beijing has increased from 57.6 % in 1980 to 86.3 % in 2013. Rapid urbanization has an adverse impact on the urban rainfall-runoff processes, which may result in the increase of urban flood risk. In the present study, the major purpose is to investigate the impact of land use/cover change on hydrological processes. The intensive human activities, such as the increase of impervious area, changes of river network morphology, construction of drainage system and water transfer, were considered in this study. Landsat TM images were adopted to monitor urbanization process based on Urban Land-use Index (ULI). The SWMM model considering different urbanized scenarios and anthropogenic disturbance was developed. The measured streamflow data was used for model calibration and validation. Precipitation with different return periods was taken as model input to analyse the changes of flood characteristics under different urbanized scenarios. The results indicated that SWMM provided a good estimation for storms under different urbanized scenarios. The volume of surface runoff after urbanization was 3.5 times greater than that before urbanization; the coefficient of runoff changed from 0.12 to 0.41, and the ratio of infiltration decreased from 88 to 60 %. After urbanization, the time of overland flow concentration increased while the time of river concentration decreased; the peak time did not show much difference in this study. It was found that the peak flow of 20-year return-period after urbanization is greater than that of 100-year return-period before urbanization. The amplification effect of urbanization on flood is significant, resulting in an increase of the flooding risk. These effects are especially noticeable for extreme precipitation. The results in this study will provide technical support for the planning and management of urban storm water and the evaluation on Low Impact Development (LID) measures.

1 Introduction

Urbanization is an important index to reflect the development level of a country, but it also enhances the interaction between human society and the environment. The unbalance of the water conservancy facilities transformation and the rapid urban development resulted in considerable urban water problems in China. From 2008 to 2010, different kinds of floods and inundation disasters occurred in 62 % of cities in China. For example, the heavy rainfall on 21 July 2012 in Beijing, resulted in 63 severe urban inundations and claimed the lives of 79 people.

Urbanization increases the risk of flooding due to increased peak discharge and volume, and decreased time to peak that restricted the development of cities (Nirupama and Simonovic, 2007; Saghafian et al., 2008). In highly urbanized areas, over half of rainfall becomes surface runoff, and deep infiltration is only a small fraction of natural situation (Chester et al., 1998). However, some researchers did not found significant changes in runoff coefficients in an urbanized catchment (Brun and Band, 2000; Chang, 2003). Rainfall-runoff process is known to be related to complex factors in urban catchment, such as land use/cover, river network morphology, construction of drainage system and water transfer. The complexity of underlying surface, the uncertainty of anthropogenic disturbance, and the lack of high-quality datasets for calibration and validation may limit our research for the rainfall-runoff processes in an urban catchment. Thus, there is an urgent need to develop a hydrological model to identify the impact on urbanization with a support of long-term monitoring data in a typical urbanized catchment.

A variety of hydrological models have been developed and applied to simulate rainfall-runoff processes in urban area, including SWMM, MIKE, HSPF, STORM and INFOWORKS. For instance, Guan et al. (2015) used the SWMM model to simulate a series of scenarios in a developing urban catchment. Koudelak and West (2008) and Peng et al. (2015) adopted the InfoWorks model to simulate sewerage network flow in the city of Latvia and China, respectively. Guo et al. (2013) and Alam et al. (2014) used MIKE model to assess flood hazard combined sewer system in urban areas.

In this study, SWMM model considering different urbanized scenarios and anthropogenic disturbance was developed on the basis of remote sensing image and survey. The measured streamflow data was used for model calibration and validation. Precipitation with different return periods was taken as model input to analyse the changes of rainfall-runoff process under different urbanized scenarios.

2 Study area description

Dahongmen catchment (Fig. 1) is located in the upstream of Liangshui River basin in Beijing, between 39°48′–39°55′ N, 116°9′–116°24′ E and with an area of 131 km^2. The terrain in the catchment shows a downward trend from the western mountains to the eastern plains. The annual average precipitation is 522.4 mm and 80 % of precipitation occurs during the period from June to September. Dahongmen catchment, which has experienced a rapid urbanization process over the past two decades, has gradually become a transport hub and core zone of Beijing metropolitan. Floods and inundation frequently occur in this area, which was further intensified after rapid urbanization. For the purpose to solve this problem, a set of anthropogenic measures including water transfer, river and sewer system, and morphology changes were taken. However, these measures further changed the local rainfall runoff processes.

3 Model and data description

3.1 SWMM model

The EPA Storm Water Management Model (SWMM) is a dynamic rainfall-runoff simulation model used for single event or long-term (continuous) simulation of runoff quantity and quality from primarily urban areas. It was first developed in 1971 and continued to be widely used throughout the world for planning, analysis and design related to storm water, combined sewers, and other hydraulic structures in urban areas. The main components of SWMM include surface runoff, infiltration, surface concentration, channel and pipe concentration and LID controls. Further technical details of the model are given by Rossman et al. (2015).

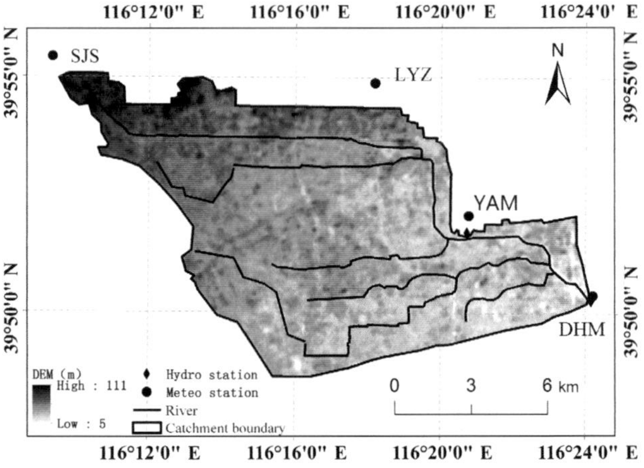

Figure 1. Location, river systems, and hydrometeorological stations in the Dahongmen catchment.

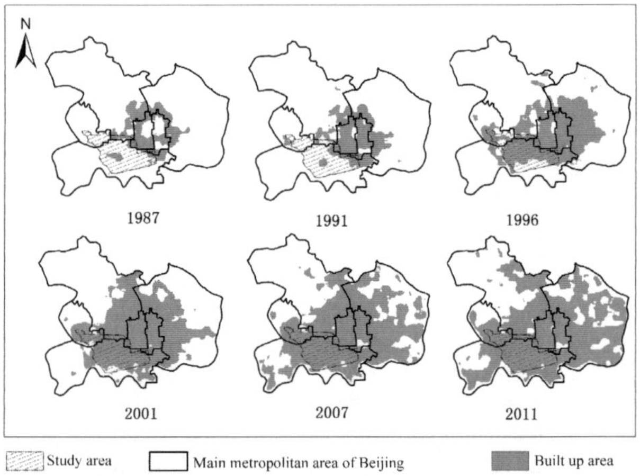

Figure 2. Urbanization processes in the Dahongmen catchment.

3.2 Data description

In this study, urban land-use index (ULI) was used to monitor the urbanization process of Dahongmen catchment in Beijing (Fig. 2). ULI is an effective index automatically extracting from urban land uses with Landsat 7 image. It reduces the impact of low density vegetation on extraction accuracy. According to Fig. 2, the study area has experienced rapid urbanization process from 1987 to 2011, and the proportion of impervious area in SWMM model is set based on ULI index.

Digital Elevation Model (DEM) and sewer system map were provided by the National Aeronautics and Space Ad-

Table 1. Results of parameter identification.

Parameters in SWMM		Before urbanization	After urbanization
Manning coefficient	Pervious zone	0.43	0.36
	Impervious zone	0.07	0.04
	Channels	0.025	0.048
	Pipeline	–	0.02
Horton coefficient	Maximum infiltration	124.6	142.4
	minimum infiltration	74.7	89.6
	Decay coefficient	11.4	23.2

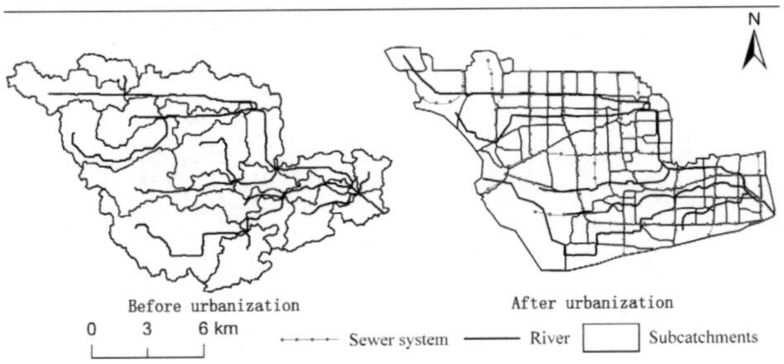

Figure 3. Model development in different urbanization cases.

ministration (NASA, ASTER GDEM) and Beijing Municipal Institute of City Planning and Design. Hourly precipitation series at 4 stations (Fig. 1) and streamflow data at DHM station from 1980 to 2012 were obtained from the Hydrographic Station of Beijing. Hourly inflow data of YAM station were obtained from the Liangshui River Basin Authority.

4 Model development

4.1 Catchment subdivision

Catchment subdivision should reflect the urbanization process and has great impacts on the rainfall runoff simulation. Before urbanization, DEM and D8 algorithm were used to extract the subcatchment (Fig. 3). But for highly developed urban area, catchment was divided by design drainage areas provided by the Beijing Municipal Institute of City Planning and Design.

4.2 Model accuracy evaluation

According to the standard for hydrological information and hydrological forecasting in China, Nash–Sutcliffe efficiency coefficient (R_{NS}), relative error of flood peak discharge (RE_p) and absolute error of flood peak appearance (AE_T) were used for measuring the accuracy of model performance.

$$R_{NS} = 1 - \frac{\sum_{i=1}^{N} \left(q_t^{obs} - q_t^{sim}\right)^2}{\sum_{i=1}^{N} \left(q_t^{obs} - \overline{q}^{obs}\right)^2}, \tag{1}$$

where $q_t^{obs}$is observed discharge sequence, N is the number of observed discharge data and $\overline{q}^{obs}$ is the average of the observed discharge.

$$RE_p = \frac{\left|q_p^{obs} - q_p^{sim}\right|}{q_p^{obs}} \times 100\,\%, \tag{2}$$

where q_p^{obs} is the observed peak flow, q_p^{sim} is the simulated peak flow.

$$AE_T = T_p^{obs} - T_p^{sim}, \tag{3}$$

where T_p^{obs} is the observed occurrence time of flood peak, $T_p^{sim}$is the simulated occurrence of flood peak.

5 Results analysis

5.1 Model calibration and validation

Most of the parameters in SWMM model have its specific physical meaning, which can be measured theoretically. Due to the limitation of data, these parameters are usually determined by empirical method or optimization method. In this study, the genetic algorithm (GA) was adopted to optimize the sensitive parameters in SWMM model, and Nash–Sutcliffe efficiency (R_{NS}) was used as the objective function. Results of parameter identification are given in Table 1.

Manning coefficient in both pervious and impervious area decreased after urbanization, it shows that the concentration time of overland flow declined. However, the increase in manning roughness coefficient of channel suggested that the concentration time of river raised after urbanization. Horton infiltration coefficient reflects the infiltration capacity of pervious zone, which changes little after urbanization. Error statistics of simulation results and flood hygrograph are shown in Table 2, Figs. 4 and 5.

Table 2. Error statistics of simulation results.

		Calibration		Validation	
Before urbanization	Storm	19810703	19830619	19850702	19870813
	R_{NS}	0.62	0.61	0.83	0.64
	RE_p	3 %	3 %	6 %	15 %
	AE_T	−1	0	0	+1
After urbanization	Storm	20110623	20120721	20110814	20110726
	R_{NS}	0.88	0.95	0.84	0.60
	RE_p	2 %	2 %	11 %	4 %
	AE_T	0	0	+1	0

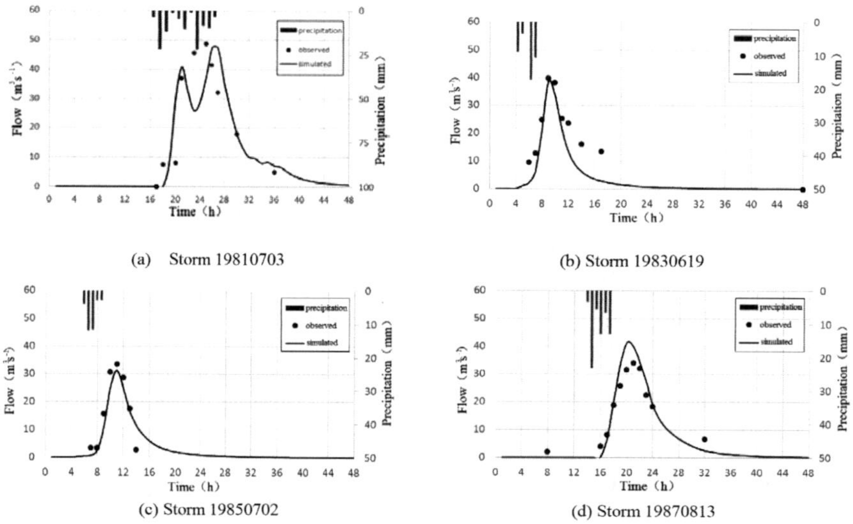

Figure 4. Simulation of flood before urbanization.

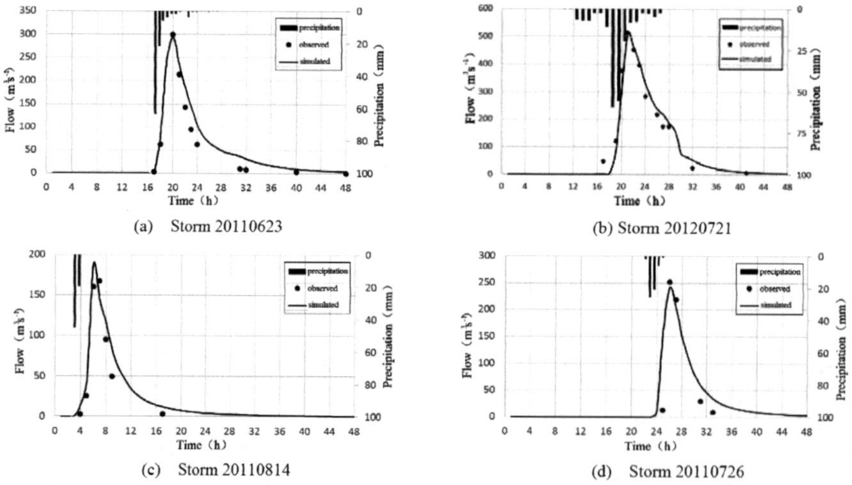

Figure 5. Simulation of flood after urbanization.

Table 3. Changes of rainfall-runoff processes in different scenarios.

Hydrological Various	$P = 100\,\%$		$P = 20\,\%$		$P = 5\,\%$		$P = 2\,\%$		$P = 1\,\%$	
	before	after	before	after	before	after	before	after	before	after
Precipitation (mm)	47.7	47.7	150.4	150.4	260.3	260.3	335.9	335.9	395.0	395.0
Runoff (mm)	5.5	14.6	17.9	66.3	30.3	109.8	40.0	136.3	48.6	161.1
Runoff-Coefficient	0.12	0.31	0.12	0.44	0.12	0.42	0.12	0.41	0.12	0.41
Peak flow ($m^3\,s^{-1}$)	30.6	64.3	118.7	337.4	208.5	438.6	298.8	526.7	384.9	612.8
Peak time (h)	3 h	5 h	2 h	3 h	2 h	2 h	2 h	2 h	2 h	2 h

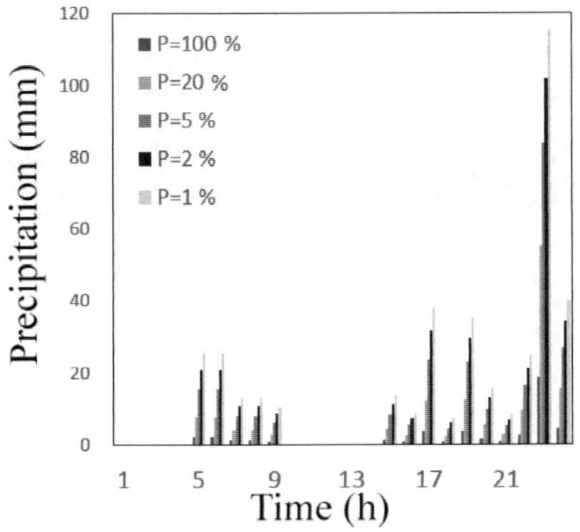

Figure 6. 24-h rainfall processes at different frequency.

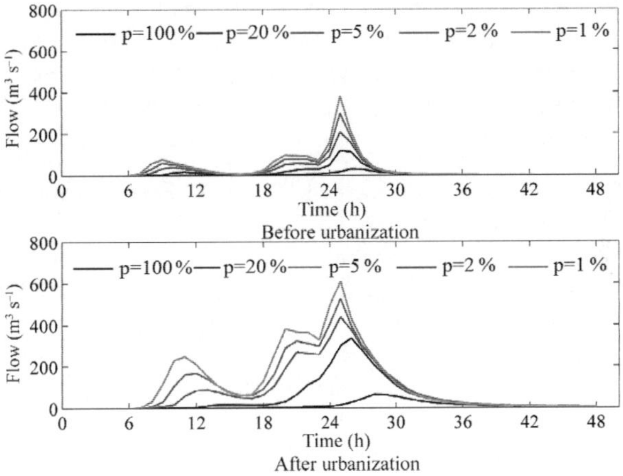

Figure 7. Changes of floods in different urbanization scenarios.

As can be seen, Nash–Sutcliffe efficiency is greater than 0.6 and the relative error of the peak flow is smaller than 15 % in both calibration and validation periods, which means that SWMM provided a good estimation both pre- and post-urbanization. Moreover, the simulation results after urbanization are much better than before, particularly on the occurrence time of flood peak.

5.2 Rainstorm scenario analysis

According to the hydrological handbook in Beijing, the design storms with different return periods ($p = 100\,\%$, $p = 20\,\%$, $p = 5\,\%$, $p = 2\,\%$, $p = 1\,\%$) were calculated based on design rainfall formula and the 24-h rainfall distribution. The results of design storms are given in Fig. 6. Changes of floods in different urbanization scenarios were simulated with these input precipitation data and the results are shown in Fig. 7 and Table 3.

According to Table 3, the urbanization process has a significant amplification effect on runoff generation. Taking 100-year return-period ($p = 1\,\%$) as an example, the surface runoff after urbanization is 395 mm, which is 3.55 times than that before urbanization. The runoff coefficient increased from 0.12 to 0.41, and the infiltration after urbanization covers 65 % of that before urbanization; however, the peak flow after urbanization is 1.68 times than that before urbanization. Therefore, the amplification effect of urbanization on low-frequency flood is obvious, resulting in an increase of the flooding risk in urban area.

However, the flood peak did not appear earlier, and even delayed on high-frequency precipitation ($p = 100\,\%$ and $p = 20\,\%$). Field investigation found that there are many artificial river barrages (such as bridges, gates or embankments) which delayed the time of channel concentration. On the other hand, the runoff flowed along the artificial path after urbanization, which makes it faster but longer than that before urbanization, and hence influenced the peak time. It is also found that the peak flow of 20-year return-period after urbanization is higher than that of 100-year return-period before urbanization, which puts forward new requirements and challenges for the transformation of urban facilities and urban water management.

6 Conclusions

In this study, the SWMM model considering different urbanized scenarios and anthropogenic disturbance was successfully applied to simulate the rainfall-runoff process in Dahongmen catchment, Beijing. The measured streamflow data was used for model calibration and validation. Precipitation with different return periods was used as model input to analyse the changes of flood characteristics under different urbanized scenarios. Major conclusions can be summarized as follows:

1. SWMM model has good performance for the simulation of rainfall-runoff process before and after urbanization. Nash–Sutcliffe efficiency is greater than 0.6, and the relative error of flood peak discharge is smaller than 15 % both during calibration and validation periods.

2. The volume of surface runoff after urbanization was 3.5 times greater than that before urbanization; the coefficient of runoff changed from 0.12 to 0.41, and the ratio of infiltration decreased from 88 to 60 %. The amplification effect of urbanization for low-frequency precipitation is more obvious and the peak flow of 20-year return-period (5 %) after urbanization is higher than that of 100-year return-period (1 %) before urbanization, leading to a higher flood risk in urban area.

3. The occurrence time of flood peak shows little change after urbanization. The concentration of overland flow accelerates and the river concentration had grown over time after urbanization. Therefore, the occurrence time of flood peak did not appear in advance or even delayed on high-frequency precipitation ($p = 100\,\%$ and $p = 20\,\%$), mainly due to the growing number of artificial river barrages in the study area.

Acknowledgements. This study was financially supported by the key project of Beijing Natural Science Foundation (8141003).

References

Alam, S., Willems, P., and Alam, M.: Comparative Assessment of Urban Flood Risks Due to Urbanization and Climate Change in the Turnhout Valley of Belgium, J. Adv. Res., 3, 14–23, 2014.

Brun, S. E. and Band, L. E.: Simulating runoff behavior in an urbanizing watershed, Comput. Environ. Urban, 24, 5–22, 2000.

Chang, H.: Basin Hydrologic Response to Changes in Climate and Land Use: the Conestoga River Basin, Pennsylvania, Phys. Geogr., 24, 222–247, 2003.

Chester, L., Arnold, J. C., and James, G.: Impervious Surface Coverage: The Emergence of a Key Environmental Indicator, J. Am. Plann. Assoc., 62, 243–258, 1998.

Guan, M., Sillanpää, N., and Koivusalo, H.,: Modelling and assessment of hydrological changes in a developing urban catchment, Hydrol. Process., 29, 2880–2894, 2015.

Guo, F., Hanfei, Q. U., Zeng, H., Cong, P., and Geng, X.: Flood hazard forecast of Pajiang River flood storage and detention basin based on MIKE21, J. Nat. Disasters, 22, 144–152, 2013.

Koudelak, P. and West, S.: Sewerage network modelling in Latvia, use of InfoWorks CS and Storm Water Management Model 5 in Liepaja city, Water Environ. J., 22, 81–87, 2007.

Nirupama, N. and Simonovic, S. P.: Increase of Flood Risk due to Urbanization: A Canadian Example, Nat. Hazards, 40, 25–41, 2007.

Peng, H. Q., Liu, Y., Wang, H. W., and Ma, L. M.: Assessment of the service performance of drainage system and transformation of pipeline network based on urban combined sewer system model, Environ. Sci. Pollut. R., 22, 15712–15721, 2015.

Rossman, L. A.: Storm Water Management Model User's Manual Version 5.0, EPA United States Environmental Protection Agency, 2015.

Saghafian, B., Farazjoo, H., Bozorgy, B., and Yazdandoost, F.: Flood Intensification due to Changes in Land Use, Water Resour. Manag., 22, 1051–1067, 2008.

Proc. IAHS, 373, 13–17, 2016
proc-iahs.net/373/13/2016/
doi:10.5194/piahs-373-13-2016

Open Access
PIAHS

Quantification of resilience to water scarcity, a dynamic measure in time and space

S. P. Simonovic and R. Arunkumar

Department of Civil and Environmental Engineering,
University of Western Ontario, London – N6A 5B9, Canada

Correspondence to: S. P. Simonovic (simonovic@uwo.ca)

Published: 12 May 2016

Abstract. There are practical links between water resources management, climate change adaptation and sustainable development leading to reduction of water scarcity risk and re-enforcing resilience as a new development paradigm. Water scarcity, due to the global change (population growth, land use change and climate change), is of serious concern since it can cause loss of human lives and serious damage to the economy of a region. Unfortunately, in many regions of the world, water scarcity is, and will be unavoidable in the near future. As the scarcity is increasing, at the same time it erodes resilience, therefore global change has a magnifying effect on water scarcity risk. In the past, standard water resources management planning considered arrangements for prevention, mitigation, preparedness and recovery, as well as response. However, over the last ten years substantial progress has been made in establishing the role of resilience in sustainable development. Dynamic resilience is considered as a novel measure that provides for better understanding of temporal and spatial dynamics of water scarcity. In this context, a water scarcity is seen as a disturbance in a complex physical-socio-economic system. Resilience is commonly used as a measure to assess the ability of a system to respond and recover from a failure. However, the time independent static resilience without consideration of variability in space does not provide sufficient insight into system's ability to respond and recover from the failure state and was mostly used as a damage avoidance measure. This paper provides an original systems framework for quantification of resilience. The framework is based on the definition of resilience as the ability of physical and socio-economic systems to absorb disturbance while still being able to continue functioning. The disturbance depends on spatial and temporal perspectives and direct interaction between impacts of disturbance (social, health, economic, and other) and adaptive capacity of the system to absorb disturbance. Utility of the dynamic resilience is demonstrated through a single-purpose reservoir operation subject to different failure (water scarcity) scenarios. The reservoir operation is simulated using the system dynamics (SD) feedback-based object-oriented simulation approach.

1 Introduction

Risk, resilience, and vulnerability are the fundamental characteristics that defines the state of a system and are widely used to assess the performance of the system. Among these indices, resilience refers to the systems capability to recover from a failure. The concept of resilience was first introduced by Holling (1973) in ecological systems and defined as the measure of ability of the system to absorb changes and still persist with same basic structure when subjected to stress. Further, Hashimoto et al. (1982) extended the application of resilience to water resources system. They defined resilience as the measure that describes how quickly a system will likely to recover or bounce back from failure once failure has occurred. Over the period of time, the concept of resilience has been implemented in various domains. Thus, there are various definitions of resilience in different fields. Bruneau et al. (2003) reported in the context of earthquake engineering that the measure of system resilience should show the reduced failure probability, reduced consequences and reduced time to recovery. Haimes (2009) defined resilience as the ability of the system to withstand a major disruption within acceptable degradation parameters and to recover within an acceptable time and composite costs and risks. Vugrin et al. (2010) defined resilience as the ability of a system to efficiently reduce both the magnitude and duration of the deviation of system performance from the targeted system performance levels. Ayyub (2014) mentioned resilience as the ability of the system to return to a stable state after a perturbation. Based on the concepts of Bruneau et al. (2003), various dynamic measures were reported for assessing the system's ability to bounce back from a failure state, as seen above.

Though, various approaches were developed for estimating the resilience of the system based on the concepts of Bruneau et al. (2003), most of the approaches are estimating the resilience as a time independent measure that do not provide much insight about the recovery capability of the system over time. Also, most of these measures are domain specific and a number of issues must be taken into consideration while applying in other domains. The time independent static resilience is merely an abstract attribute of the system and does not completely describe the state of the system composed of vectors specific to any sub-state. Thus, the time independent static resilience measures are practically ineffective for efficient planning and developing appropriate system recovery strategies from a failure. Simonovic and Peck (2013) first developed a framework to quantify the resilience as a dynamic measure through system dynamics (SD) simulation approach and demonstrated the concept for the coastal urban flooding caused due to climate change. The developed framework considers the economic, social, organizational, health and physical impacts of climate change.

The main objective of this study is implementation of the framework developed by Simonovic and Peck (2013) for quantifying the resilience to a water scarcity (deficit) scenario with consideration of variability in time and space. The implementation is illustrated using simulation of reservoir operation for irrigation. The rest of the paper is organized in the following manner. The quantitative space-time dynamic resilience measure (STDRM) is presented in the next section. The presentation of the SD simulation of STDRM for a single-purpose reservoir operation is then presented. Results of the reservoir simulation and its space time dynamic resilience are then discussed.

2 Space-time dynamic resilience

The quantitative resilience measure, first introduced by Simonovic and Peck (2013) following Cutter et al. (2008), has two qualities: inherent (functions well during non-disturbance periods); and adaptive (flexibility in response during disturbance events) and can be applied to physical environment (built and natural), social systems, governance network (institutions and organizations), and economic systems (metabolic flows). An original space-time dynamic resilience measure (STDRM) of Simonovic and Peck (2013) is

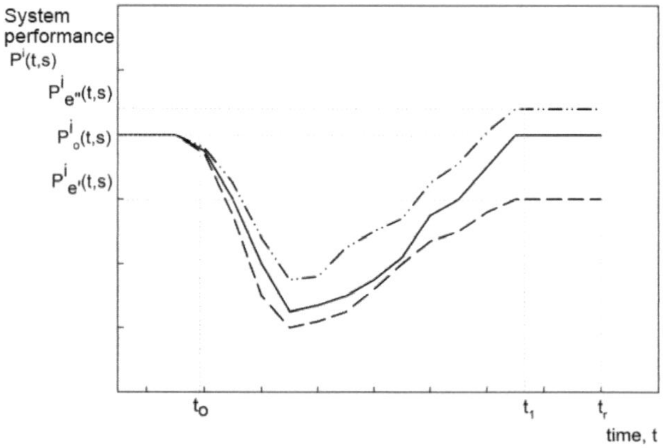

Figure 1. System performance.

designed to capture the relationships between the main components of resilience; one that is theoretically grounded in systems approach, open to empirical testing, and one that can be applied to address real-world problems in various domains.

STDRM is based on two basic concepts: level of system performance and adaptive capacity. They together define resilience. The level of system performance integrates various impacts (i) of system disturbance. The following impacts (units of resilience (ρ^i)) can be considered: physical, health, economic, social and organizational, but the general measure is not limited to them. Measure of system performance $P^i(t,s)$ for each impact (i) is expressed in the impact units (for example physical impact may be length [km] of road being inundated; and so on). This approach is based on the notion that an impact, $P^i(t,s)$ which varies with time and location in space, defines a particular resilience component of a system under consideration, see Fig. 1 adapted from Simonovic and Peck (2013). The area between the initial performance line $P_0^i(t,s)$, and performance line $P^i(t,s)$ represents the loss of system performance, and the area under the performance line $P^i(t,s)$ represent the system resilience ($\rho^i(t,s)$). In Fig. 1, t_0 denotes the beginning of the disturbance, t_1 the end of disturbance and t_r the end of the recovery period.

In mathematical form, the loss of resilience for impacts (i) represents the area under the performance graph between the beginning of the system disruption event at time (t_0) and the end of the disruption recovery process at time (t_r). Changes in system performance can be represented mathematically as:

$$\rho^i(t,s) = \int_{t_0}^{t} \left[P_0^i - P^i(\tau,s)\right] d\tau \text{ where } t \in [t_0, t_r] \tag{1}$$

When performance of the system does not deteriorate due to disruption, $P_o^i(t,s) = P^i(t,s)$ for an impact (i), the loss of resilience is 0 (i.e. the system is in the same state as at the

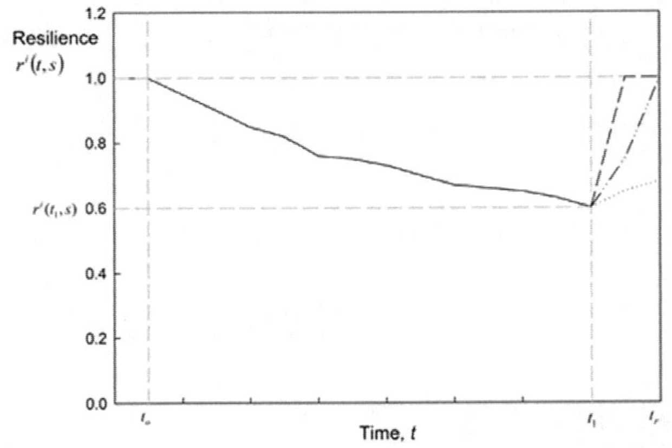

Figure 2. System resilience.

beginning of disruption). When all of system performance is lost, $P^i(t,s) = 0$, the loss of resilience is at the maximum value. The system resilience, $r^i(t,s)$ is calculated as follows:

$$r^i(t,s) = 1 - \left(\frac{\rho^i(t,s)}{P^i_o \times (t - t_o)}\right) \tag{2}$$

As illustrated in Fig. 1, performance of a system which is subject to a disaster event drops below the initial value and time is required to recover the loss of system performance. Disturbance to a system causes a drop in system resilience from value of 1 at t_o to some value $r^i(t_1,s)$ at time t_1, see Fig. 2. Recovery usually requires longer time than the duration of disturbance. Ideally, resilience value should return to a value of 1 at the end of the recovery period, t_r (dashed line in Fig. 2); and the faster the recovery, the better. The system resilience (over all impacts (i)) is calculated using:

$$R(t,s) = \left\{\prod_{i=1}^{M} r^i(t,s)\right\}^{\frac{1}{M}} \tag{3}$$

where M is total number of impacts.

The calculation of STDRM for each impact (i) is done at each location (s) by solving the following differential equation:

$$\frac{\partial \rho^i(t)}{\partial t} = \mathrm{AC}^i(t) - P^i(t) \tag{4}$$

where AC^i represents adaptive capacity with respect to impact i.

The STDRM integrates resilience types, dimensions and properties by solving for each point in space (s):

$$\frac{\partial R(t)}{\partial t} = \mathrm{AC}(t) - \prod_i P^i(t) \tag{5}$$

The implementation of the presented framework is proceeding by using system dynamics simulation approach together with spatial analysis software.

Space-time dynamic resilience to water scarcity

In the present study, the space-time dynamic resilience of a simple single-purpose reservoir has been quantified using SD simulation approach. To achieve this, the operations of the single-purpose reservoir and its dynamic resilience is modelled using Vensim package – a SD simulation software (Ventana Systems, 2005).

Reservoir simulation model

The simple single-purpose reservoir simulation model consists of the continuity equation and a set of operational constraints. The continuity equation is expressed as:

$$S_t = S_{t-1} + I_t - \mathrm{IR}_t - O_t - \mathrm{SP}_t \tag{6}$$

where S_t is the final storage at the time period t; S_{t-1} is the initial storage in the reservoir; I_t is the inflow during the time period t; IR_t is the total irrigation release from the reservoir during the time period t; O_t are the losses from the reservoir (evaporation and other leakage losses); and SP_t is the spill from the reservoir during the time period t. The system constraints, reservoir operating rules and the release decisions are captured using IF-THEN-ELSE statements in the simulation model. If the available storage is greater than the irrigation demand, then the actual demand is released; else the available storage is released. Then, the releases to the individual fields are made sequentially based on their demand.

The system performance for the reservoir irrigation $(P_{t,s})$ for each individual field s at time t is expressed as the ratio of actual release made for irrigation and the demand during the time t:

$$P_{t,s} = \frac{\mathrm{IR}_{t,s}}{\mathrm{IR}^{\mathrm{demand}}_{t,s}} \tag{7}$$

where $\mathrm{IR}^{\mathrm{demand}}_{t,s}$ is the irrigation demand during the time period t for the field s. This performance measure is used for quantifying the space-time dynamic resilience.

3 An illustrative case study

Data required for the simulation are taken from Arunkumar and Jothiprakash (2012). Ten years of historical daily inflow observed at the dam site was collected and used for simulation. Analysis of inflow data shows that the reservoir is highly intermittent in nature and receives inflow only during the monsoon season. Inflow during the non-monsoon season is highly negligible. The reservoir supplies irrigation water to the command area at the downstream through the lift irrigation scheme. More details about the reservoir system can be found in Arunkumar and Jothiprakash (2012).

The illustrative model is set to include 100 individual fields to be irrigated by the reservoir. All individual fields are assumed to have equal area and same crop, therefore the

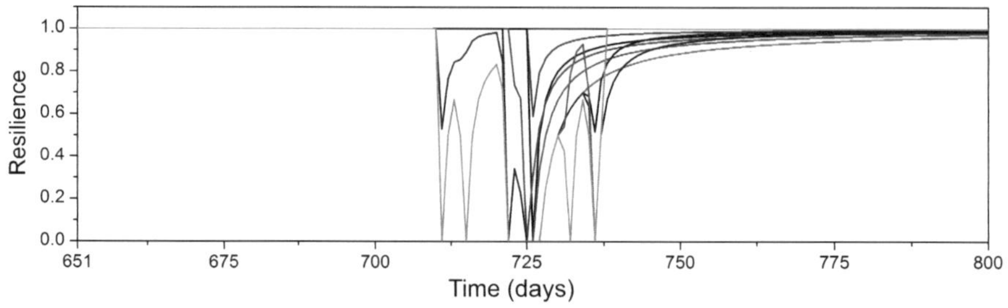

Figure 3. Temporal variation of resilience to irrigation water scarcity.

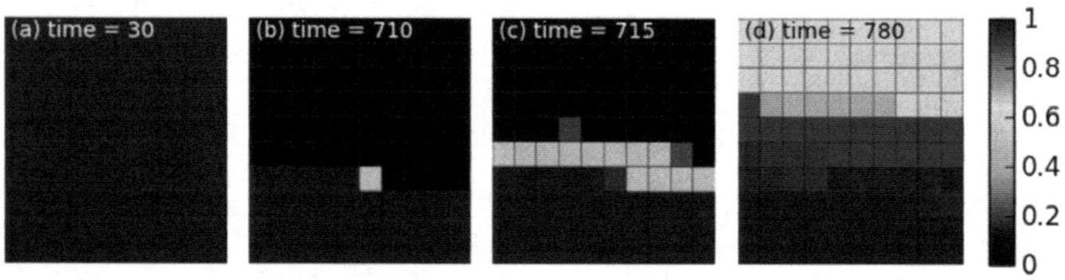

Figure 4. Spatial variation of resilience to irrigation water scarcity.

all fields have same irrigation demand. Water from the reservoir is delivered according to the distance from the reservoir, closer fields get the water first. However, it is be noted that the demand of individual fields may vary depending on crop type, plot size, etc. and release schedule can be modified to accommodate any other order of priority. In the illustrative case, the releases are made sequentially starting from field 1, which is closest to the reservoir finishing with filed 100 being furthest away from the reservoir. Thus fields which are closer to reservoir get irrigated frequently and with increase in distance from the reservoir, irrigation frequency decreases. In event of inadequate water availability, all the fields may not be irrigated to their full demand. In such case, the available water is released to the fields closer to the reservoir. The system performance is estimated individually for each field (using Eq. 7).

The computational procedure integrates (a) reservoir operation and space time dynamic resilience models simulation using system dynamics approach in Vensim and (b) spatial irrigation release distribution model in Python. Integrated system provides for space-time dynamic resilience calculation of single-purpose reservoir operations.

Simulation experiment

The reservoir operation is simulated for 10 years using observed inflow and specified daily irrigation demand. Both, the inflow and the irrigation demand vary significantly on daily time scale. The reservoir receives predominant inflow during the monsoon season. However, the irrigation demand during monsoon season is insignificant due to the availability of rainfall. The irrigation demand is very high during non-monsoon seasons, but water availability is low. Thus, the system is prone to failure during non-monsoon season. Hence, the capability to recover from failure due to water scarcity is studied using SD approach.

Simulation results

The dynamic resilience of all 100 fields for the time period 651 to 800 is shown in Fig. 3. It is observed that initially all fields receive sufficient amount of water resulting in high resilience for most of the time. When the water availability is not sufficient to meet the demands, failure occurs. Then, the resilience of individual fields vary significantly and drops due to either partial satisfaction of the demand or no satisfaction. During the failure periods, due to lower amount of water being available, not all the fields receive the full demand. Only the fields near the reservoir receive the full demand, one field receives partial demand and most of the fields do not receive any water. Hence, there is variation in performance value and resilience with time and in space. It is observed from the Fig. 3 that the rate of recovery of induvial fields also vary. Some fields have higher rate of recovery and hence their resilience value suddenly increases from zero. These fields are closer to the reservoir, which are more resilient when failure happens. Fields far away from the reservoir exhibit much slower recovery and are less resilient.

The spatial dynamic resilience of all 100 individual fields is shown in Fig. 4 for four selected time periods. Figure 4a shows resilience of all fields at early part of the simulation period, day 30. During the initial periods, irrigation demand of all individual fields is fully met and therefore they have high resilience value. However, during non-monsoon season, due to less inflow in to the reservoir, irrigation demand are satisfied for only few fields. Due to the simple assumptions that the water from the reservoir is delivered in accordance to the distance (first the closets and so on) some fields closer to the reservoir get their demand fully satisfied whereas fields further away do not. Figure 4b shows the spatial distribution of resilience at day 710. The fields closer to the reservoir always receive sufficient amount of water and therefore show high resilience. Most of the fields further away fail to receive the irrigation releases and hence their resilience drops to zero. Figure 4c shows day 715 where some of the fields already recovered from failure and their resilience increased. Figure 4d shows day 780 when almost all fields are receiving the water for irrigation and their resilience is on the rise. As it can be seen from Fig. 4, the resilience of individual fields varies significantly. This is due to difference in the performance value and failure time of each field. Thus, their rate of recovery also varies. Due to our assumption, the fields closer to the reservoir have always higher resilience then fields further away.

4 Discussion

In this study, the concept of space-time dynamic resilience is applied to reservoir operation for the water scarcity scenario. The developed framework of the dynamic resilience is based on the original work of Simonovic and Peck (2013). Irrigation is the reservoir purpose used to measure the performance of the reservoir operation. Less reservoir inflow and insufficient storage resulted in changes in system performance and dynamic resilience. Temporal and spatial presentation of resilience can be used as the mechanism to investigate various adaptation options by repeating the simulations for various scenarios. In this way space-time resilience can provide for more efficient and better informed decision making process.

References

Arunkumar, R. and Jothiprakash, V.: Optimal Reservoir Operation for Hydropower Generation using Non-linear Programming Model, J. Inst. Eng. (India) Ser. A, 93, 111–120, doi:10.1007/s40030-012-0013-8, 2012.

Ayyub, B. M.: Systems Resilience for Multihazard Environments: Definition, Metrics and Valuation for Decision Making, Risk Anal., 34, 340–355, doi:10.1111/risa.12093, 2014.

Bruneau, M., Chang, S. E., Eguchi, R. T., Lee, G. C., O'Rourke, T. D., Reinhorn, A. M., Shinozuka, M., Tierney, K., Wallace, W. A., and von Winterfeldt, D.: A Framework to Quantitatively Assess and Enhance the Seismic Resilience of Communities, Earthq. Spectra, 19, 733–752, doi:10.1193/1.1623497, 2003.

Cutter, S. L., Barnes, I., Berry, M., Burton, C., Evans, E., and Tate, E.: A place-based model for understanding community resilience to natural disasters, Global Environ. Chang., 18, 598–606, doi:10.1016/j.gloenvcha.2008.07.013, 2008.

Haimes, Y. Y.: On the Definition of Resilience in Systems, Risk Anal., 29, 498–501, doi:10.1111/j.1539-6924.2009.01216.x, 2009.

Hashimoto, T., Stedinger, J. R., and Loucks, D.: Reliability, Resiliency, and Vulnerability Criteria for Water Resource System Performance Evaluation, Water Resour. Res., 18, 14, doi:10.1029/WR018i001p00014, 1982.

Holling, C. S.: Resilience and Stability of Ecological Systems, Annu. Rev. Ecol. Syst., 4, 1–23, doi:10.1146/annurev.es.04.110173.000245, 1973.

Simonovic, S. P. and Peck, A.: Dynamic Resilience to Climate Change Caused Natural Disasters in Coastal Megacities Quantification Framework, Br. J. Environ. Clim. Chang., 3, 378–401, doi:10.9734/BJECC/2013/2504, 2013.

Ventana Systems: Vensim User's Guide. Ventana Systems, Inc., Belmont, MA, USA, 2005.

Vugrin, E. D., Warren, D. E., Ehlen, M. A., and Camphouse, R. C.: A Framework for Assessing the Resilience of Infrastructure and Economic Systems, in: Sustainable and Resilient Critical Infrastructure Systems, edited by: Gopalakrishnan, K. and Peeta, S., Springer Berlin Heidelberg, Berlin, Germany, 77–116, doi:10.1007/978-3-642-11405-2_3, 2010.

Proc. IAHS, 373, 19–24, 2016
proc-iahs.net/373/19/2016/
doi:10.5194/piahs-373-19-2016

Scaling issues in multi-criteria evaluation of combinations of measures for integrated river basin management

Jörg Dietrich

Institute of Water Resources Management, Hydrology and Agricultural Hydraulic Engineering, Leibniz Universität Hannover, Hannover, 30167, Germany

Correspondence to: Jörg Dietrich (dietrich@iww.uni-hannover.de)

Published: 12 May 2016

Abstract. In integrated river basin management, measures for reaching the environmental objectives can be evaluated at different scales, and according to multiple criteria of different nature (e.g. ecological, economic, social). Decision makers, including responsible authorities and stakeholders, follow different interests regarding criteria and scales. With a bottom up approach, the multi criteria assessment could produce a different outcome than with a top down approach. The first assigns more power to the local community, which is a common principle of IWRM. On the other hand, the development of an overall catchment strategy could potentially make use of synergetic effects of the measures, which fulfils the cost efficiency requirement at the basin scale but compromises local interests. Within a joint research project for the 5500 km^2 Werra river basin in central Germany, measures have been planned to reach environmental objectives of the European Water Framework directive (WFD) regarding ecological continuity and nutrient loads. The main criteria for the evaluation of the measures were costs of implementation, reduction of nutrients, ecological benefit and social acceptance. The multi-criteria evaluation of the catchment strategies showed compensation between positive and negative performance of criteria within the catchment, which in the end reduced the discriminative power of the different strategies. Furthermore, benefit criteria are partially computed for the whole basin only. Both ecological continuity and nutrient load show upstream-downstream effects in opposite direction. The principles of "polluter pays" and "overall cost efficiency" can be followed for the reduction of nutrient losses when financial compensations between upstream and downstream users are made, similar to concepts of emission trading.

1 Introduction

The Werra river basin is situated in central Germany within the upper part of the Weser catchment. Before German reunification it was divided by the inner-German borderline. The main industry in the catchment is potash mining, associated with a high salt load of the Werra River. Like for many German rivers, the morphological conditions of the river courses and the ecological continuity were affected before implementation of the European Water Framework Directive (WFD). Agricultural land use dominates in the North-Eastern area of the catchment. In former Eastern Germany many dispersed settlements were not connected to the public sewer system and were often not equipped with decentralized wastewater treatment. While the degree of connection was 98 % in Hessen, it was 48 % in the Thuringian part of the Werra catchment in 2001. As a consequence, the nutrient load of the catchment was high compared to the relatively extensive land use and the low population density. The ecological community was degraded in several water bodies, showing a good ecological status according to the AQEM assessment system (Hering et al., 2004) only in upstream regions of the Thuringian Forest.

For the implementation of the WFD, an exemplary river basin management plan (RBMP) was elaborated by an interdisciplinary research team, supported by local water authorities (Dietrich and Schumann, 2006). The RBMP provided several alternative strategies for the catchment, which were prepared for a final decision procedure supported by a multi criteria decision support system (Dietrich et al., 2007). Within this paper we focus on spatial aspects of measures for the improvement of the hydro-morphological conditions and the reduction of nutrient loads from point sources and diffuse sources (for a detailed description see Dietrich and Funke, 2009).

One of the challenges in spatial decision analysis is the spatial aggregation of criteria. For an RBMP, measures are located throughout the catchment area. The criteria for the individual measures can be aggregated in space to get an overall multi-criteria assessment of alternative combinations for the RBMP. This technique was applied in the widely used MULINO-DSS (Giupponi et al., 2002). Alternatively the multi-criteria analysis (MCA) can be applied for each of the locations separately, and then the outcome of the MCA is aggregated in space. Both pathways of aggregation of criteria and space can lead to different overall results (Herwijnen and Rietveld, 1999). The first path better represents the characteristics of the basin, whereas the second path allows different preference structures for the smaller sub-units, hence better represents the local situation. By aggregating criteria, positive and negative effects can be smoothened, with the consequence of reduced distinctive character of the alternatives. This kind of spatial compensation can be addressed by introducing additional criteria as Nijssen and Schumann (2014) showed for flood risk management. In this study, we present a strategic combination approach, which includes a criterion for social acceptance of the measures in order to represent the stakeholders' preference for the local measures.

2 Werra catchment diagnosis and integrated planning of measures for the WFD implementation

2.1 Morphological state and nutrient emissions

The ecological assessment with AQEM showed significant deviations from the species composition, which could be expected for the types of water bodies in that catchment. The salt load of the lower Werra River was not subject of the investigations even if it was known that it is one of the causes of ecological degradation for the affected water bodies. Apart from this, morphological deficits in most river courses (Fig. 1) were identified as a major problem to address in river basin management (Dietrich and Schumann, 2006), hence in the implementation of the Water Framework Directive. The morphological deficits include the riparian and river bed structure, but also numerous structures from groundsills to reservoir dams which disturb or prevent fish migration. Also the overall saprobial state (Fig. 1), as well as nitrate and phosphorus concentrations were found to be beyond the levels which support a good ecological state according to the WFD. The quantitative investigation of the nutrient cycle was done with a chain of models, combining an agricultural production model to compute nutrient losses from agricultural areas, a point source emission model for sewage treatment, and a coupled SWAT-RWQM1 model to simulate nutrient turnover and transport at catchment scale. The emissions of nitrogen and phosphorus from point and non-point sources show an uneven distribution over the catchment, closely related with urban land use in the case of point sources (Fig. 2) and agricultural land use in the case of diffuse (non-point) sources (Fig. 3).

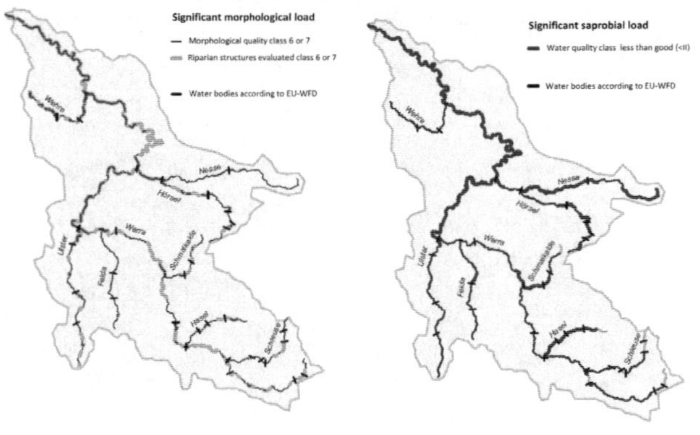

Figure 1. Significant morphological alterations (left) and significant saprobial load (right), indicating priority areas for measures (changed from Dietrich and Schumann, 2006).

2.2 Development of alternative environmental measures

The objective of river basin management according to the WFD is to reach a good ecological state of all water bodies by 2015, with some possible exemptions e.g. for heavily modified water bodies or due to long lasting sanitation or disproportionate costs. The WFD gives a framework for the development and 6-yearly update of river basin management plans (RBMP). The RBMP collects all measures, which were decided by the respective bodies. Within the Werra project, an exemplary RBMP was developed to address the environmental issues of the catchment that were introduced in Sect. 2.1. Different from the formal and final WFD RBMP, in this paper we provide alternative solutions for the selection phase of the decision process, which means that we present not a single solution but alternative measures, which follow the same overall objective. The following types of measures were considered and then designed for the water bodies in order to fulfil the objectives of the WFD:

- Improvement of the morphological conditions of the river
  - Ecological continuity by removing barriers or building fish passes
  - Creation of riparian buffer strips
  - Plantation of natural woods along the rivers
  - Removing bank reinforcement

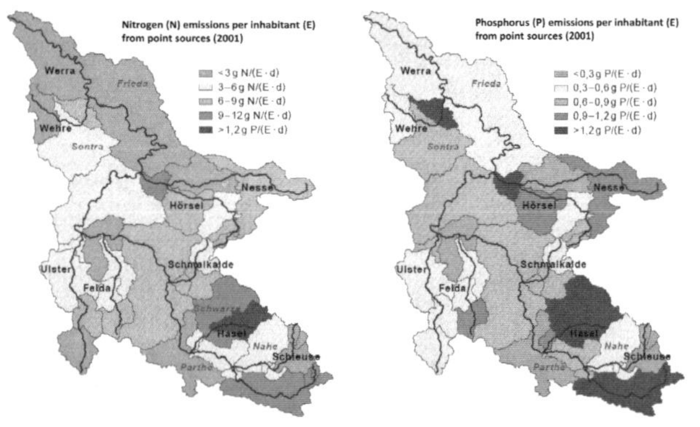

Figure 2. Nitrogen emissions (left) and Phosphorus emissions (right) from point sources, indicating priority areas for measures (changed from Dietrich and Schumann, 2006).

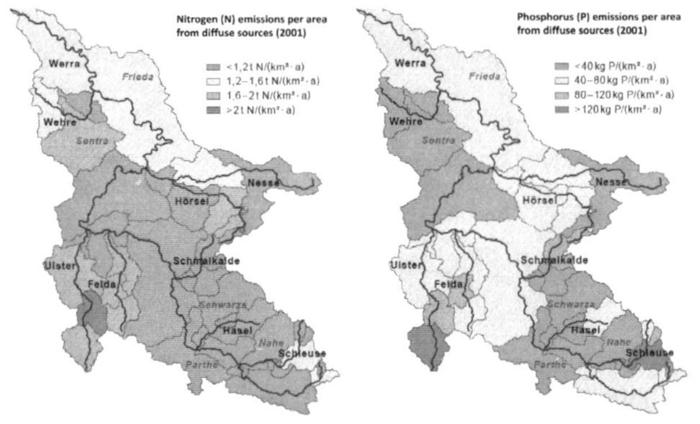

Figure 3. Nitrogen emissions (left) and Phosphorus emissions (right) from diffuse sources, indicating priority areas for measures (changed from Dietrich and Schumann, 2006).

- Removing of canalization
- Reduction of nutrient pollution from diffuse sources
  - Conversion of arable land to permanent grassland
  - Reduction of fertilizer use
  - Optimization of crop rotation
- Reduction of nutrient pollution from point sources
  - Connect dispersed settlements to sewage system
  - New stage of expansion of treatment plants
  - Increase capacity of sewage treatment plants
  - Construct new sewage treatment plants

2.3 Multi-criteria assessment of measures

All measures were evaluated with the following methods and criteria (Dietrich and Schumann, 2006):

Ecology: the WFD formulates aspiration levels for ecological criteria. If all combinations of measures for the RBMP can reach the objectives, there is no degree of freedom, which justifies an ecological decision criterion. Nevertheless, overfulfillment of the ecological status (very good status instead of good) provides an additional value and could be formulated as criterion. Furthermore, making use of WFD exemptions reduces the ecological value of the measures, which again justifies a decision criterion. In this study, all measures were planned to reach the aspiration levels only, and exemptions were negotiated separately. Thus, we do not investigate purely ecological criteria for the spatial aggregation issue. Ecological consequences of the implementation of measures were included in the ecological benefit analysis, which is human centred and expressed in monetary units.

Ecological benefit: the total economic value (TEV, Turner et al., 2003) includes use-values and non-use values of natural systems. All future values were discounted. We assumed a project lifetime of 20 years.

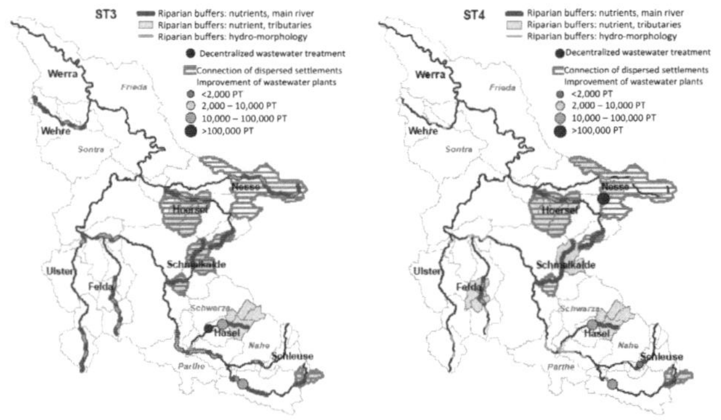

Figure 4. Priority areas for actions within the two strategies focussing on polluter pays principle (ST3, left) and cost-efficiency (ST4, right). PT is the total sewage plant capacity of population equivalent (changed from Dietrich and Schumann, 2006, ST1 and ST2 are shown in Dietrich and Funke, 2009).

Costs: the calculation of the costs of measures is based on literature values. All future costs were discounted.

Reduction of nutrient loads: the effect of measures targeting point sources and diffuse sources was simulated with a model chain agro-economy – eco-hydrology (SWAT) – water quality (RWQM). This included the implementation of typical crop rotations and application of organic and inorganic fertilizer and the change of crop management according to the measures as described in Sect. 2.2. We evaluated the reduction of the total mass of nutrients (N and P) within the catchment and the fulfilment of the requirements of a good ecological state within the water bodies.

Social acceptance: the Werra research project could not perform a complete participatory planning and decision making process. Nevertheless, the management of conflicts between stakeholders and the acceptance of measures among stakeholders should be regarded in the exemplary RBMP. Thus, a "cooperation index" was developed as a concept to represent the preference structure of stakeholders as a so-

Table 1. Categories and scales of the ecological benefit criterion (changed from Dietrich and Schumann, 2006).

Effects of measures	Conserving/improving biodiversity	Improving ecological quality of rivers	Recreation
Benefit category	Non-use value, indirect use value	Use value; Indirect use value	Direct use value
Scale	Water body, catchment	Single measure, water body, catchment	Water body, catchment
Evaluation technique	Benefit transfer	Replacement costs	Benefit transfer
Unit	Willingness to pay in € per household	Replacement value in € per ha riparian zone	Willingness to pay in € per visit * number of visitors

cial criterion within the decision process, and up to the highest level of aggregation. Based on a dynamic actor network analysis, the index is computed from questionnaires obtained from a representative group of stakeholders of different sectors (tourism, agriculture, nature protection, fishers). These expressed their general acceptance of the measures listed in Sect. 2 before measures were planned in detail. The cooperation index incorporates (a) the degree of being affected by potential measures; (b) the acceptance of the potential measures; (c) the relevance of the affected uses in the region and (d) the question of who will bear the costs (Hirschfeld et al., 2005; Dietrich and Schumann, 2006).

3 Combination of measures to catchment scale strategies

The final result of the project's planning are several alternative combinations of measures for the RBMP, which can be used as a decision matrix for multi-criteria evaluation and computation of a ranking based on preferences for the different criteria. This final matrix is computed for the entire Werra catchment. The aggregation of criteria from locations (single measures) via water bodies and their contributing catchments up to the catchment scale was complex and hat to be treated differently for the different criteria. For that reason, the pathway of aggregating criteria first was not possible. We decided to build combinations of measures according to different principles of strategic planning and policy making. Thus we called the final alternatives "strategies".

The aggregation of the criteria introduced in Sect. 2.3 faced the following issues:

Ecology: morphological riverbed improvement was mostly assessed local for single measures (creating or improving habitat structures), but there can be additional ecological effects at larger scale by habitat connectivity. The ecological continuity is very important for long distance travelling fishes. Therefore measures are most effective from downstream to upstream, whereas single measures in the middle of the catchment have reduced value when downstream connectivity is not given.

Ecological benefit: the TEV calculation includes components, which could not be obtained at the scale of single measures or water bodies, in particular by applying the benefit transfer from other studies (Table 1). This includes a superadditive benefit for developing the whole basin into a good ecological status.

Costs were attributed to single measures and aggregated by summation.

Reduction of nutrient loads: the transport of nutrients in the river network can have basin wide effects of measures. Upstream measures are more effective because they also reduce the inflow load of downstream areas. Here, the aggregation can not only count the criteria at the locations where the measures take place. With the model chain, it was possible to calculate basin wide effects of measures. Due to complex interactions in nutrient conversion and retention processes, the effects of upstream measures on downstream regions cannot be quantified separately for single measures. Thus the model performs the aggregation, and the final catchment outflow was taken as aggregated consequence of the different strategies for nutrient reduction.

Social acceptance: for getting a representative number of questionnaires, this kind of analysis could not be done separately for the single measures or water bodies. Even if the index value could be assigned for the measures within a water body, the cooperation potential cannot be related to the sub-set of stakeholders living in the respective water body sub-catchment.

The complexity of the problem does not allow a spatial multi-criteria aggregation at smaller scales than the overall catchment. Otherwise, much more detailed studies had to be performed regarding the ecological benefit and the social criteria. Furthermore, a decomposition of the upstream – downstream effects of nutrient reduction had to be done. As a consequence, we performed a coordinated catchment strategy development. This follows the following principles:

Table 2. Four catchment management strategies and respective decision criteria (from Dietrich and Funke, 2009; Dietrich and Schumann, 2006).

Strategy	Cost (mill. €) point sources/non-point sources/total incl. morphology	Total phosphorus reduction (t/a)	Ecological benefit (mill. €)	Cooperation index (polluter bears costs, higher values show more conflicts)	Number of water bodies with extended need for monitoring
ST1	9.2/2.8/56.0	32.0	104.3	2.07	10
ST2	6.3/13.9/64.2	30.1	112.1	1.73	10
ST3	7.0/8.1/102.3	33.6	118.1	2.73	11
ST4	4.8/6.6/55.6	32.6	127.2	1.80	13

ST1: first reduce point sources, then diffuse sources – the idea is a lower cost and better predictability of the consequences of measures at point sources;

ST2: first reduce diffuse sources, then point sources – the idea is to make use of combined beneficial effects from reducing diffuse sources by hydro-morphological structures like riparian buffers;

ST3: polluter oriented distribution of measures – the idea is to strictly follow the "polluter pays" principle;

ST4: most cost efficient allocation of measures – the idea is an economic optimization of the overall RBMP.

All the four basic strategies were computed and all criterion values were calculated with the respective methods. For the ecological benefit, the willingness to pay for biodiversity was calculated with a declining value. The measures for ecological continuity prefer the removal of structures where possible. Table 2 shows the results of the overall assessment.

The polluter oriented strategy ST3 does not only show the highest costs, but also the highest conflict potential because farmers expressed negative about the planned measures (Table 2). The optimized strategy ST4 is marginally cheaper than ST1, but shows better ecological benefit due to high valued riparian buffers. But, ecologists estimated that 13 instead of 10 resp. 11 water bodies need extended monitoring due to a marginal fulfilment of the ecological objectives, which (under uncertainty) can lead to the need for additional measures.

4 Conclusions

The results of the simulation and aggregation of criteria highlight problems in following the "polluter pays" principle and the WFD requirement of overall "cost efficiency of the program of measures" for the RBMP at the same time. A decomposition of larger scale measures and the redistribution of costs for measures with basin wide effects could be done by concepts like emission trading for nutrients. Then, the cost recovery happens at the polluters, but the spatial aggregation effects of nutrient reduction can be utilized in the best way.

Further work will be done in comparing different aggregation methods and different MCA methods. For very large basins, the study could be designed differently – e.g. the Werra basin is one part of the Weser basin, and the Fulda basin and the middle Weser and lower Weser sub-basins could be assessed separately. Then, the four larger parts of the whole basin could be aggregated in both ways (first space or first criteria).

Acknowledgements. This work is based on results of the joint research project "River Basin Management of the Werra River" (principal investigator: Andreas Schumann, Bochum). The German Federal Ministry of Education and Research (Bundesministerium für Bildung und Forschung, BMBF, FKZ 0330211) is acknowledged for funding of the project and the book publication of the results. We would like to thank all those involved in the original project for their collaborative efforts and for sharing data, knowledge and results.

References

Dietrich, J. and Schumann, A. H. (Eds.): Werkzeuge für das integrierte Flussgebietsmanagement - Ergebnisse der Fallstudie Werra, Konzepte für die nachhaltige Entwicklung einer Flusslandschaft Bd. 7, Weissensee-Verlag, Berlin, 470 S., 2006.

Dietrich, J., Schumann, A. H., and Lotov, A. V.: Workflow oriented participatory decision support for integrated river basin planning, in: Topics on System Analysis and Integrated Water Resource Management (IWRM), edited by: Castelletti, A. and Soncini Sessa, R., Elsevier, 2007.

Dietrich, J. and Funke, M.: Integrated catchment modelling within a strategic planning and decision making process: Werra case study, J. Phys. Chem. Earth, 34, 580–588, doi:10.1016/j.pce.2008.11.001, 2009.

Giupponi, C., Mysiak, J., Fassio, A., and Cogan, V.: Towards a spatial decision support system for water resource management: MULINO-DSS 1st release. Proceedings 5th AGILE Conference on Geographic Information Science, Palma (Spain), 25–27 April, 2002.

Hering, D., Moog, O., Sandin, L., and Verdonschot, P. F. M.: Overview and application of the AQEM assessment system, Hydrobiologia, 516, 1–20, 2004.

Herwijnen, v. M. and Rietveld, P.: Spatial dimensions in multicriteria analysis, in Thill, J.-C.: Spatial multicriteria decision making and analysis, Ashgate, 77–102, 1999.

Hirschfeld, J., Dehnhardt, A., and Dietrich, J.: Socioeconomic analysis within an interdisciplinary spatial decision support system for an integrated management of the Werra River Basin, Limnologica, 35, 234–244, 2005.

Nijssen, D. and Schumann, A. H.: Aggregating spatially explicit criteria: avoiding spatial compensation, International Journal of River Basin Management, 12, 87–98, 2014.

Turner, R. K., Paavola, J., Cooper, P., Farber, S., Jessamy, V., and Georgiou, S.: Valuing Nature: lessons learned and future research directions, Ecol. Econom., 46, 493–511, 2003.

Proc. IAHS, 373, 25–29, 2016
proc-iahs.net/373/25/2016/
doi:10.5194/piahs-373-25-2016

Open Access
PIAHS

Sustainable land and water management of River Oases along the Tarim River

Markus Disse

Chair of Hydrology and River Basin Management, Technical University of Munich (TUM), 80333 Munich, Germany

Correspondence to: Markus Disse (markus.disse@tum.de)

Published: 12 May 2016

Abstract. The Tarim Basin in Xinjiang province in northwest China is characterized by a hyper arid climate. Climate change and a strong increase in agricultural land use are major challenges for sustainable water management. The largest competition for water resources exists between irrigated fields and natural riparian vegetation, which is dependent on seasonal flooding of the Tarim River. In addition to numerous water management measures implemented by the Chinese government, the Sino-German project SuMaRiO (Sustainable Management of River Oases along the Tarim River) provided a decision support system based on ecosystem services for the Chinese stakeholders. This tool will help to implement sustainable land and water management measures in the next 5-year plan.

1 Project overview

In consequence of the rapid economic development in China, the cotton production in the Xinjiang Autonomous Region has risen sharply in recent decades. The farming industry and the natural vegetation are thus in conflict over the very limited water resources.

1.1 The Tarim Basin

The Tarim Basin is an intercontinental region with extreme arid climate. The Tarim River gets its water mainly from the meltwater from snow and glaciers of the surrounding mountains (see Fig. 1). In the Tarim Basin the annual rainfall can reach about 70 mm and the potential evaporation is about 2000 mm year^{-1}. People settled already in ancient times along the Tarim River, resulting in a series of oases forming the historical Silk Road. Today the oases are greatly increased mainly by cotton fields. The character of the landscape and the socio-economic situation of the population has changed considerably. Due to the high water demand of irrigation farming more and more water has been extracted from the river and the groundwater. This led to the fact that the Tarim River no longer reached its terminal lake (Taitema Lake) and dried up halfway. Secondly, the water is contaminated with salt, fertilizers and pesticides, which impacts not only on the floodplain forests and river ecology but also on human health.

1.2 SuMaRiO

Large areas of agricultural soils have become unusable through salinization, the river lining flood-plain vegetation has vastly receded, and important Ecosystem Functions/Ecosystem Services (ESF/ESS) such as providing uncontaminated irrigation water or attenuating dust and sand storms by vegetation have been severely decreased, or completely lost. The BMBF (German Federal Ministry of Education and Research) funded the interdisciplinary project SuMaRiO which aims to consider the complete land and water management system with its ESS/ESF in an integrated way taking into account diverse problem perceptions of the stakeholders.

The overall goal of SuMaRiO is to support oasis management along the Tarim River under conditions of climatic and societal changes by:

- developing methods for analysing ESF/ESS, and integrating them into land and water management of oases areas and floodplain forests;

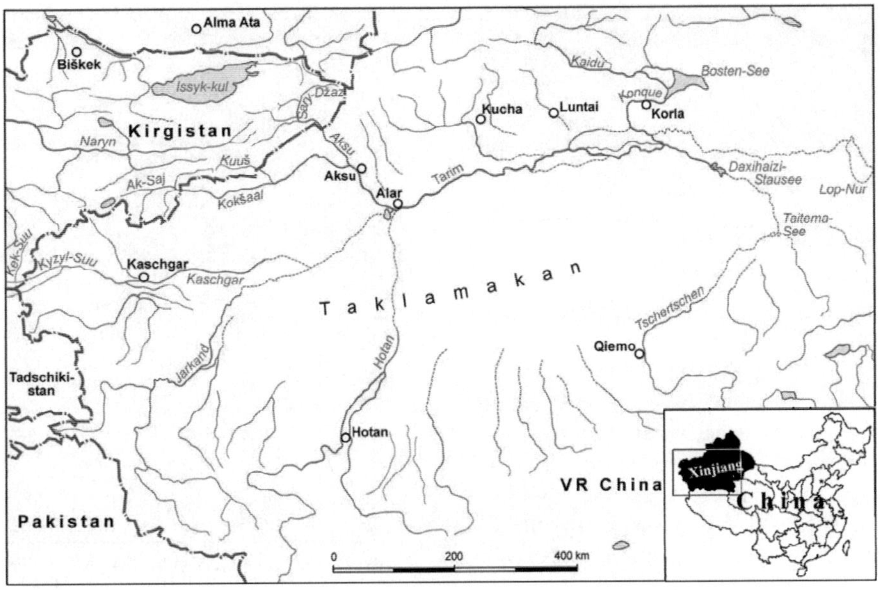

Figure 1. Location of the study area in Xinjiang/China.

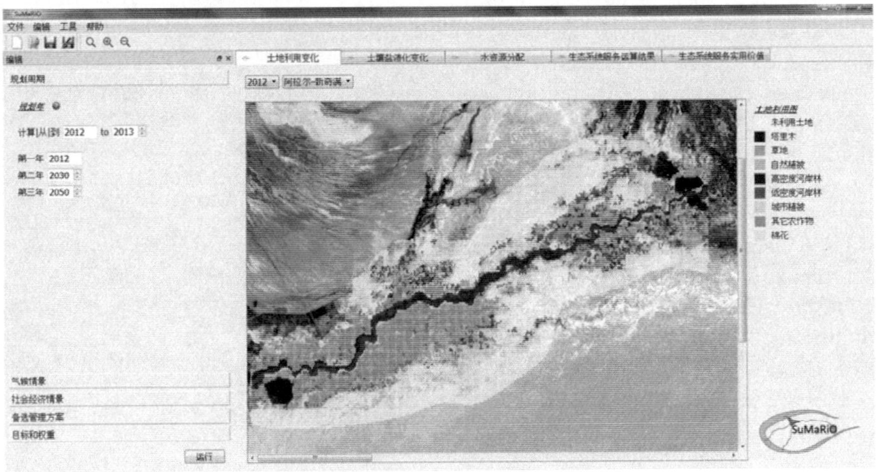

Figure 2. Exemplary representation of the Chinese user interface of the Decision Support System (DSS).

- involving stakeholders in the research process to integrate their knowledge and problem perceptions into the scientific process;
- developing tools with Chinese decision makers that demonstrate the ecological and socio-economic consequences of their decisions in a changing world;
- introducing participatory approaches into the development of sustainable management structures;
- jointly identifying options for optimizing economic, ecological, and societal utilities; and
- implementation of sustainable land management strategies.

This article focuses on the hydrological modelling of the water balance and the structure of the decision support system (DSS). The implementation phase will begin in 2016 when the Chinese partners will get training in the use of the developed tools.

1.3 Decision Support System (DSS) as a water and land management tool for the Tarim Basin

The demand for irrigation water for cotton cultivation exceeds especially in spring the discharge of the Tarim River.

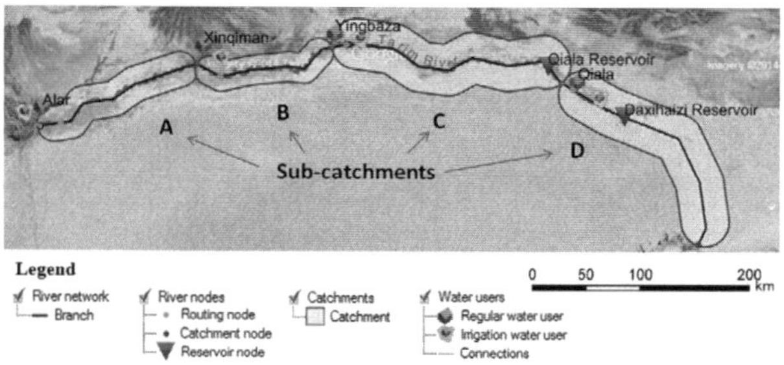

Figure 3. Aggregated sub-basins of the model MIKE HYDRO.

Therefore, the poplar forests (tugai) have declined significantly in recent decades and are endangered in the lower river. The so-called "green corridor" represents a (vitally) important barrier to the silting of the desert oases and the adjacent infrastructure. Other ecosystem services of the Tugai forests are biodiversity, climate regulation, carbon sink and tourism.

A decision support system (DSS) was designed and programmed to evaluate and compare different ESS indicators. The Potsdam Institute for Climate Impact Research (PIK) simulated different climate scenarios with the model SWIM and provided the discharge of Aksu, Hotan and Yarkant Rivers to their confluence in Alar (see Fig. 1) as input for the DSS. The user of the DSS can select one out of four climate scenarios and thus initiate the simulation for the corresponding hydrograph and associated evapotranspiration along the Tarim River.

To bridge the gap between science and practice, the DSS must be understood, accepted and used by Chinese stakeholders and decision makers. Therefore, the graphical user interface should be intuitive to use and in addition to English also available in Chinese (see Fig. 2).

2 Hydrological modelling

The study area for the water allocation model MIKE HYDRO begins in Alar and ends on Taitema Lake (see Fig. 1). MIKE HYDRO is a deterministic, semi-distributive and physically based model system that can be used for the simulation of water consumption, actual evapotranspiration, irrigation management, crop yields and plant growth. For large-scale modelling, the Tarim River has been divided into four main sections, which were marked by discharge gauges (see Fig. 3): Alar Xinqiman (A), Xinqiman-Yingbaza (B), Yingbaza-Qiala (C), and Qiala-Taitema Lake (D). In each of these sub-regions, the results are aggregated for water consumption and biomass production and passed to the DSS.

After calibration of MIKE HYDRO various water-saving scenarios were simulated (Yu et al., 2015). Important results are shown in Fig. 4 and Table 1. From Fig. 4 it is obvious that the plant available water, which can be determined from usable field capacity and effective root length for most agricultural crops must not be filled up to 70 %. For cotton a deficit irrigation up to about 40 % TAW is acceptable without having severe yield losses. Therefore, this type of irrigation should be promoted in Xinjiang. A second way to save water for agriculture is the irrigation technology "drip irrigation under plastic mulch" (DIUM). Table 1 shows a water saving potential of up to 40 %, mainly due to prevention of (unproductive) soil evaporation by covering the soil with plastic wrap ("mulch").

The results of MIKE HYDRO – supplemented by scenarios of water losses along irrigation canals – were finally passed as simplified regression relationships to the DSS. In this way it could be ensured that the impact of land management scenarios is based on a solid model and data base. The loss of accuracy by the transfer of the model results to the DSS are negligible. If necessary, additional MIKE HYDRO scenario calculations can supplement the database of the DSS.

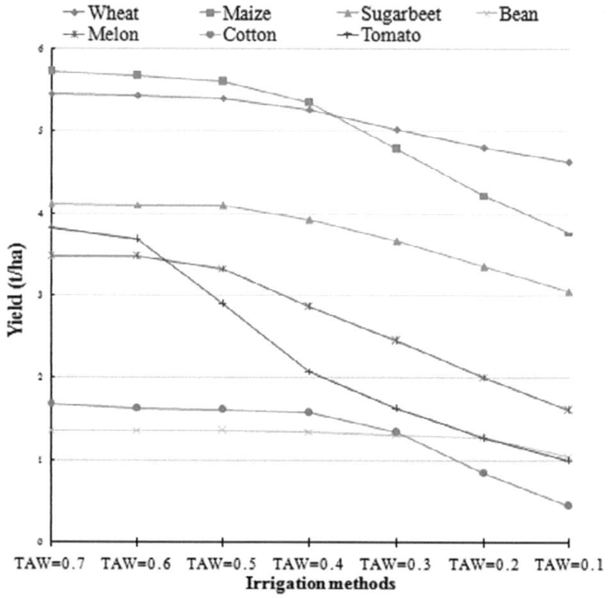

Figure 4. Irrigation scenarios depending on the plant available water (TAW). The average crop yield are based on a 3-year simulation period (see also Doorenbos and Kassam, 1979).

Table 1. Results of five different scenarios DIUM (Drip Irrigation Under Mulch) with SL = spray loss, WF = Wetting Fraction, WS = Water Saving.

% DIUM	% SL	% WF	% WS
10	46	91	6
30	38	73	17
50	30	55	25
70	22	37	32
100	10	10	40

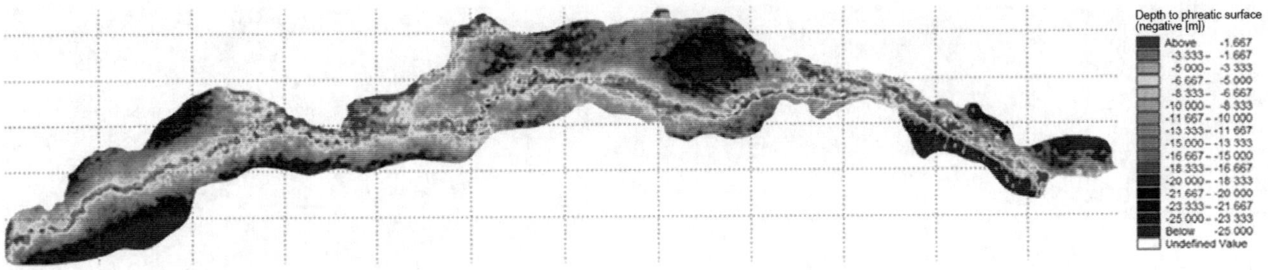

Figure 5. Simulated groundwater levels along the Tarim.

Table 2. ESS Indicators.

Ecosystem	Ecosystem Service	ESS Indicator
Agriculture	Provisioning services	Cotton Production [million t] Production of fruits [million t] Production of other crops [million t] Income of the farmer [million CNY]
Riparian Forest	Provisioning services Regulating Service	Biomass production [million t] Dust filtering through floodplain forests [kg] Control of sand drift by riparian forests [million t] Wind control [damping in % in 2 m height] Carbon storage [million t]
	Supporting Services	Main species [number]
Grassland	Provisioning services	Production of Apocynum [million t] Production of reeds [t]

Table 3. Socio-economic indicators in the DSS.

Production costs for cotton, fruits and other agricultural products	[CNY ha^{-1}]
Sales revenue for cotton, fruits and other agricultural products	[CNY ha^{-1}]

Table 4. Management indicators for the design of management measures in the DSS.

Proportion of drip irrigation to the entire irrigation area	[%]
Subsidy for drip irrigation	[CNY ha^{-1}]
Proportion of water allocation for households and commercial/industrial	[%]
Proportion of the flooded area for natural vegetation	[%]

Table 5. Land use types.

Fallow land	Grass	Reed
Light riparian vegetation Cotton	Dense riparian vegetation other agricultural cash crops	Urban vegetation

Since the effect of the respective scenarios has a direct impact on the groundwater situation along the Tarim River, the internal calculation of groundwater levels is indispensable (Keilholz et al., 2015). The DSS should be usable for subsequent users without a license, so that the software can be easily distributed and freely accessed to both public authorities and for educational institutions.

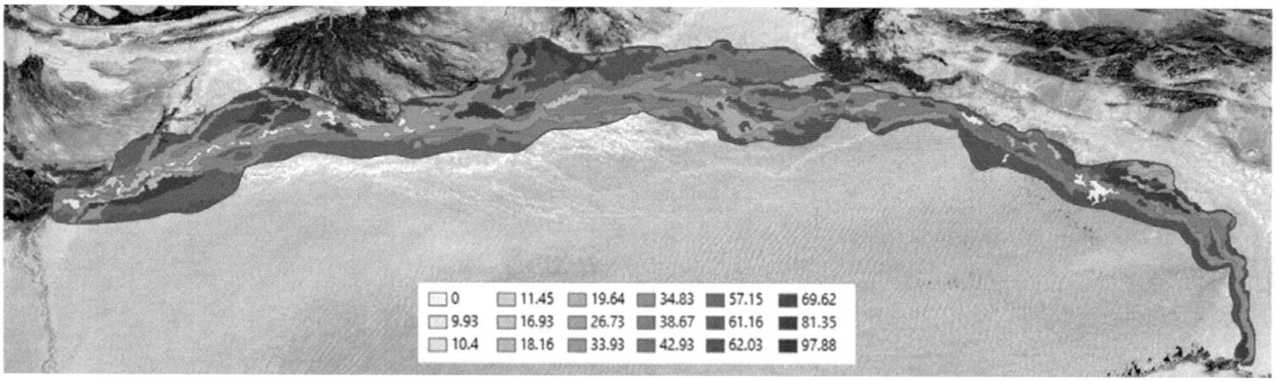

Figure 6. Distribution of the salt concentration in the soil in ECe [$\mathrm{mS\,cm^{-1}}$].

One common open-source code for simulating groundwater movement is the model MODFLOW. Due to its modular structure and the grid-based calculation, it was used for calculating the groundwater distribution. The resulting groundwater levels of MODFLOW due to artificial flooding of the Lower Tarim River and other management practices are coupled with the DSS with a spatial resolution of 500 m × 500 m (see Fig. 5).

3 Implementation of the Decision Support System (DSS)

Based on these two models and the contribution of the other work blocks within the project, the DSS could be designed and implemented.

The DSS is an indicator based tool that enables stakeholders and decision-makers to evaluate the consequences of their actions. Among selected scenarios of future climate and socio-economic development, the impact of planned management measures is determined by quantitative and semi-quantitative methods based on ESS indicators (Table 2). All indicators were conducted by SuMaRiO workshops (DSS workshops, stakeholder workshops) and has been classified as relevant. The following Tables 3–5 show examples of the required input data (i.e. management data) which formed the core for the evaluation of the management alternatives.

The impact on the ESS indicators and the available discharge along the Tarim River for each scenario year is calculated with the help of multivariate functions, lookup tables or fuzzy rules. The output is presented in tables and graphs. Based on these results and the objectives assigned to the ESS indicators by the user, a degree of target achievement between 0 and 1 is calculated, indicating the relative target fulfilment by each measure for each ESS indicator. Additionally, the DSS displays for each year to what extent there was an improvement/deterioration of the utility value compared to the first planning year. Based on this differentiated outputs the trade-offs between the ESS indicators can be analysed. Furthermore, it calculates the change in land use and soil salinization for each grid cell for the last year of the simulated time period (see Fig. 6).

The DSS with the coupled MODFLOW groundwater model will be delivered to the Chinese partners free of charge. In the coming 5-year plan (starting 2016) three pilot areas have been defined in Xinjiang. The DSS will be tested and further developed in these areas under the supervision of the Chinese Academy of Science (Xinjiang Institute of Ecology and Geography). Generally, the DSS is applicable for other regions, too, and is free available on request.

Acknowledgements. We thank the German Federal Ministry of Education and Research for the funding of the SuMaRiO consortium as part of the Sustainable Land Management Program (FKZ: 01LL0918L). We also thank our partners from the Xinjiang Institute of Ecology and Geography, Chinese Academy of Sciences who helped with our fieldwork and provided the basic data and information. And of course all the other Chinese and German SuMaRiO colleagues should be mentioned for SuMaRiO works only as a team.

References

Doorenbos, J. and Kassam, A. H.: Yield Response to Water, Irrigation and Drainage Paper, FAO, Rome, Italy, 33, p. 257, 1979.

Keilholz, P., Disse, M., and Halik, Ü.: Effects of Land Use and Climate Change on Groundwater and Ecosystems at the Middle Reaches of the Tarim River Using the MIKE SHE Integrated Hydrological Model, Water, 7, 3040–3056, doi:10.3390/w7063040, 2015.

Yu, Y., Disse, M., Yu, R., Yu, G., Sun, L., Huttner, P., and Rumbaur, C.: Large-Scale Hydrological Modeling and Decision-Making for Agricultural Water Consumption and Allocation in the Main Stem Tarim River, China, Water, 7, 2821–2839, doi:10.3390/w7062821, 2015.

Proc. IAHS, 373, 31–35, 2016
proc-iahs.net/373/31/2016/
doi:10.5194/piahs-373-31-2016

Managing saltwater intrusion in coastal arid regions and its societal implications for agriculture

Jens Grundmann[1], **Ayisha Al-Khatri**[1,2], **and Niels Schütze**[1]

[1]Institute of Hydrology and Meteorology, Technische Universität Dresden, Dresden, 01062, Germany
[2]Ministry of Regional Municipalities and Water Resources, P.O. Box 2575, Postal Code 112, Ruwi, Sultanate of Oman

Correspondence to: Jens Grundmann (jens.grundmann@tu-dresden.de)

Published: 12 May 2016

Abstract. Coastal aquifers in arid and semiarid regions are particularly at risk due to intrusion of salty marine water. Since groundwater is predominantly used in irrigated agriculture, its excessive pumping – above the natural rate of replenishment – strengthen the intrusion process. Using this increasingly saline water for irrigation, leads to a destruction of valuable agricultural resources and the economic basis of farmers and their communities. The limitation of resources (water and soil) in these regions requires a societal adaptation and change in behaviour as well as the development of appropriate management strategies for a transition towards stable and sustainable future hydrosystem states. Besides a description of the system dynamics and the spatial consequences of adaptation on the resources availability, the contribution combines results of an empirical survey with stakeholders and physically based modelling of the groundwater-agriculture hydrosystem interactions. This includes an analysis of stakeholders' (farmers and decision makers) behaviour and opinions regarding several management interventions aiming on water demand and water resources management as well as the thinking of decision makers how farmers will behave. In this context, the technical counter measures to manage the saltwater intrusion by simulating different groundwater pumping strategies and scenarios are evaluated from the economic and social point of view and if the spatial variability of the aquifer's hydrogeology is taken into consideration. The study is exemplarily investigated for the south Batinah region in the Sultanate of Oman, which is affected by saltwater intrusion into a coastal aquifer system due to excessive groundwater withdrawal for irrigated agriculture.

1 Introduction

Sustainable water management strategies are required to meet future water demands considering the predicted changes e.g. for population growth and climate as well as the diverse availability of water resources worldwide (WWAP, 2015). Since environmental systems are intensively used by human interventions, societal, environmental, and economic aspects needs to consider for evaluating appropriate management options for regional development (see IWRM definition, GWP-TAC, 2000). As a prerequisite, fundamental knowledge about linkages and feedbacks of the various system components is needed to portray the complexity of systems and the outcome of future scenarios adequately (Montanari et al., 2013). This includes considering different sectors and resources with their relevant players and stakeholders (Hering and Ingold, 2012). Each of them having different interests and mostly contradicting objectives which have to be formulated in a multi-objective, multiple-decision maker water management problem. To handle this complexity for planning of water management options and actions participatory approaches are recommended following a systematic procedure based on an appropriate system of well-defined objectives, criteria and indicators for evaluating, discussing, and negotiating different water management options (Soncini-Sessa et al., 2007).

The contribution focuses on water management of coastal (semi-) arid regions, as an example of water a management problem under limited resources. Besides a description of the system dynamics and the spatial consequences of adap-

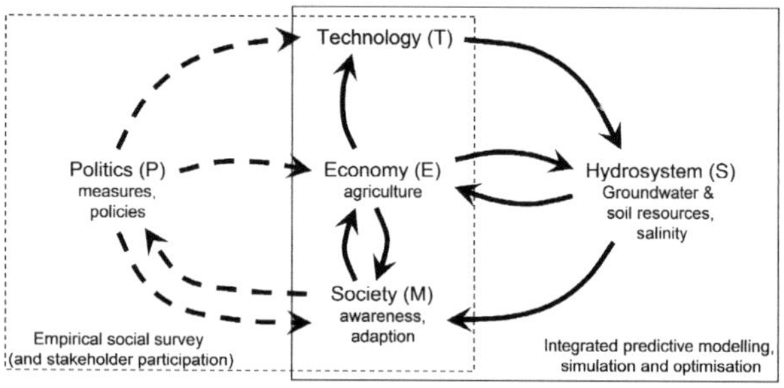

Figure 1. Schematic of system dynamics for managing saltwater intrusion in agricultural coastal plain.

tation on the resources availability, the contribution seeks to determine selected indicators for social, environmental and economic criteria in order to evaluate different water management options and policies. Results are obtained from empirical social surveys about stakeholders' behaviour and opinions on possible management interventions and an integrated predictive physically based modelling system of the groundwater-agriculture interactions. The study is exemplarily investigated for the south Batinah region in the Sultanate of Oman, which is affected by saltwater intrusion into a coastal aquifer system due to excessive groundwater withdrawal for irrigated agriculture.

2 Methods

2.1 Conceptualizing societal adjustment to saltwater intrusion

The principles of societal adaptation on saltwater intrusion are illustrated in Fig. 1. The linkages between the system components agricultural economy (E), hydrosystem (S), technology (T), society (M) and politics (P) are displayed by lines with arrows showing their direction. The adaptation as a consequence of saltwater intrusion is illustrated by solid lines. Due to an increased application of technology (T) for groundwater pumping as well as inefficient agricultural practices (E) from a society (M) showing less awareness about resources limitation, the groundwater-soil-hydrosystem (S) is damaged by salinity which feeds back to agriculture (E) by reduced crop yields and society (M) which loses its connection to water of good quality. Management feedbacks are illustrated by dashed lines. They are triggered by the request of the society (M) to politics (P), which may respond by different actions focusing/supporting on technology (T), agricultural practices (E), and society's behaviour (M), through e.g. technological countermeasures, agricultural incentives or education. The system components and their linkages are assessed and evaluated by different types of investigation and modelling (see rectangular boxes in Fig. 1). The responses of the system components on management interventions can be described by several criteria characterising the system states in the modelling system as well as the acceptance of interventions in the society represented by concerned stakeholders (e.g. decision maker (P), farmers (M)). Selected criteria as well as the methods of their determination are described in the following.

2.2 Integrated predictive modelling, simulation and optimisation

Physically based modelling of bio-physical processes is applied to portray the interactions between the agriculture and the groundwater-soil hydrosystem (S) and to obtain predictions of unknown future system states. Groundwater flow phenomena introduced by different geological units and water quality problems caused by saltwater intrusion are simulated using a density dependent transient groundwater model (Kolditz et al., 2012). Water abstraction in terms of water amount and quality (i.e. salinity of water) for agricultural production is realized by several vertical wells positioned at different distances from the sea to represent farm locations. The impact of management interventions represented by groundwater pumping schemes on the groundwater system is indicated by changes of salinity concentration s and water levels h on x_k observation points between the end t_n and the initial state t_1 of the simulation period. It is calculated by a sustainability index SI (Eq. 1), where a small value shows a high stability of the aquifer system.

$$\mathrm{SI}(t_n) = \sum_{k=1}^{\mathrm{end}} \frac{\mathrm{abs}(s(t_1, x_k) - s(t_n, x_k))}{s_{\max}} + \sum_{k=1}^{\mathrm{end}} \frac{\mathrm{abs}(h(t_1, x_k) - h(t_n, x_k))}{h_{\max}} \quad (1)$$

The adaption of the society (M) on the water quality is addressed by implementation of a moving inner boundary condition in the numerical density dependent groundwater model, which adjusts the locations for groundwater abstraction according to the position of the saltwater intrusion front controlled by thresholds of relative chloride concentration. The adaption process is performed for each management cycle within transient model simulations and allows for considering feedbacks with the society (M) by moving agricultural farms more inland or towards the sea if more fertile soils at the coast could be recovered.

The behaviour of agricultural farms and its economy (E) is simulated by a database of crop-water-production functions which is constructed by an extension of the OCCASION methodology (Schütze and Schmitz, 2010; Schütze et al., 2012) presented in Grundmann et al. (2012). It describes the relationship between the applied amount of water, its salinity and the produced yield for a given combination of crop/soil/climate under the assumption of optimal water application considering climate uncertainty. Assuming a rational behaviour of the farmers the Profit from agricultural

Table 1. Selected results of a social survey for potential management interventions.

Intervention measures	Farmers (mean)	DM's (mean)	Farmers (SD)	DM's (SD)	p value	Direction	System impacted
Introducing water quotas	3.47	1.88	1.391	1.023	0	D	E
Introducing water quotas with subsidies in form of equipments for modern irrigation systems	2.75	1.7	1.309	0.779	0	D	E, T
Use of treated wastewater for agricultural use, if it is available and the quality is acceptable	2.17	1.61	1.092	0.834	0.001	R	T
Encourage the farmer to reduce the withdrawal of groundwater by guidance and training	2.31	1.63	0.974	0.756	0	D	M
Implementation of centralized well fields which provides water in a good quality to farmers	2.42	2.27	1.193	1.162	0.458	R	T
Convince the farmer to change the type of crops to ones with lower crop water requirements	2.48	2.03	1.084	1.058	0.017	D	E
Encourage farmers to improve irrigation methods	2.02	1.45	0.968	0.61	0	D	T
Construction of more desalination plants for brackish and seawater for use in irrigation	2.14	3.09	1.082	1.311	0	R	T
Increase the effectiveness of water use by public awareness	1.55	1.46	0.561	0.636	0.424	D	M
Introduce water prices for pumped groundwater	3.92	2.48	1.276	1.133	0	D	E
Forming water managers groups	1.88	1.91	0.864	0.9	0.819	D	M, E
Stop all agricultural activities in the coastal zone	3.28	3.79	1.315	1.32	0.029	–	E, M
Leave the system as it is (No action)	4.09	4.19	1.205	1.305	0.649	R	–

production is calculated (Eq. 2), where CI_j and CP are fixed and variable costs for irrigated agriculture and groundwater pumping respectively, Pr_j the current prices for the cultivated crops $j = 1, \ldots, m$ which are produced from the acreage L_j for the cultivation period $i = 1, \ldots, n$, and Y_j the crop yield.

$$\mathrm{Profit} = \sum_{i=1}^{n} \left[\left(\sum_{j=1}^{m} \mathrm{Pr}_j Y_j(t_i) L(t_i) - \mathrm{CI}_j(t_i) \right) - \mathrm{CP}(t_i) \right] \quad (2)$$

Technology (T) is represented on the water abstraction side by different types of pumps having different economic and hydraulic behaviour (pump characteristic curves). On the water consumption side, different irrigation technologies are considered via their irrigation efficiencies and production costs.

The described simulator is used in a simulation–optimisation framework to find optimal solutions for water management considering contradicting objectives. Two main principles are applied to achieve a fast and reliable operation of the system: (1) customized tuned simulation-optimization based on surrogate models, which emulate the behaviour of the process models and (2) the decomposition of complex optimization problems into smaller, independently solved problems (Grundmann et al., 2014).

2.3 Analysis of stakeholder's behaviour and opinions

One of the most appropriate tools to explore the opinions of stakeholders in a domain is a social survey by distributing questionnaires and face to face interviews. After identification of relevant stakeholders e.g. water professionals, farmers from the study area and decision makers from different organizations, a pre-test survey is performed to collect stakeholder's ideas for possible management interventions. Afterwards questionnaires are designed and a social survey is performed to collect a combination of environmental, social and economic data as well as opinions regarding several management interventions from stakeholders. The obtained data are analysed statistically for each group separately by calculating primary statistic measures like mean and standard deviation. Applying these statistics on the list of management interventions indicates the degree of acceptance of an intervention by the group. Differences between opinions of groups are examined by an independent sample t test. If calculated p values are next to zero, significant differences between opinions of groups existing, which indicate a low implementation potential for the intervention measure from the social point of view (M, P).

3 Results and discussion

Selected results are presented for a study performed in the south Al-Batinah region of Oman. The region is affected by saltwater intrusion into an alluvial coastal aquifer system due to uncontrolled, excessive groundwater withdrawal for irrigated agriculture by numerous small scaled farms located in different distances from the sea. The marine saltwater intrusion process is ongoing since many years and led to a continuously increasing number of abandoned farms and a damage of valuable soil resources for agriculture.

A social survey was performed for collecting data and opinions from more than 130 stakeholders in the study area (Al-Khatri et al., 2014). Stakeholders were split into 2 groups; (1) farmers from the region and (2) decision makers and experts. Amongst other issues they were asked to rate several management options and interventions between 1 (strongly agree) and 5 (strongly disagree). Selected results for some interventions are presented in Table 1 by their mean, standard deviation (SD) and p values. Furthermore, it is indicated in the table if the management options and interventions either focus on water demand side (D) measures to reduce water consumption and use the resources more efficiently, or on water resources side (R) measures to increase the availability of water. Additionally, the system component (see Fig. 1) impacted by the measure is addressed. Results show that the need of improvement and implementing new management strategies is supported by all groups of stakeholders. In most cases, farmers are more likely to interventions of increasing water availability (R), while decision makers (DM's) were more likely to demand management (D). Opinions of farmers are mostly more diverse than DM's opinions shown by SD values. Results showing the functionality of the simulation-optimisation framework can be obtained from Grundmann et al. (2014).

4 Conclusion and outlook

The paper deals with a management problem under limited resources availability (water and soil) and quality, illustrated on the situation of arid coastal regions infected by marine saltwater intrusion and its implications on agriculture and society. A simulation-optimisation framework of bio-physical processes allows for multi-objective optimization of sustainable water resources and profitable agricultural management on farm and regional scale by providing optimal groundwater abstraction schemes and an appropriate choice of crop patterns regarding their salinity tolerance. Furthermore, trade-offs between contradicting objectives can be explored (Grundmann et al., 2013) and help to understand system behaviour and the impact of different management strategies on the society. Working with stakeholders by using empirical social surveys show the implementation potential of management interventions and foster a participatory process for supporting decision makers to take more informed decisions. Statistical analysis of stakeholder's responses helps to identify contradicting opinions between and within groups. Understanding the drivers behind these opinions offers the opportunity to trigger the acceptance of management interventions, which is part of our ongoing research. Due to complexity of processes and linkages in coastal systems there is no single solution which fits all constraints and requirements. Therefore, future work is focused on supporting management decisions by the development of indicators and criteria to evaluate management options preferably within a participatory stakeholder dialogue as well as to develop methods for identifying and scheduling appropriate combinations of measures considering different response times of system components.

Acknowledgements. The authors wish to thank the Ministry of Regional Municipalities and Water Resources of the Sultanate of Oman for supporting the joint Omani-German IWRM-APPM initiative.

References

Al-Khatri, A., Grundmann, J., v. d. Weth, R., Schütze, N., and Lennartz, F.: Analysis of Stakeholder's Characteristics for IWRM Implementation and its Application in the Batinah Region of Oman, 11th WSTA Gulf Water Conference, Muscat, Oman, 20–22 October, 2014.

Kolditz, O., Bauer, S., Bilke, L., Böttcher, N., Delfs, J. O., Fischer, T., Görke, U. J., Kalbacher, T., Kosakowski, G., McDermott, C. I., Park, C. H., Radu, F., Rink, K., Shao, H., Shao, H. B., Sun, F., Sun, Y. Y., Singh, A. K., Taron, J., Walther, M., Wang,W., Watanabe, N., Wu, Y., Xie, M., Xu, W., and Zehner, B.: OpenGeoSys: an open-source initiative for numerical simulation of thermohydro-mechanical/chemical (THM/C) processes in porous media, Environ. Earth Sci., 67, 589–599, doi:10.1007/s12665-012-1546-x, 2012.

Global Water Partnership Technical Advisory Committee (GWP-TAC): Integrated Water Resources Management, TAC Background papers, No. 4, Global Water Partnership, 2000.

Grundmann, J., Schütze, N., Schmitz, G.-H., and Al-Shaqsi, S.: Towards an integrated arid zone water management using simulation based optimisation, Environ. Earth Sci., 65, 1381–1394, doi:10.1007/s12665-011-1253-z, 2012.

Grundmann, J., Schütze, N., and Lennartz, F.: Sustainable management of a coupled groundwater-agriculture hydrosystem using multi-criteria simulation based optimisation, Water Sci. Technol., 67, 689–698, doi:10.2166/wst.2012.602, 2013.

Grundmann, J., Schütze, N., and Heck, V.: Optimal integrated management of groundwater resources and irrigated agriculture in arid coastal regions, Proc. IAHS, 364, 216–221, doi:10.5194/piahs-364-216-2014, 2014.

Hering, J. G. and Ingold, K. M.: Water resources management: what should be integrated?, Science, 336, 1234–1235, 2012.

Montanari, A., Young, G., Savenije, H. H. G., Hughes, D., Wagener, T., Ren, L. L., Koutsoyiannis, D., Cudennec, C., Toth, E., Grimaldi, S., Blöschl, G., Sivapalan, M., Beven, K., Gupta, H.,

Hipsey, M., Schaefli, B., Arheimer, B., Boegh, E., Schymanski, S. J., Di Baldassarre, G., Yu, B., Hubert, P., Huang, Y., Schumann, A., Post, D. A., Srinivasan, V., Harman, C., Thompson, S., Rogger, M., Viglione, A., McMillan, H., Characklis, G., Pang, Z., and Belyaev, V.: "Panta Rhei – Everything Flows": Change in hydrology and society – The IAHS Scientific Decade 2013–2022, Hydrol. Sci. J., 58, 1256–1275, 2013.

Schütze, N. and Schmitz, G. H.: OCCASION: A new Planning Tool for Optimal Climate Change Adaption Strategies in Irrigation, J. Irrig. Drain. E., 136, 836–846, doi:10.1061/(ASCE)IR.1943-4774.0000266, 2010.

Schütze, N., Kloss, S., Lennartz, F., Al Bakri, A., and Schmitz, G. H.: Optimal planning and operation of irrigation systems under water resource constraints in Oman considering climatic uncertainty, Environ. Earth Sci., 65, 1511–1521, 2012.

Soncini-Sessa, R., Weber, E., and Castelletti, A.: Integrated and participatory water resources management-theory, Elsevier, Amsterdam, 582 pp., 2007.

WWAP (United Nations World Water Assessment Programme): The United Nations World Water Development Report 2015: Water for a Sustainable World, Paris, UNESCO, 2015.

Proc. IAHS, 373, 37–43, 2016
proc-iahs.net/373/37/2016/
doi:10.5194/piahs-373-37-2016

Open Access
PIAHS

Local irrigation systems, regional hydrological problems and the demand for overarching solutions at the example of an irrigation system in the P.R. of China

David Nijssen[1], **Andreas H. Schumann**[2], **and Bertram Monninkhoff**[3]

[1]Federal Institute of Hydrology, 56002 Koblenz, Germany
[2]Ruhr-University, 44801 Bochum, Germany
[3]DHI-WASY GmbH, 12489 Berlin, Germany

Correspondence to: David Nijssen (nijssen@bafg.de)

Published: 12 May 2016

Abstract. The utilization of groundwater for irrigation purposes becomes problematic if groundwater recharge decreases through climate variability. Nevertheless, the degree of groundwater utilization for irrigation increases significantly in dry periods, when the amount of green water is strongly limited. With an increasing gap between water demand and supply, new water management activities are started, which are mostly directed to increase the supply, often by overuse of local resources. In many cases such local activities results in their summarization in side-effects, which worsen the hydrological conditions throughout a region. Step by step the spatial scale of water management measures has to be extended in such cases by implementation of water transfer systems. In this contribution this general scale problem of water management is discussed at the example of an agricultural region in the Province of Shandong (P.R. of China). The local irrigation systems and the options to increase the water supply at the local scale (e.g. by waste water reuse) are discussed as well as regional measures e.g. reservoirs or barrages in rivers to increase the groundwater recharge. For this purpose, several socio-economic and hydrological models were combined. It is shown how a change of water policy towards a demand management requires a new approach to spatial aspects. Here the question arises, how hydrological most effective measures can be allocated within a region. In the case study, a reduction of agricultural irrigation and a change of the crop structure would be essential to improve the groundwater conditions, which are impaired by ongoing sea-water intrusions. A model hierarchy, which is needed to answer such problems not only from the hydrological point of view, but also considering their socio-economic feasibility, are presented.

1 Introduction

Water resources management can be implemented on a wide range of legal, socio-economic and technical-geographical levels. The large variety of possible measures and their direct and indirect implications poses specific problems concerning their comparison. Often the solution of importing water from neighbouring regions and even over long distances seems to be a "deus ex machina" to solve local and regional water problems that had previously seemed impossible to solve. The question arises if regional water problems are really unsolvable or if the conditions to solve them are only too complex to determine a feasible solution. Like the sources of water supply, the different aspects of water demand have to be analyzed in a holistic way. To objectively compare water management measures to each other, their effects have to be commensurable. Therefore, comparing benefits of water utilizations across sectors should only be performed as an initial and very rough orientation aid and not as a final decision instrument as for instance in Addams et al. (2009). Cubic meters of saved water across sectors are not commensurable: their possible merits vary from a localized few hundred cubic meters reduction in irrigation water losses (Brouwer, 2005; Bjornlund, 2009) over city-wide thousands cubic meters of rainwater collection (Patel and Sah, 2008), to inter basin water transfers of millions of cubic meters of water (Ghassemi and White, 2007; Cai, 2008); from replacing water faucets in households (Governing Board of the National Research Council, 2005; National Development and Reform Commission, 2005) over province-wide crop changes (Bishop et al., 2010) to high-tech nuclear desalination plants (El-Dessouky and Ettouney, 2002; Wang et al., 2003). To compare these measures and their variations in implementation extent with each other according to their effect on the specific local hydrological conditions, a common denominator has to be found. This is especially important when the water demand and supply display a strong local variability. The water balance of a region can describe water problems that water managers are faced with on a larger scale, but on a local scale this instrument has its limitations. In this paper, spatial aspects are considered in three ways: IWRM measures are compared at the river basin scale to show the options to balance water demand with supply. The conditions for the applicability of the most effective measures at the local scale are analyzed in the second step. In the last phase the allocation of such measures within the river basin under consideration of their spatial efficiency is discussed.

2 Methodology

2.1 Objectively comparing IWRM measures on the river basin scale

The recommended spatial dimension of integrated water resources management (IWRM) is the river basin. A water balance, which combines the hydrological conditions of the natural system with its man-made alterations, is a useful tool to visualize the interlinkages of the hydrological and social system and determine the extent of human interventions. It has to quantify water flows between different sources and sinks caused by water consumption of different users within, as well as in and out of the study area. Besides being a useful tool to gain insight in the hydrological functioning of the catchment basin, the water balance can clearly point out the deficits or misbalances in water fluxes. The goal of IWRM measures is to alleviate the deficits in this sector. This can achieved through a wide range of measures, as the IWRM tool palette is multifarious. However, not all measures will have the same impact on the identified deficits.

The water balance can be expanded from a static, one-time inventory of quantities and fluxes to a dynamic system, where components or elements can be changed according to predicted effects of IWRM measures. Especially if the fluxes in the water balance are based on robust hydro(geo)logical models, the standard water balance is upgraded to a dynamic meta-model. A dynamic water balance can translate the effect of water saving measures across different components. Even if IWRM measures can save large quantities of water in one component of the water system seems to be promising, their actual effect has to be assessed with regard to the limiting, or misbalanced components and fluxes of the water balance. Integrating the IWRM measures in a dynamic water balance based on an ensemble of compartments and fluxes based on interactively updatable units, allows a testing and quantifying of all measures uniformly according to their impacts on specific targets like groundwater increase/decrease.

2.2 Specifying the feasibility of selected IWRM measures on a local scale

Once the optimal IWRM strategy to alleviate the water stress for the specific hydrosystem in the catchment basin is determined, the next steps are to substantiate and even optimize this strategy for the sector under consideration. The first step contains an assessment of the efficiencies of different measures to improve the hydrological conditions in the basin in general, the second one the determination of their feasibility as well as the framing conditions for implementation. This means that the rough level on which different IWRM measures were compared over the entire catchment basin and across all sectors, has to be specified in more detail. Optimally both, measures selection and quantification, should be re-evaluated on a sub-basin level and compared to the overall goal. If the overall goal is still not achieved, inter-basin measures should be evaluated, starting in the upstream direction.

2.3 Allocation of IWRM measures within a river basin

In a third level, the spatial allocation of most effective measures has to be specified. For this purpose, the spatial distribution of hydrological characteristics and the impact of feasible water management measures have to be combined to find most efficient locations for the implementation of local measures within the river basin.

3 Application

3.1 Study site

With insufficient water resources to meet rising water consumption in northern and eastern China, over-withdrawal of both surface and groundwater is widespread (Jiang, 2009) and becoming one of the most significant limiting factors affecting sustainable development in this region (Xia et al., 2007). Shandong Peninsula, in the east of the North China Plain, is facing a particularly grim situation with an average total of only 357 m^3 fresh water per capita (Wang et al., 2003), these freshwater resources are thus only 1/24-th of the world average (The World Bank, 2010). Caused by high rainfall intensities, along with an expansion of agriculture, the high silt content of rivers is reducing retention basin capacity by almost 6 Mio m^3 every year (Sun et al., 1998). The Huangshui river basin in Shandong (see Fig. 1) is typical for the region, with water demands exceeding the water resources in average by about 25 % (Geiger, 2005), thus combining all of

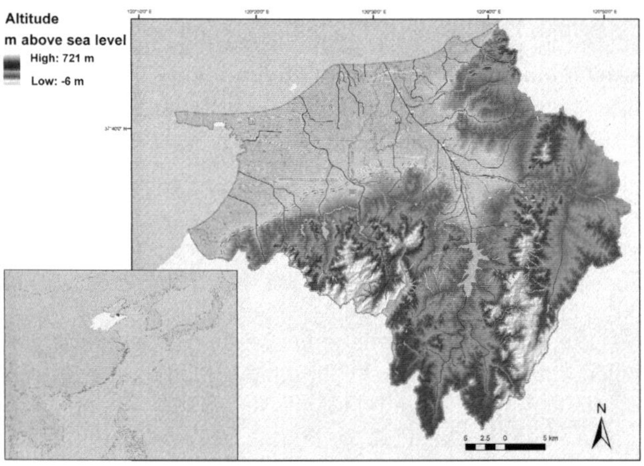

Figure 1. Location and altitude of the Huangshui river basin.

the problems that occur throughout Shandong, like groundwater depletion, dry rivers and sea water intrusion.

3.2 Assessments of IWRM measures at the regional scale

3.2.1 List of IWRM Measures

An extensive list of IWRM related management measures was collected by literature review and interviews with local and foreign IWRM experts. A water management measure was mainly defined as an approach to increase the availability of water resources compared to the current water intensity of demand and currently available supply. The details of the technical measures in the measure catalogue are compiled in an Excel workbook. Extent estimates and other calculations for many of the measures have been done on separate sheets within the same workbook. The IWRM measures were categorized in two ways; first according to water management fields or sectors, and then according to both their position in the management hierarchy as well as their effect on water management goals. More extensive Information about this catalogue can be found in Kaden and Geiger (2014).

3.2.2 Hydrological modelling

A water balance meta-model was set up to specify the relevance of water utilizations and their impact on the regional hydrological conditions. The three hydrological building blocks of the meta model concept consist of the simulation packages SIWA (Wirsing et al., 2010), WBalMo (Kaltofen et al., 2008), MIKE11 and FEFLOW.

SIWA is a 1-D empirical soil water model based on land-use, soil, slope and groundwater depth data. The model calculates spatially distributed monthly groundwater recharge and direct runoff rates. The input parameters for SIWA have been derived from the land-use, soil type map and groundwater depth of the year 2000. In total approx. 38 000 polygons with 1900 hydrotopes were generated. The average rainfall was derived for approx. 20 stations from the period 1960–2008 within the Huangshui River Basin. The evaporation could be derived from the pan-evaporation measurements made at WangWu reservoir station (Xu, 2006). Pan evaporation data were available on a monthly basis for the period 1980–2007.

To simulate the hydrological processes within the rivers and reservoirs a MIKE11 model has been set up. The software offers to simulate unsteady flow in river networks as well as looped networks using an implicit finite difference scheme. In total 25 rivers have been modeled with a total length of approx. 327 km. Within the model 13 weirs have been integrated, among the largest is the Wangwu reservoir.

Most important input data for the vertical discretization in FEFLOW© were 17 borehole descriptions as well as 2 longitudinal profiles and a detailed geological map (1 : 50 000). The conditions for transient calibration could be obtained from 31 groundwater observation points from the period 2000–2007. The river system was implemented as line-in elements in the mesh. The horizontal mesh consists of 34 547 elements and 17 651 nodes. In vertical direction 10 layers represent the geological characteristics within the model domain. The recharge could be obtained from the SIWA model. The final input parameter for the model is the extraction, both for agricultural, industrial and domestic use. Information for the years 2000–2007 was available per land use for each individual district. The main calibration period for the numerical models supporting the water balance are was 2000–2007. Two examples, which demonstrate the fit between observed and simulated water levels are shown in Fig. 2.

3.3 Considerations of the water demand within the water balance meta-model

A very important interface between hydrological conditions and human water use is agriculture. Falkenmark (1999) argued effectively for a broadening of the water sector's attention from the conventional "blue" water in aquifers and rivers to include also the evapotranspiration flow of "green" water stored in the unsaturated upper soil zone and involved in rain fed biomass production in agriculture, forestry, and natural vegetation systems. In the water balance of the Huangshuihe River, this demarcation of blue and green water is also integrated and visualised in blue and green colours. About 26.8 % of the rainfall amount in the Huangshui-basin is converted to blue water, which is in line with global averages. The green/blue water determination has been an extremely useful illustrative concept in many situations where the role of land use in water resources management needs to be highlighted.

The input data of the meta-model water balance can be changed according to different study areas. They are categorized in three categories:

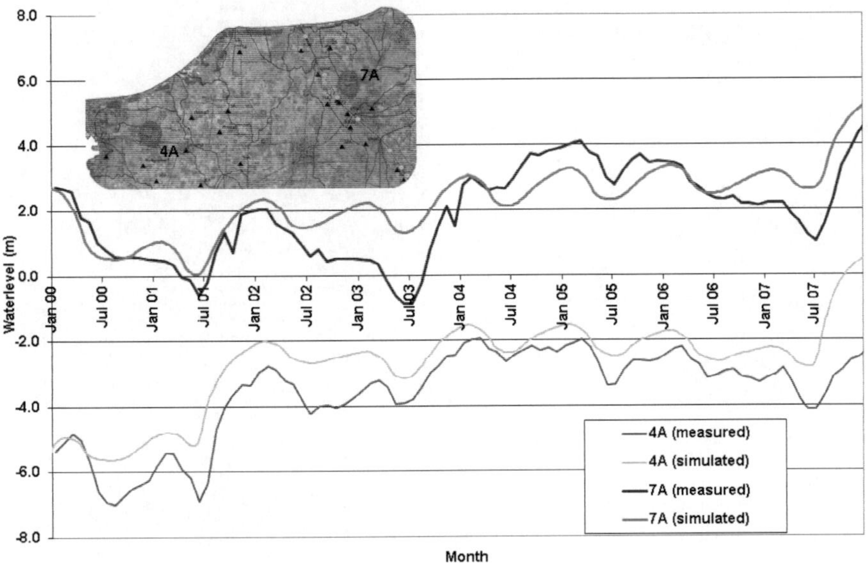

Figure 2. Measured and simulated results for two groundwater monitoring stations.

- Base parameters, like land-use categories, precipitation and PET, where mainly long-time changes are considered
- Natural balance, partitioning keys, properties of the local topography, and land-uses like runoff, percolation and outflow
- Anthropogenic balance, municipal, industrial and agricultural water needs and their respective extraction and discharge sources.

The resulting water balance is visualized as a flowchart (see Fig. 3).

Decision makers can explore the effects on the water balance of changing (decreasing, increasing) needs, proportions in the extraction sources, etc. This allows an improved understanding of water related problematic beyond each individual sector and throughout the entire water cycle. The different steering variables can be accessed through a user interface. For instance, the base parameters display the current amounts of water use, precipitation, evapotranspiration (including their increase by irrigation) and water storage. Changing e.g. the agricultural area automatically changes the related water balance factors, the supply from ground- and surface water, the transpiration and the groundwater balance. This way, rough ideas about possible trends like increases in municipal water demand, decreases in agricultural areas, etc. can be explored. Also trends in decreasing precipitation or increasing radiation can very roughly be simulated. This allows the decision maker to develop a feeling of the magnitudes and effects of, for instance, increasing surface water use in the agricultural, municipal or industrial sectors instead of groundwater use.

As can be seen from Fig. 3, the element with a yearly structural deficit of 14 million m^3 $year^{-1}$ is the groundwater component. To alleviate this unbalance in the hydrosystem, the IWRM measures are to be compared according to their effects on the recharging the groundwater deficit.

3.4 Translating the IWRM measure to a common denominator to estimate their local feasibility

Since the interactive water balance is to "norm" the effects of the measure catalogue, the balance was opted to be programmed in the same software package, Microsoft Excel©. The measures were dimensioned in million m^3 $years^{-1}$. Measures have to be integrated in three levels. First, the measures have to be implemented. Not all measures can be implemented simultaneously, or can reach their optimum when combined with others. Each measure was attributed a Boolean operator that can be switched to active or not. Secondly, the measures have a certain impact magnitude that is estimated, as the average value between minimum and maximum. Using intervals, the amounts of water that are shifted from one or more compartments to one or more other compartments are estimated according to the implemented measure. The third level of integration is allowing the variables to vary between these respective minima and maxima and finding the optimum accordingly. A routine was written that, using the dynamic water balance algorithms evaluates all measures and their ranges individually and in combination and writes out their costs and benefits, in this case groundwater quantities. In order to explore the end results, a GUI was created that allowed de-selection of individual parameters, in/exclusion of categories, the maximum available investment

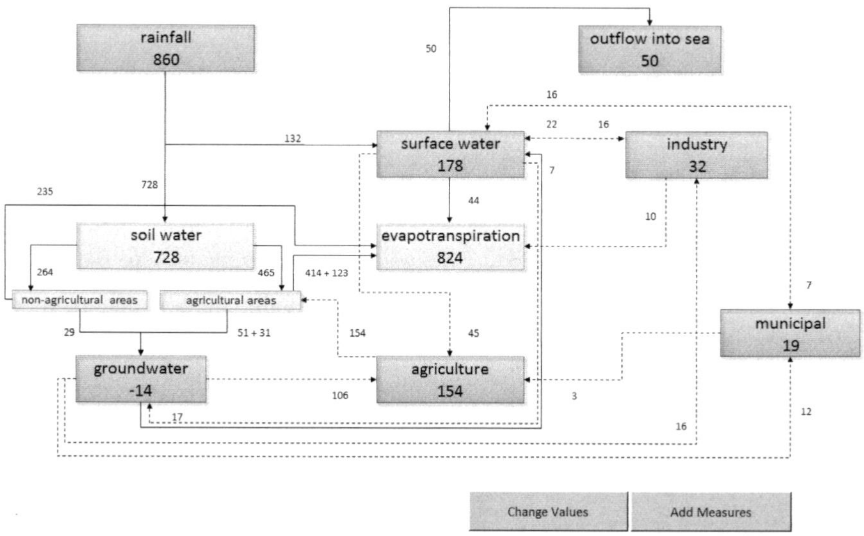

Figure 3. The water balance meta-model depicted as a flowchart with the main components and water flows.

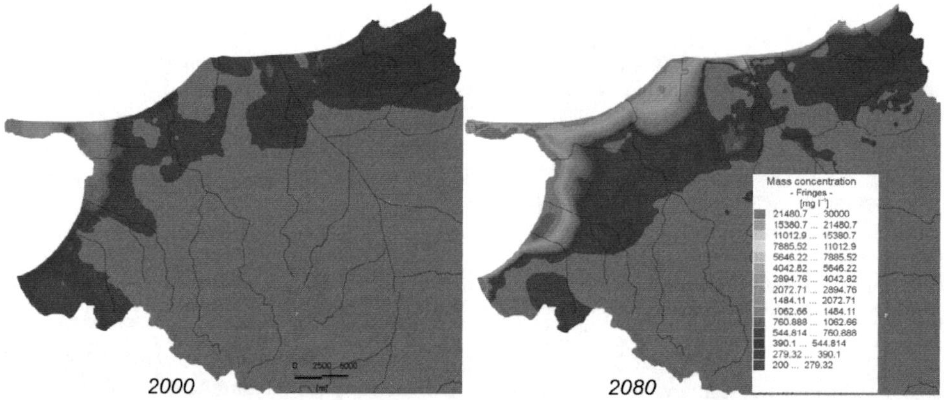

Figure 4. Saltwater intrusion at the start of simulations (year 2000) and after 80 years.

amount and the minimum required effect upon the most relevant compartment.

3.5 Specifying the selected IWRM measure on a local scale

After the objective selection of the optimal IWRM measure (here the reduction of the agricultural water consumption), feasible strategies to change it under consideration of socio-economic effects have to be specified. Details for this step are presented in Nijssen (2013) and Schumann and Nijssen (2014). The local dimensions and extents are to be modelled based on the recharge and run-off model (SIWA) and the water allocation model (WBalMo).

3.6 Allocation of measures within the river basin

As result of the previous analysis, the reduction of agricultural consumption has to be allocated to the parts of the river basin, where such measures would be most efficient. As the sea water intrusion is a creeping problem, which results from the overutilization of groundwater for irrigation and an increasing deficit of groundwater recharge, a more detailed analysis of the long-term development of groundwater conditions became necessary. To analyse the saltwater intrusion caused by the observed over-extraction, a long-term simulation was performed for about 80 years. This resulted in an alarming saltwater and groundwater level distribution at the end of the simulation (Fig. 4).

To specify feasible strategies to prevent this development, a model-based optimal allocation of the measure "reduction of the agricultural water consumption" was estimated. This

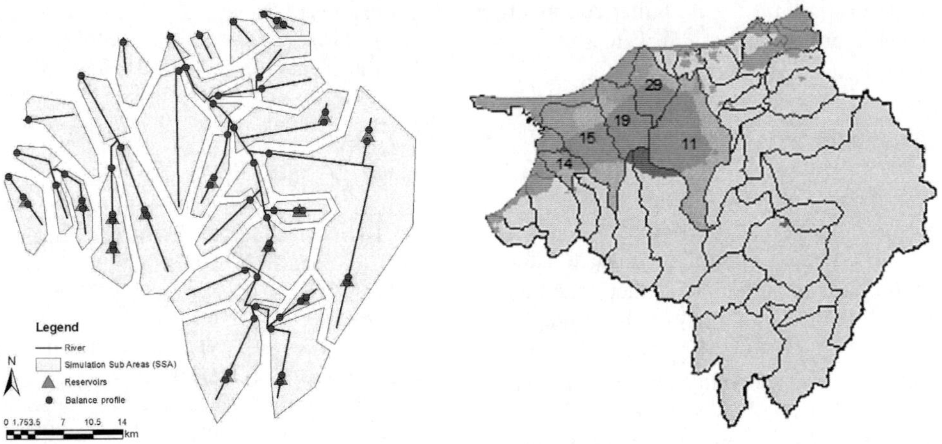

Figure 5. Left: Subdivision of the basin into water balance units, connected by river links, Right: Areas where the water extraction has to be reduced urgently (dark grey); the areas in orange show strong saltwater intrusion (mass concentration >400 mg L^{-1}, after 40 years of simulation without intervention); the areas in blue indicate a groundwater depression cone (hydraulic heads ≤3 m, after 40 years of simulation without intervention).

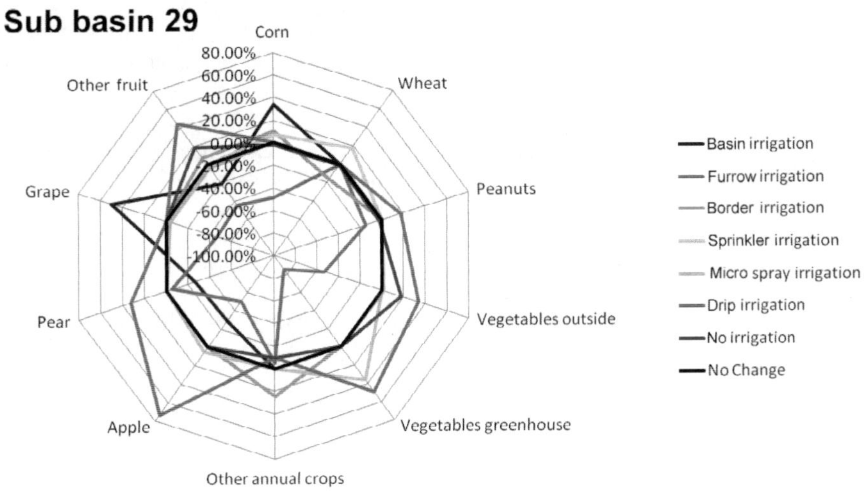

Figure 6. Proposed irrigation changes for sub basin 29, optimised to reduce the irrigation water demand whilst retaining the overall basin yield, according to local soil properties.

has to be done under consideration of the local flow conditions. For this purpose the river basin was subdivided into sub-areas (Fig. 5) which were described in their water balance by the model WBalMo. The computed groundwater recharge was the input into the model FEFLOW. Both models were used in combination to estimate the effects of with several scenarios of reductions of the agricultural water consumption with different spatial focusses. In Fig. 5 (right) the hotspot of groundwater problems, a groundwater depression cone after 40 years of simulations, is located in the subbasins 11, 14, 19, 15, 19 and 29.

Several scenarios were simulated with the water balance model WBalMo and the groundwater model FEFLOW (Business as usual, Climate change, several agricultural developments with different strategies to reduce the water demand for agriculture). Under consideration of the maximum of feasible reductions of irrigation water demand, which were estimated under point 3.4, strategies to reduce the irrigation water demand for all hot-spot basins were derived (Fig. 6). It became evident that the stabilization of groundwater conditions could not reached by these localized measures. Thus the neighbouring upstream sub-basins were also considered. The impacts of the increase of groundwater recharge by changed agriculture upstream were analysed with the hydrological models to find feasible but effective measures. The final result of the combined model approach showed that for the Huangshui river basin, irrigation changes, crop changes, and the implementation of water pricing on top of irrigation changes, rank as the top three measures. Other promising measures for the local hydrological system were swale infiltration and dam construction.

4 Summary and conclusions

Spatial aspects are a basic problem of water management planning. Local interventions have very different regional efficiencies. Thus, the allocation of hydrological most effective local measures to improve the regional water balance is a non-trivial issue. Different steps of analyses are needed to characterize the decision space considering the hydrological effects and the socio-economic feasibility of measures. The local impacts on the water balance have to be assessed in the regional context. Efficient measures in the regional context have to be allocated in such a way, that their local feasibility is accompanied by with a regional effectivity.

References

Addams, L., Boccaletti, G., Kerlin, G., and Stuchtey, M. M.: Charting our Water Future, Economic frameworks to inform decision making, 2030 Water Resources Group, McKinsey and Company, 2009.

Bishop, C. D., Curtis, K. R., and Kim, M.-K.: Conserving water in arid regions: Exploring the economic feasibility of alternative crops, J. Agr. Syst., 103, 535–542, 2010.

Bjornlund, H.: Incentives and Instruments for Sustainable Irrigation, WIT Transactions on State-of-the-art in Science and Engineering 37, 240 pp., ISBN: 978-1-84564-406-2, 2009.

Brouwer, C.: Irrigation Water Management: Irrigation Methods, Irrigation water management: training manuals 6, FAO, Rome, 2005.

Cai, X.: Water stress, water transfer and social equity in Northern China-Implications for policy reforms, J. Environ. Manage., 87, 14–25, 2008.

El-Dessouky, H. T. and Ettouney, H. E.: Fundamentals of Salt Water Desalination, Elsevier Science B.V., 2002.

Falkenmark, M.: Forward to the future: a conceptual framework for water dependence, AMBIO, 28, 356–361, 1999.

Geiger, W. F.: Cost-effective measures for flood control and groundwater recharge in coastal catchments. Research Project Founded by BMBF, Förderkennzeichen: CHN 04/015, Bonn, 2005.

Ghassemi, F. and White, I.: Inter-Basin Water Transfer, Case Studies from Australia, United States, Canada, China and India, Cambridge University Press, 2007.

Governing Board of the National Research Council: Water Conservation, Reuse and Recycling Proceedings of an Iranian-American Workshop, Inter-academy workshop on water resources management, conservation, and recycling, Tunis, 2005.

Jiang, Y.: China's water scarcity, J. Environ. Manage., 90, 3185–3196, 2009.

Kaltofen, M., Koch, H., and Schramm, M.: Water management strategies in the Spree region upstream of Berlin, in: Integrated Analysis of the impacts of Global Change on Envronment and Society in the Elbe basin, edited by: Wechsung, F., Kaden, S., Behrendt, H., and Klöcking, B., Weißensee Verlag, Berlin, ISBN: 978-3-89998-145-2, 2008.

Kaden, S. and Geiger, W. (Eds.): Overall-effective measures for sustainable water resources management in the coastal area of Shandong Province, PR China, Concepts for Sustainable River Landscape Development, Vol. 10, 346 pp., ISBN: 978-3-89998-212-1, 2014.

National Development and Reform Commission (NDRC): China Water Conservation Technology Policy Outline, Announcement No. 17, 2005.

Nijssen, D.: Improving Spatiality in Decision Making for River Basin Management, Schriftenreihe Hydrologie – Wasserwirtschaft 27, RUB, Ruhr-Universität Bochum, 147 pp., 2013.

Patel, A. S. and Sah, D. L.: Water Management – Conservation, Harvesting and Artificial Recharge, New Age Int., New Delhi, 2008.

Schumann, A. and Nijssen, D.: Shortage and surplus of water in the socio-hydrological context, Proc. IAHS, 364, 292–298, doi:10.5194/piahs-364-292-2014, 2014.

Sun, Z., Song, C., Pang, H., and Li, W.: Eco-restoration engineering and techniques in the Muyu reservoir watershed in Shandong, P.R. of China. Ecol. Eng., 11, 209–219, 1998.

The World Bank: Development and Climate Change, World Development Report, Washington, ISBN: 978-0-8213-7987-5, 2010.

Xia, J., Zhang, L., Liu, C., and Yu, J.: Towards better water security in North China, Water Resour. Manag., 21, 233–247, doi:10.1007/s11269-006-9051-1, 2007.

Xu, Ch.-Y.: Analysis of spatial distribution and temporal trend of reference evapotranspiration and pan evaporation in Changjiang (Yangtze River) catchment, J. Hydr., 327, 81–93, 2006.

Wang, Y., Zhang, Y., and Zheng, W.: Water resources and potential of seawater desalination in Shandong peninsula, Desalination, 157, 269–276, 2003.

Wirsing, T., Deinlein, W., Hofmann, B., Maier, M., and Roth, K.: Modellierung des Bodenwasserhaushalts als Prognose-Instrument zur praxisnahen Abschätzung der Umweltauswirkungen einer Grundwasserabsenkung, Hydrogeologie, Fachberichte, gwf-Wasser|Abwasser, 2010.

Proc. IAHS, 373, 45–49, 2016
proc-iahs.net/373/45/2016/
doi:10.5194/piahs-373-45-2016

Reservoir impacts downstream in highly regulated river basins: the Ebro delta and the Guadalquivir estuary in Spain

María J. Polo[1], Albert Rovira[2], Darío García-Contreras[1], Eva Contreras[1], Agustín Millares[3], Cristina Aguilar[3], and Miguel A. Losada[3]

[1]Andalusian Institute of Earth System Research, University of Cordoba, Córdoba, Spain
[2]Aquatic Ecosystems, IRTA, Sant Carles de la Rápita, Tarragona, Spain
[3]Andalusian Institute of Earth System Research, University of Granada, Granada, Spain

Correspondence to: María J. Polo (mjpolo@uco.es)

Published: 12 May 2016

Abstract. Regulation by reservoirs affects both the freshwater regime and the sediment delivery at the area downstream, and may have a significant impact on water quality in the final transitional water bodies. Spain is one the countries with more water storage capacity by reservoirs in the world. Dense reservoir networks can be found in most of the hydrographic basins, especially in the central and southern regions. The spatial redistribution of the seasonal and annual water storage in reservoirs for irrigation and urban supply, mainly, has resulted in significant changes of water flow and sediment load regimes, together with a fostered development of soil and water uses, with environmental impacts downstream and higher vulnerability of these areas to the sea level rise and drought occurrence. This work shows these effects in the Guadalquivir and the Ebro River basins, two of the largest regulated areas in Spain. The results show a 71 % decrease of the annual freshwater input to the Guadalquivir River estuary during 1930–2014, an increase of 420 % of the irrigated area upstream the estuary, and suspended sediment loads up to 1000 % the initial levels. In the Ebro River delta, the annual water yield has decreased over a 30 % but, on the contrary, the big reservoirs are located in the main stream, and the sediment load has decreased a 99 %, resulting in a delta coastal regression up to 10 m per year and the massive presence of macrophytes in the lower river. Adaptive actions proposed to face these impacts in a sea level rise scenario are also analyzed.

1 Introduction

Highly regulated catchments can be found over the world, being Spain one of the countries with more water storage capacity. The decrease of flood events and the storage of water results in significant decreases of both the freshwater and sediment supply to the final transitional waters, with morphological, water-quality related and ecological impacts, especially enhanced by the current sea level rise. The social benefits of dam construction may be hidden by these environmental effects that finally affect the economy and the social welfare (Morris and Fan, 2000).

This work shows the current regime of freshwater flow and suspended sediments in both the Ebro and the Guadalquivir River basins, in the context of their Hydrological Plans, and some of the adaptation actions proposed at each site.

2 The Ebro River basin: current conditions

The Ebro River has a contributing area of approximately 85 900 km^2 with a total river length of 970 km. It is the biggest catchment in Spain and one of the most European regulated rivers, with three big in-stream reservoirs in the lower parts of the river, draining to the Mediterranean Sea through a deltaic formation, the Ebro Delta (Fig. 1). The basin goes through nine different regions in Spain, with different competences on water uses. The allocation of water

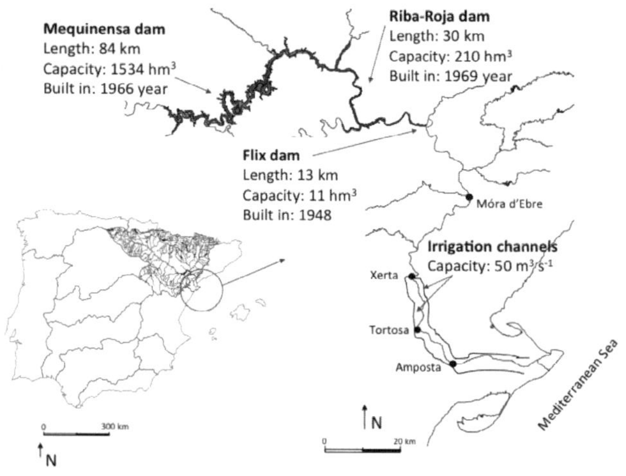

Figure 1. The Ebro River basin in Spain. Location and drainage network (left), the final reach of the river and the Ebro Delta in the Mediterranean Sea (right) with location of dams and main irrigation channels.

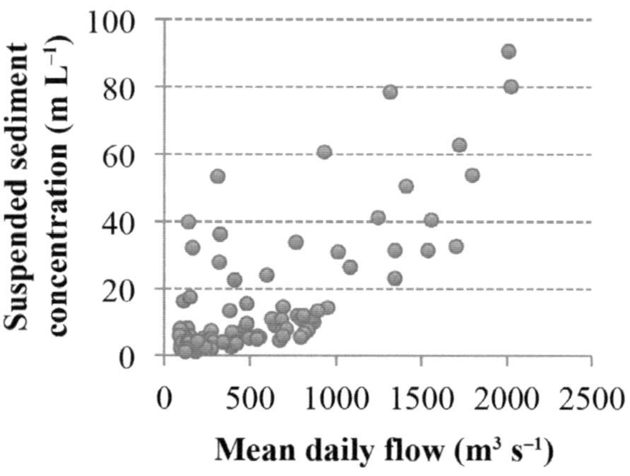

Figure 2. Rating curve of water flow-suspended sediment concentration at Tortosa (Fig. 1) in the lower Ebro River, obtained from data during March 2008–December 2015.

resources from the basin is a source of conflicts, not only due to the water uses within and upstream from each region, but also because of the established water uses recently approved by the Spanish Government in the new Ebro River Basin Plan.

The mean annual average precipitation in the catchment, with strong snow influence, ranges from over 2000 mm in the Pyrenees to less than 400 mm in the internal arid areas, with mean annual average temperature of 12.5 °C but strong local gradients across the river basin (www.chebro.es, last access: 30 January 2016). The annual runoff is highly variable as usually found in Mediterranean catchments, but it has dramatically decreased during the 20th century due to increasing water uses and the river regulation from reservoirs (Gallart and Llorens, 2004). The mean annual flow at Tortosa (Fig. 1), close to the Mediterranean Sea, for the period 1981–2010 is $289\,m^3\,s^{-1}$ (Rovira et al., 2015), value well below the estimated $464\,km^3\,s^{-1}$ under natural regime conditions given by the current hydrological plan (www.chebro.es). Water resources are allocated to irrigation, hydropower production, and urban uses. Up to 190 reservoirs, most of them built during the 1950–1975 period, regulate 57 % of the mean annual water yield (Batalla et al., 2004), regulate the river flow with a total impoundment capacity of $7524\,hm^3$ ($0.088\,hm^3\,km^{-2}$).

The reservoir network and the existing dams close to the delta capture most of the sediment loads from the catchment (Rovira et al., 2014), which adds to the effects of the current trend of sea level rise; both processes and the loss of freshwater delivery affect the subsidence dynamics of the delta, as many authors state (Ibáñez et al., 2010). The final in-stream system of dams formed by the Mequinensa-Ribaroja and Flix reservoirs (see Fig. 1) virtually trap all sediments transported by the Ebro River mainstream and the Segre-Cinca tributaries. In consequence, the total sediment load debouched to the Ebro delta and the Mediterranean Sea has dramatically dropped from approximately 20–30 million $t\,yr^{-1}$ at the end of the 19th century (Varela et al., 1986; Ibáñez et al., 1996), to 0.99 million $t\,yr^{-1}$ at present (Rovira et al., 2015). Figure 2 shows the relationship between the mean water discharge and suspended sediment concentration at Tortosa (Fig. 1), located approximately 40 km upstream from the delta and 15 km from the estuarine zone, obtained from daily data during 2008–2015. The values are well below the reported concentrations during floods at the beginning of the past century, which ranged between 1 and $10\,g\,L^{-1}$ (Gorría, 1877; Ibáñez et al., 1996).

The loss of sediment supply has resulted in a coastal regression greater than 10 m per year at the river mouth, where 150 ha of wetland were lost between 1957 and 2000. This sediment deficit is augmented by the reduction in the average elevation or height of the Delta, due to the sea level rise and the natural subsidence affecting the Delta (Ibáñez et al., 2010). Hence, around 50 % of the Delta's surface area could be under the sea level at the end of the 21st century (Alvarado-Aguilar et al., 2012). In addition, changes in the Ebro fluvial regime (i.e. annual water yield reduction, decrease on the magnitude and frequency of medium and large floods; Batalla et al., 2004), has lead to the explosion of macrophytes along the lower Ebro River (Ibáñez et al., 2012) with a tremendous ecological impact.

3 The Guadalquivir River basin: current conditions

The Guadalquivir River has a contributing catchment of approximately $57\,400\,km^2$ and a length of 657 km. It is one of the biggest catchment in Spain and one of the most regulated, with 60 reservoirs distributed throughout the tributaries of

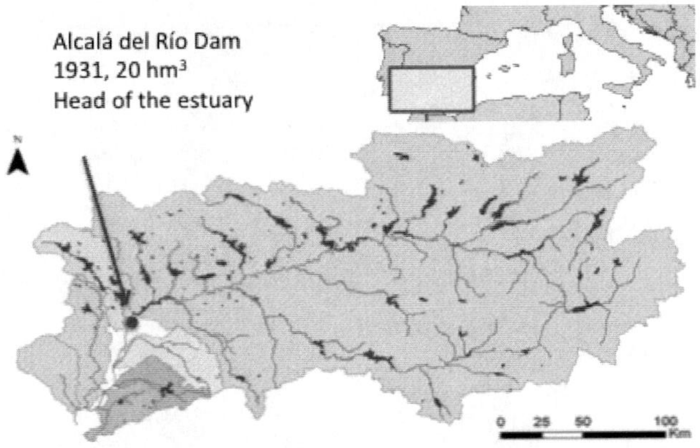

Figure 3. The Guadalquivir River basin in Spain draining to the Atlantic Ocean. Drainage network and reservoirs (dark blue small areas) upstream the estuary (grey area); the red point locates the Alcalá del Río dam, at the head of the estuary. The coloured sub-basins drain directly to the estuary area.

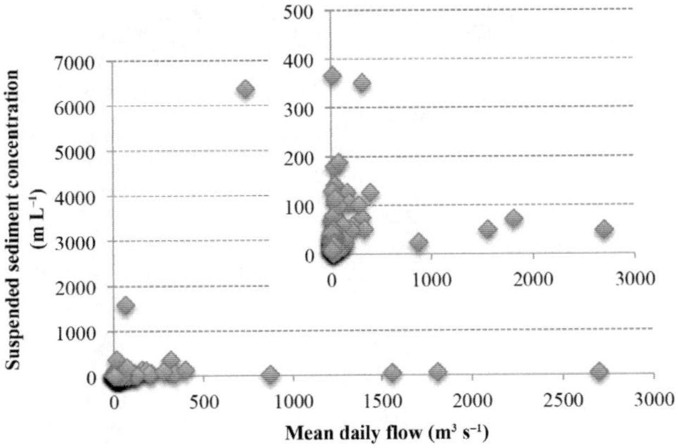

Figure 4. Rating curve water flow-suspended sediment concentration downstream the Alcalá del Río Dam (Fig. 3) in the head of the Guadalquivir Estuary, obtained from data during 1993–2007; a zoom for the interval 0–500 $mg\,L^{-1}$ is included for clarity.

the river, draining to the Atlantic Ocean through an estuarine area, the Guadalquivir Estuary (Fig. 3). The basin belongs to three different regions in Spain but up to a 90 % of its area is in Andalusia, and the allocation of water resources from the basin is mainly for irrigation (80 % of the water resource volume).

The mean annual average precipitation in the catchment, with a slight snow influence at its headwaters, is higher than 2000 mm in mountainous areas throughout the basin, and lower than 300 mm in the eastern arid areas, with mean annual average temperature of 16.9 °C and strong local gradients, especially at the East, where snow and semiarid areas coexist (www.chguadalquivir.es, last access: 30 January 2016). The annual runoff is also highly variable, but its significant decrease during the 20th century is associated to the increase of irrigated areas (Contreras, 2012).

Most of the reservoirs were built during the 1930–1970 period and the 1970–1990 period (equivalent to the 50 and 75 % of the total capacity; Contreras, 2012), regulate the river flow with a global capacity of 8101 hm^3 (0.1411 $hm^3\,km^{-2}$). The mean annual flow at Alcalá del Río (Fig. 3), at the head of the estuary, was estimated as 129 $m^3\,s^{-1}$ during the period 1931–1980, and as 63 $m^3\,s^{-1}$ after this period (Contreras, 2012); these values, however, strongly contrast with the maximum flows during flooding periods, that reached 5400 and 6700 $m^3\,s^{-1}$ in Cordoba and Seville, respectively, in February 1963 (the historical maximum during the 20th century), and reduced to values over 3000 $m^3\,s^{-1}$ after the building of dams afterwards (www.chguadalquivir.es).

The high regulation per unit area in the catchment has also led to a high trapping efficiency of the sediments; however, the lack of in-stream reservoirs in the Guadalquivir River (the influence of the Alcalá del Río dam with just 20 hm^3 is negligible) and the cropping system used downstream dams result in significant suspended sediments concentration in the river. Figure 4 shows the relationship between the mean daily flow delivered by Alcalá del Río and the suspended sediment concentration during 1993-2007.

However, significantly higher concentrations are usually found along the estuary, in which the dramatic decrease of freshwater delivery from the catchment has altered the sediment dynamics and budget, two turbidity maximum areas found (Díez-Minguito et al., 2012), and tidal dynamics influencing both salinity and sediments dynamics (Díez-Minguito et al., 2013, 2014), with turbidity events that may persist over weeks after a flooding event. Contreras and Polo (2012) reported values up to 14 and 9 $g\,L^{-1}$ at the upper and medium-low reaches of the estuary.

4 Adaptation actions: some reflections

Both the Ebro Delta and the Guadalquivir Estuary suffer strong morphological changes due to the decrease of freshwater from their catchments and the trapping effect of their reservoir networks. However, despite the differences from the marine dynamics between both systems (Mediterranean Sea and Atlantic Ocean), some additional factors produce the opposite effects. In the Ebro River, the presence of in-stream reservoirs in the lower parts of the river basin results in a very low suspended-sediment concentration and a low sediment delivery to the delta. In the Guadalquivir River, the lack of reservoirs in the main stream results in high suspended-sediment concentrations along its pathway that is significantly increased in the estuary due to the very low freshwater delivery from the catchment, and the usual dominance under such conditions of the tidal processes. Figures 2 and 4 show the different order of magnitude of the current concentrations of suspended sediments in each one of both cases.

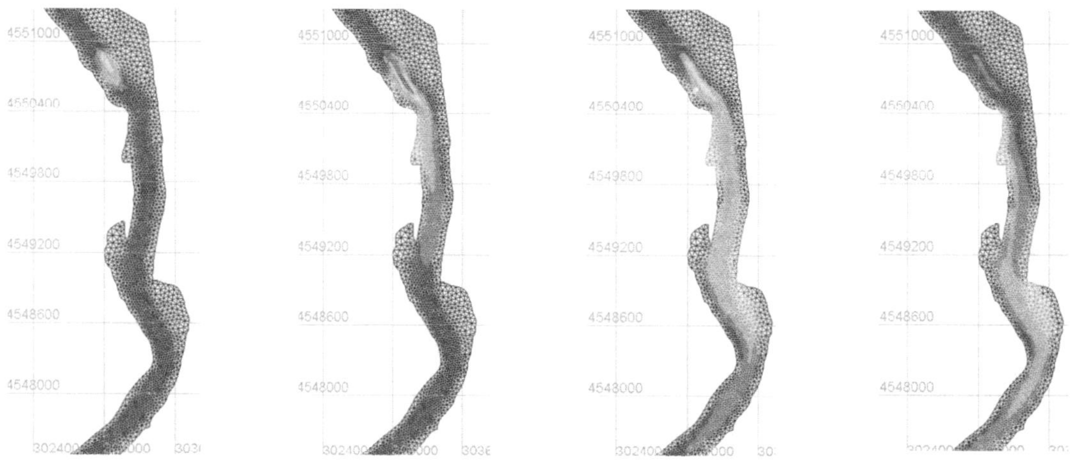

Figure 5. 2-dimensional simulation of the time evolution (from left to right) of the suspended sediment plume generated by a controlled injection at Mora d'Ebre (see Fig. 1) in the final reach of the Ebro River; permanent flow conditions of 100 $m^3 s^{-1}$ and a triangular sedimentograph for the load with maximum value of 50 $mg L^{-1}$.

In the projected actions to mitigate some of the effects of the water use upstream, the Guadalquivir River Authority proposes a transfer of water from upstream Alcalá del Río to the rice farmers in the estuary, who have serious damages from the higher salinity during the medium and low water regime. Some actions are also intended for environmental objectives.

In the Ebro River, the planned water uses have recently arisen conflicts and protests from the neighboring areas of the delta and Catalonia in general. As to the dramatic sediment deficit, the on-going LIFE project, EBRO-ADMICLIM, is developing pilot actions of sediment injections in the lower reach of the river to cope with the trapping by reservoirs (http://www.lifeebroadmiclim.eu/es/, last access: 8 April 2016). Figure 5 shows one example of the simulations performed to design these injections within the project.

However, no adaptation focus is included in the hydrological plans to decrease water use in the catchments. Despite the specific processes leading to the environmental affections in both the Ebro delta and the Guadalquivir estuary, and the warming period and increasing sea level rise, undoubtedly an increase of the freshwater delivery to these systems would mitigate the imbalances and make any other action much more efficient. When will society assume the excessive consumption of water in these regions? Adaptation should be an optimal point between adapting the environment to us, and adapting us to the environment.

5 Conclusions

The work presents two highly regulated but different catchments in Spain and the main general impacts of their reservoir networks. The different type of network, with or without in-stream reservoirs, has a different impact on the sediment delivery to the river mouth. Despite their differences in the tidal dynamics, the significant decrease of freshwater suffered by both of them greatly affects the riverine, and deltaic/estuarine morphologies. Adaptation actions must foresee a decrease in the water use upstream throughout the catchment, if an efficient adaptation plan is to be developed. Specific actions such as study of the pilot injections studied in the EBRO-ADMICLIM project are needed to cope with the future conditions, and science and engineering can jointly contribute to a rigorous decision-making framework that provides support for unpopular but necessary measurements.

Author contributions. María J. Polo collected the information about the Guadalquivir River Basin, designed the experiments simulated in the Ebro River, and prepared the manuscript; A. Rovira collected the information about the Ebro River basin, carried out the field campaigns at the Ebro River and collaborated in the experimental design and manuscript redaction; D. García-Contreras adapted the model code of the Ebro River simulations and performed them; E. Contreras and C. Aguilar analyzed the Guadalquivir River data; A. Millares participated in the experimental design of the Ebro River simulations; Miguel A. Losada supervised the Guadalquivir River Basin study.

Acknowledgements. This work was funded by the project LIFE EBRO-ADMICLIM (ENV/ES/001182) and the project RNM-4735-Proyecto de Excelencia-Junta de Andalucía.

References

Alvarado-Aguilar, D., Jiménez, J. A., and Nicholls, R. J.: Flood hazard and damage assessment in the Ebro delta (NW Mediterranean) to relative sea level rise, Nat. Hazards, 62, 1301–1321, 2012.

Batalla, R. J., Gómez, J. C., and Kondolf, G. M.: Reservoir-induced hydrological changes in the Ebro River Basin (NE Spain), J. Hydrol., 290, 117–136, 2004.

Contreras, E.: Influence of the fluvial discharges on the water quality of the Guadalquivir Estuary, PhD thesis, University of Cordoba, Spain, 171 pp., 2012.

Contreras, E. and Polo, M. J.: Measurement frequency and sampling spatial domains required to characterize turbidity and salinity events in the Guadalquivir estuary (Spain), Nat. Hazards Earth Syst. Sci., 12, 2581–2589, doi:10.5194/nhess-12-2581-2012, 2012.

Díez-Minguito, M., Baquerizo, A., Ortega-Sánchez, M., Navarro, G., and Losada, M. A.: Tide transformation in the Guadalquivir estuary (SW Spain) and process-based zonation, J. Geophys. Res., 117, C03019, doi:10.1029/2011JC007344, 2012.

Díez-Minguito, M., Contreras, E., Polo, M. J., and Losada, M. A.: Spatio-temporal distribution, along-channel transport, and post-riverflood recovery of salinity in the Guadalquivir estuary (SW Spain), J. Geophys. Res., 118, 2267–2278, 2013.

Díez-Minguito, M., Baquerizo, A., de Swart, H. E., and Losada, M. A.: Structure of the turbidity field in the Guadalquivir estuary: Analysis of observations and a box model approach, J. Geophys. Res., 119, 7190–7204, 2014.

Gallart, F. and Llorens, P.: Observations on land cover changes and the headwaters of the Ebro catchment, water resources in Iberian Peninsula, Phys. Chem. Earth, 29, 769–773, 2004.

Gorría, H.: Desecación de las marismas y terrenos pantanosos denominados de los alfaques, Ministerio de Agricultura, Madrid, Technical report, 1877.

Ibàñez, C., Prat, N., and Canicio, A.: Changes in the hydrology and sediment transport produced by large dams on the lower Ebro River and its estuary, Regul. River., 12, 51–62, 1996.

Ibáñez, C., Sharpe, P. J., Day, J. W., Day, J. N., and Prat, N.: Vertical accretion and relative sea level rise in the Ebro delta wetlands (Catalonia, Spain), Wetlands, 30, 979–988, 2010.

Ibáñez, C., Alcaraz, C., Caiola, N., Rovira, A., Trobajo, R., Alonso, M., Duran, C., Jiménez, P. J., Munné, A., and Prat, N.: Regime shift from phytoplankton to macrophyte dominance in a large river: top-down versus bottom-up effects, Sci. Total Environ., 416, 314–322, 2012.

Morris, G. L. and Fan, J.: Reservoir Sedimentation Handbook, McGraw-Hill Book Co., New York, 2000.

Rovira, A., Ibáñez, C., and Martín-Vide, J. P.: Suspended sediment load at the lowermost Ebro River (Catalonia), Quatern. Int., 388, 188–198, 2015.

Varela, J. M., Gallardo, A., and López de Velasco, A.: Retención de sólidos por los embalses de Mequinenza y Ribarroja. Efectos sobre los aportes al Delta del Ebro, in: El sistema integrado del Ebro, edited by: Mariño, M., Gráficas Hermes, Madrid, 203–219, 1986.

Proc. IAHS, 373, 51–55, 2016
proc-iahs.net/373/51/2016/
doi:10.5194/piahs-373-51-2016

Open Access
PIAHS

Integrated management of water resources demand and supply in irrigated agriculture from plot to regional scale

Niels Schütze and Michael Wagner

Institute of Hydrology and Meteorology, TU Dresden, 01279 Dresden, Germany

Correspondence to: Niels Schütze (niels.schuetze@tu-dresden.de)

Published: 12 May 2016

Abstract. Growing water scarcity in agriculture is an increasing problem in future in many regions of the world. Recent trends of weather extremes in Saxony, Germany also enhance drought risks for agricultural production. In addition, signals of longer and more intense drought conditions during the vegetation period can be found in future regional climate scenarios for Saxony. However, those climate predictions are associated with high uncertainty and therefore, e.g. stochastic methods are required to analyze the impact of changing climate patterns on future crop water requirements and water availability. For assessing irrigation as a measure to increase agricultural water security a generalized stochastic approach for a spatial distributed estimation of future irrigation water demand is proposed, which ensures safe yields and a high water productivity at the same time. The developed concept of stochastic crop water production functions (SCWPF) can serve as a central decision support tool for both, (i) a cost benefit analysis of farm irrigation modernization on a local scale and (ii) a regional water demand management using a multi-scale approach for modeling and implementation. The new approach is applied using the example of a case study in Saxony, which is dealing with the sustainable management of future irrigation water demands and its implementation.

1 Introduction

Arid and semi-arid areas that are intensively used for agriculture, are facing water shortage which is often intensified by an overexploitation of existing water resources. Accordingly, they show an increased sensitivity to water stress and a high vulnerability that can only be reduced by a highly efficient and foresighted water resource management practices. Same observations are less pronounced for Saxony, but signals of longer and more intense drought conditions during the vegetation period can be found in future regional climate scenarios and irrigation may become relevant as a measure to mitigate drought stress for agricultural crops in future. For assessing irrigation as a measure to increase agricu-ltural water security a generalized stochastic approach for a spatial distributed estimation of future irrigation water demand is proposed in this contribution, which ensures safe yields and a high water productivity at the same time. Recent studies, e.g. (Semenov, 2007; Brumbelow and Georgakakos, 2007), try to analyze possible impacts of climate variability and climate change on agriculture, based on process-based simulation models.

Only, a few papers (Brumbelow and Georgakakos, 2007; Schütze and Schmitz, 2010) focus on irrigation management under limited water availability which requires a deficit irrigation strategy with the target to achieve water productivity (WP) rather than maximum yield. The improvement of WP needs a good quantitative understanding of the relationship between irrigation practices and grain yield, i.e. the crop water production function (CWPF). With this knowledge, the value of each unit of water applied to a field can be estimated and compared with alternative uses within and beyond the agricultural sector. However, the estimation of WP for limited water availability is difficult on a regional level because the performance of irrigation systems depend on a number of factors such as climate conditions, grown crops, soil hydraulic characteristics and used irrigation systems. In addition, the stochastic properties of the relevant climate factors (e.g. precipitation and temperature) and of the soil properties have to be considered in a risk-based regional water manage-

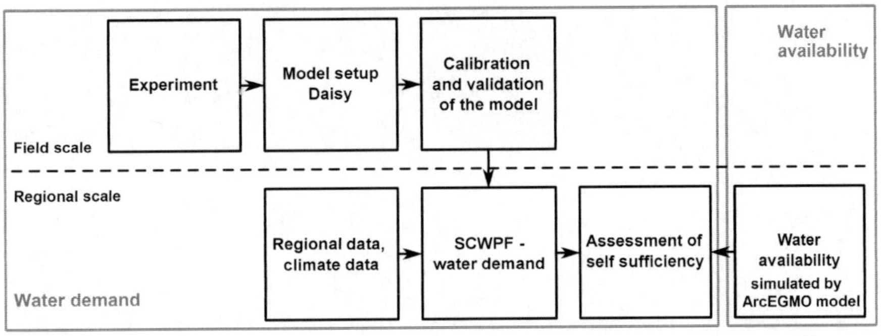

Figure 1. Work flow for estimating regional irrigation water demand from field scale simulations and regional water availability using the ArcEGMO model.

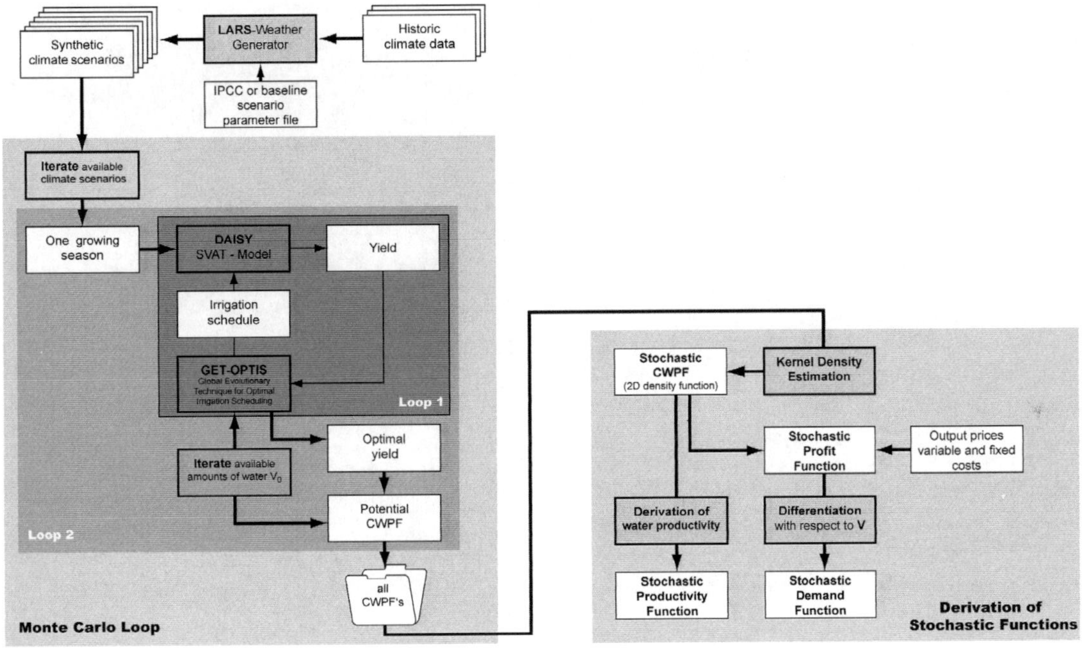

Figure 2. Layout of the stochastic framework for generating stochastic crop production functions on field scale.

ment. Thus, the objective of this study is to demonstrate in a generic framework how stochastic crop water production functions (SCWPF) generated on a field scale can serve as a tool for estimating regional water demand using an upscaling approach.

2 Overview of the stochastic framework

The work flow for estimating regional irrigation water demand from field scale experiments and simulations, shown in Fig. 1, was developed and implemented in the project "SAPHIR" (Schütze et al., 2016).

Figure 1 shows the work flow of all methods, grouped in field and regional scale. Irrigation experiments were conducted with several crops at three experimental sites in Germany. Model parameters were derived based on field irrigation experiments using the crop soil-vegetation-atmosphere-transfer (SVAT) model Daisy (Abrahamsen and Hansen, 2000). The crop model parameters and the stochastic relationship between irrigated water and crop yield (SCWPF – stochastic crop water production function) served as a basis for regional crop model simulations (see Fig. 2). The stochastic approach allows direct information of uncertainty. The corresponding crop water demand and the derived water availability result in the degree of local water self-sufficiency that was estimated for mid and northern Saxony in a real case application. For the estimation of available water resources under different climate change scenarios the hydrological model ArcEGMO was used (Schwarze et al., 2016).

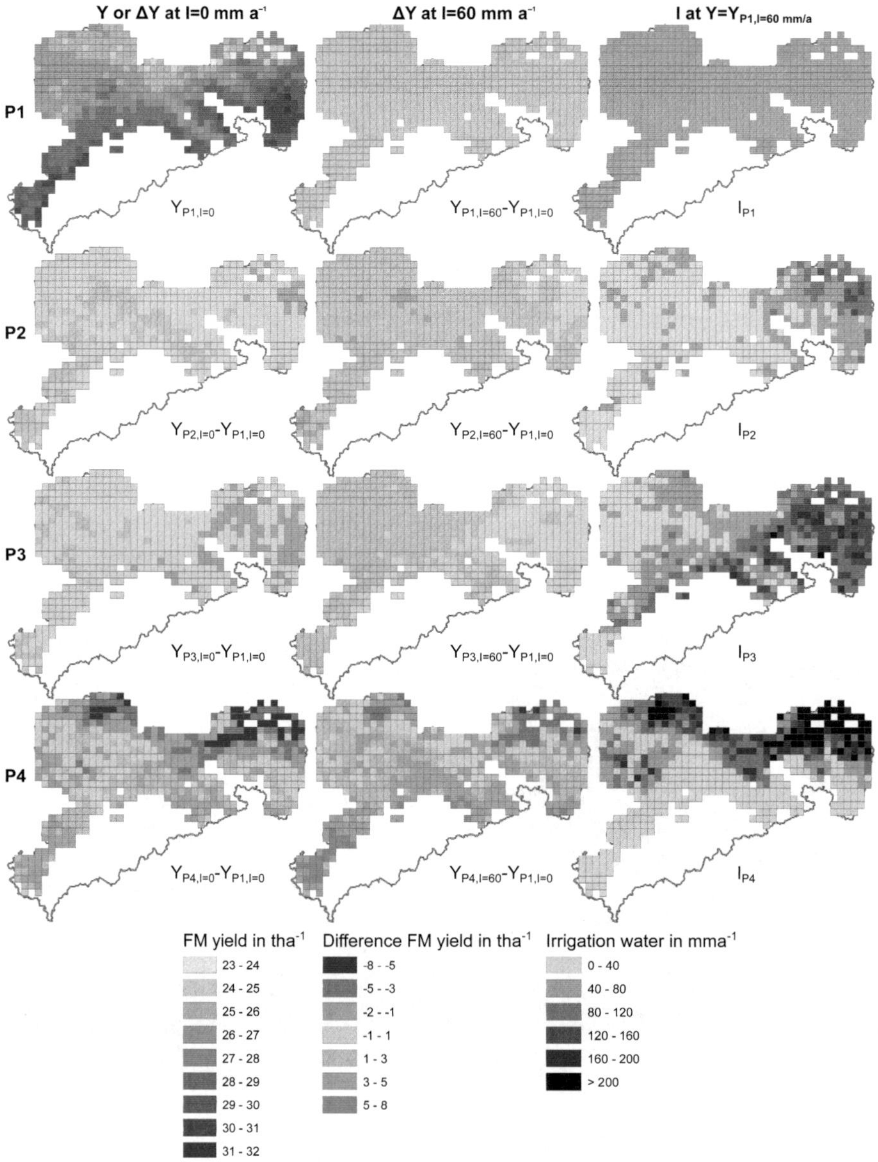

Figure 3. 1st column: regional fresh matter bean yield (1st row) and yield differences between future time periods (2nd to 4th row) with no irrigation. 2nd column: fresh matter yield differences for each time period from no irrigation to 60 mm per year irrigation. 3rd column: irrigated water amount, if crop yield shall not decrease under yield level of period P1. All results are averaged over all soils in each grid cell and medians of corresponding SCWPF (for details see Wagner et al., 2015).

The generic framework for the generation of SCWPF's comprises of an optimizer and a Monte Carlo sampler. For the optimizer two components are required to achieve reliable CWPFs for one realisation of the climate conditions (see Fig. 2, loop 1): an irrigation scheduling optimizer and a simulator of plant growth and water transport. The objective of the optimization is to maximize yield for a given climate realisation and a given amount of water for irrigation during the growing season.

Within a second loop which iterates over a range of given water volumes, a complete CWPF can be constructed.

In the third loop a required number of CWPFs is generated in order to accurately compute the statistical characteristics of the random sample of CWPFs in a non-parametric way, i.e. the resulting SCWPF is an empirical probability function.

To be useful for Monte Carlo simulations, the irrigation scheduling optimizer and the simulation model should be highly efficient and effective in finding the global maximum yield in order to generate a SCWPF within an acceptable

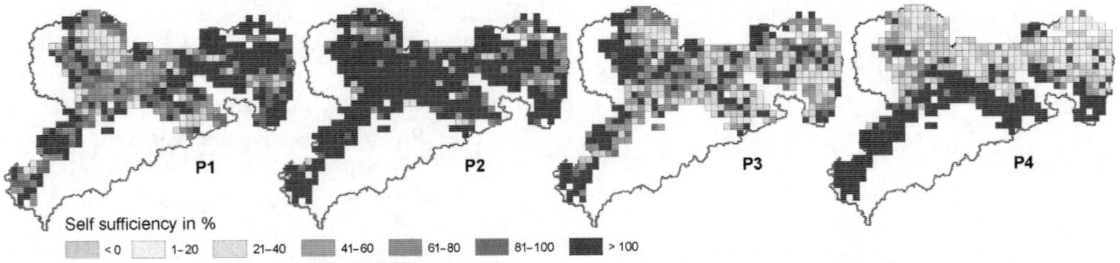

Figure 4. Predicted water self-sufficiency for common bean in relevant regions of Saxony. The water demand results from the same crop yield in all time periods as in P1 with Irrigation = 60 mm per year. The available water covers 50 % of groundwater recharge plus surface water. All results are averaged over soils in each grid cell and medians of corresponding SCWPF (for details see Wagner et al., 2015).

computation time. Finally, the generated empirical probability function SCWPF is converted into a continuously differentiable density function for the calculation of derivatives such as productivity, profit and demand function.

3 Results

For a regional analysis of future development of irrigation water demand and local self-sufficiency in Saxony the spatial resolution for estimating SCWPF depending on local climate conditions, grown crops, soil hydraulic characteristics and used irrigation systems is 5 by 5 km.

For statistical analysis, we refer to 30-year periods. 1961–1990 (P1) is assumed as recent climate, 1991–2020 (P2) is the actual situation, 2021–2050 (P3) is the near future and 2071–2100 (P4) the far future projection. For the regional analysis climate data from the statistical regional climate model WEREX V was used.

Figure 3 shows the regional development for beans in different temporal periods P1 to P4. The upper left picture displays the non-irrigated regional crop yield in P1. A decrease from south to north based on drier climate and a higher sand fraction of the upper soil is apparent. The maps illustrates developments in crop yield for P2 to P4 using the differences in non-irrigated crop yield for P2 to P4. Especially the north eastern part of Saxony suffers from minor yield decreases in P2 and P3 whereas in P4 most parts in the area of investigation show a significant decrease for larger part of the regions. Irrigation can be an appropriate measure to stabilize yields. This becomes apparent from the last column where the required irrigation water volume is shown for periods P2 to P4 which must be applied in order to achieve the crop yield of period P1. In P2 only few regions in the north east have a higher irrigation demand. However, in period P4 with a significant higher rate of evapotranspiration significant more irrigation water is required for the most regions in order to secure crop yield.

Figure 4 maps self-sufficiency in relevant regions of Saxony where agriculture is relevant. For periods P1 and P2 small deficits of self-sufficiency can be observed in mid northern Saxony.

In contrast, period P4 shows an even lower degree of self-sufficiency in the northern part, although a number of blue spots are indicating larger water reservoirs which provide additional water resources for irrigation.

4 Conclusions

In this study, a new multi-scale approach for integrated modeling and management of irrigation systems was developed, implemented and tested at a regional scale. The use of a tailor-made evolutionary optimization algorithm for optimal irrigation management with limited water supply allows the estimation of the potential performance (e.g. irrigation efficiency, economic productivity and price elasticity of water demand) of agro-hydrological units, characterized for example, by its soil, grown crops and irrigation method. To take climate variability into account the agronomic response to different levels of water application is represented by stochastic crop water production functions (SCWPF). For the region of Saxony preliminary results show that primarily the northern part of Saxony is particular affected by the drier projected future climate which can partly be compensated by irrigation measures using local water resources. The proposed multi-scale approach for integrated modeling and management of irrigation systems can easily be extended to introduce other components in the regional water management, e.g. technical systems for water storage and distribution (Müller and Schütze, 2013). Furthermore, the developed multi-scale approach is transferable to different climate conditions, e.g. it is implemented and validated for the arid region of Oman (Schütze et al., 2012).

Acknowledgements. These investigations are part of the research project "SAPHIR – Saxonian Platform for High Performance Irrigation" funded by the EU ESF "Nachwuchsforschergruppen" program under grant no. 100098204.

References

Abrahamsen, P. and Hansen, S.: Daisy: an open soil-crop-atmosphere system model, Environ. Modell. Softw., 15, 313–330, 2000.

Brumbelow, K. and Georgakakos, A.: Consideration of climate variability and change in agricultural water resources planning, J. Water Res. Pl.-ASCE, 133, 275–285, 2007.

Müller, R. and Schütze, N.: Improving the future performance and reliability of multi- reservoir systems by multi-objective optimization, in: IAHS-IAPSO-IASPEI Assembly, edited by: Schumann, A. H. and Belyaev, V., Gothenburg, Sweden, vol. 362 of Red Book, 24–32, 2013.

Schütze, N. and Schmitz, G. H.: OCCASION: A new Planning Tool for Optimal Climate Change Adaption Strategies in Irrigation, J. Irrig. Drain. E.-ASCE, 136, 836–846, doi:10.1061/(ASCE)IR.1943-4774.0000266, 2010.

Schütze, N., Kloss, S., Lennartz, F., Al Bakri, A., and Schmitz, G. H.: Optimal planning and operation of irrigation systems under water resource constraints in Oman considering climatic uncertainty, Environmental Earth Sciences, 65, 1511–1521, 2012.

Schütze, N., Stange, P., Griessbach, U., Röhm, P., and Wagner, M.: Integrated modelling and optimisation of irrigation systems at the field and catchment scale with limited water resources, Hydrologie und Wasserbewirtschaftung, 60, 7–21, 2016.

Schwarze, R., Dröge, W., Wagner, M., Spitzer, S., Maleska, V., and Kuhn, K.: Research on possible impacts of climate change on the water balance in Saxony – approach, analysis of the status quo, data and parameter model, model test, Hydrologie und Wasserbewirtschaftung, 60, 53–71, 2016.

Semenov, M. A.: Development of high-resolution UKCIP02-based climate change scenarios in the UK, Agr. Forest Meteorol., 144, 127–138, 2007.

Wagner, M., Seidel, S. J., and Schütze, N.: Irrigation water demand of common bean on field and regional scale under varying climatic conditions, Meteorol. Z., doi:10.1127/metz/2015/0698, 2015.

Proc. IAHS, 373, 57–60, 2016
proc-iahs.net/373/57/2016/
doi:10.5194/piahs-373-57-2016

Predictions in ungauged basins – where do we stand?

Günter Blöschl

Institute of Hydraulic Engineering and Water Resources Management, Vienna University of Technology, Karlsplatz 13/222, 1040 Vienna, Austria

Correspondence to: Günter Blöschl (bloeschl@hydro.tuwien.ac.at)

Published: 12 May 2016

Abstract. The spatial dimensions of water management heavily rely on accurate hydrological estimates in the landscape. This has exactly been the focus of the Predictions in Ungauged Basins (PUB) initiative of the IAHS. The initiative has significantly advanced the science by furthering process understanding and estimation methods, and by synthesising the knowledge across processes, places and scales. Ongoing research on PUB is increasingly treating water management as an intrinsic part of the hydrological cycle and is developing processes understanding and methods to account for the feedbacks between humans and water in the landscape.

1 The Prediction in Ungauged Basins (PUB) problem

Accurate estimates of stream runoff and other hydrologic quantities are needed for numerous purposes of water resources planning and management, in particular if one is interested in the spatial dimension. The most accurate way of obtaining such estimates at any one location is to measure them for an extended period of time. However, often this is not possible for financial or logistic reasons, or simply because one is interested in the future evolution of the hydrological variables. The alternative therefore is to estimate them from measurements at other locations in the region and transfer them, in some way, by modelling methods.

Methods for estimating runoff at ungauged locations such as the Rational Method and the index-flood method have been in practical use for a long time. However, there are a range of other hydrological characteristics for which estimates are needed in ungauged catchments, including low flows, rainfall and evaporation. Also, for many purposes, empirical methods do not suffice and process understanding needs to be invoked in order to make predictions that are reliable for a diverse set of hydrological conditions.

2 Best practice recommendations for predicting runoff in ungauged basins

The international research on predictions in ungauged basins has received a major impetus through a research initiative of the International Association of Hydrological Sciences (IAHS). The initiative, known as Predictions in Ungauged Basins (PUB), was started in 2003 (Sivapalan et al., 2003) and intended to advance the research on the topic around the globe. The varied and extensive outcome of the initiative suggests that this has indeed been accomplished (see, e.g., Hrachowitz et al., 2013).

One of the achievements of the initiative was a synthesis of methods that before had been treated in a fragmented way (Blöschl et al., 2013). The synthesis proceeded along the dimensions of processes, places and scales. The synthesis across processes involves a consistent and coherent treatment of annual runoff, seasonal runoff, the flow duration curve, floods, low flows and entire hydrographs. The synthesis across places is built on the notion of similarity and draws together experience and data from numerous catchments in a region and from around the world. The synthesis across scales involves a balanced view of both upscaling methods based on laboratory equations (such as distributed models) and lumped catchment scale models (such as regional statistical relationships).

The synthesis report of the initiative concluded with best practice recommendations for predicting runoff in ungauged

Table 1. Summary of best practice recommendations (from Takeuchi et al., 2013).

Step	Recommendation
Step 1: Read the landscape	Go out to your catchment, look around, what does the landscape tell you, create a photo documentation, look at the hydrogeology, ask people about previous events, obtain global, regional and local data, map hydraulic structures and other modifications. If possible, install a stream gauge.
Step 2: Runoff signatures and processes	Analyse all runoff signatures in nearby catchments to get understanding of hydrology of the catchment beyond the signature of interest. Runoff signatures include annual runoff, seasonal runoff, flow duration curve, low flows, floods, and hydrographs.
Step 3: Process similarity and grouping	On the basis of the first two steps and process similarity measures find similar gauged catchments to assist in predicting runoff in the ungauged basin (grouping of landscape units and catchments). The similarity can be based on short term and co-evolutionary processes.
Step 4: Model	Build statistical and/or process based model for the signature of interest; regionalise the parameters from similar catchments, making advantage of a priori information, dynamic proxy data and any other information on processes, including from the other signatures; account for correlations along the stream network. There is always more information than the hydrograph – use it.
Step 5: Interpretation	Interpret the parameters of the model hydrologically and justify their values against what was learned during field trip and other data, to improve parameter choice and uncertainty estimation. Parameters are, e.g., regression coefficients and runoff model parameters.
Step 6: Uncertainty	Assess uncertainty of predicted runoff by combining error propagation methods, regional cross validation and hydrological interpretation against the backdrop of the uncertainty to be expected from comparative hydrology (Level 1 and 2). We now have a prediction of runoff signature including understanding of its credibility.
All steps	Communicate all of this in a way that it contributes to the global and national body of knowledge in hydrology, especially process knowledge.

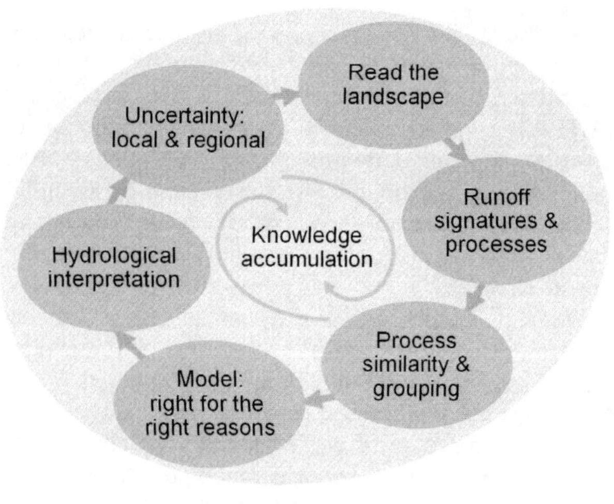

Figure 1. Best practice recommendations for predicting runoff in ungauged basins (from Takeuchi et al., 2013).

basins (see Fig 1 and Table 1). In addition to this general pattern, the exact method of estimating runoff in any one case, of course, depends on many factors, including data availability, nature of the catchment and the nature of the estimation problem.

3 Recent developments

Research on PUB has continued to flourish in the most recent years, both from a theoretical perspective and for assisting the spatial dimensions of water management. For example, flood risk management has recently evolved from a focus on individual flood protection structures to a more integrated approach at the river basin scale that considers a diversity of management measures and changes in the flood risk over time (e.g. Merz et al., 2014). New methodological developments have been aligned with the requirements of the Flood Risk Directive regarding hazard mapping, large scale interactions of floods, residual risk (flood prevention); retaining water in the landscape, linear protection measures, flood retention (flood mitigation); public participation, outreach and education (awareness); flood warning, emergency plans (preparation); and assessment of flood damage, event documentation (recovery) (Blöschl et al., 2015).

With all these new developments it is increasingly becoming clear that it is essential to treat water management as an intrinsic part of the hydrological cycle. This means it no longer suffices to make estimates for pristine catchments and correct for any human effects once the estimates have been obtained. Instead, the processes understanding and methods

Table 2. Seven steps of framing and modeling hydrological vs coupled dynamic environmental vs socio-hydrological processes (shortened from Sivapalan and Blöschl, 2015).

Step	Recommendation
Step 1: Phenomenon, domain, scale	Specify phenomenon, domain, study period, purpose of modeling. The phenomena are often defined through narratives
Step 2: Perceptual model	Causal loop diagram to conceptualise process interactions, assisted by narratives of phenomena to visualize alternative hypotheses
Step 3: Choice of state variables	Small number of variables usually of advantage. Variables should be measurable. Values are a key state variable for a long term treatment
Step 4: Causal factors that affect state variables	Causal factors can be state variables or external forcing. Known balance equations (e.g., financial budget) may assist in choice of factors (and therefore coupling)
Step 5: Functional relationships for step 4	Use of local data may require upscaling. Scaling analysis to help identify fast and slow state variables. Guidance by socio-economic data (e.g., surveys, censoring) and narratives
Step 6: Parameter estimation	Disassemble model into components and estimate parameters separately. Re-assemble model and estimate feedback parameters against emergent phenomena. Evaluate effect of model parameters on path dependence and lock-ins of model dynamics
Step 7: Model validation and uncertainty	Test component models against different streams of data. If phenomena are repeatable in space or time, test model against similar situations at different places or in different time periods. If phenomena are not repeatable, no full validation is possible. Sources of uncertainty may include non-optimum behaviour of humans.

need to account for the feedbacks between humans and water in the landscape.

This brings to the fore the time dependency of hydrological variables in the landscape. From a spatial problem one then progresses to a spatio-temporal problem. While much of the spatial PUB work, such as the best practice recommendations, are still valid under such a framework, an extension is required to allow for the temporal dynamics of the system. One appealing approach of doing this is to frame the coupled human-nature system as a dynamical system, characterized by interactions of fast and slow time scales and feedbacks between environmental and social processes. These interactions give rise to emergent phenomena such as:

- Critical transitions, path dependence, multiple equilibria, lock-in situations

- Levee effect, adaptation to change, system collapse due to resource depletion

Di Baldassarre et al. (2015) discuss such emergent phenomena for the specific case of floods.

When framing such processes there is almost always the issue of data scarcity, i.e. this is a genuine predictions in ungauged basin (PUB) problem. While there are numerous similarities with the PUB problem there also additional challenges due to the coupled nature of the processes involved and the specific characteristics of social processes, such as context dependence. Sivapalan and Blöschl (2015) proposed seven steps of framing such coupled processes that account for the specific characteristics (Table 2). An important part of the framing is to take advantage of narratives representing the phenomena of human-water interactions. The phenomena may be generic (any of the broad class of process causalities that manifest in several places, e.g., the levee effect, irrigation efficiency paradox), or they could be place based, reflecting the unique characteristics of a place and its water history.

Depending on whether one is interested in place based or generic phenomena, the model complexity may vary from complex, comprehensive models that account for much spatial and process detail to stylized models that are typically lumped and much more parsimonious in terms of their process representation. Sivapalan and Blöschl (2015) highlighted the pros and cons of the different model types. The comprehensive models have the advantage that they can be readily applied to real world management problems in specific places, but the modeling effort may be larger and it may be very difficult to specify all the required parameters in a realistic way. Also, because of the complexity, it is very challenging to understand the overall system behavior for different parameter combinations and to reveal long-term, large-scale phenomena. On the other hand, the stylized models have the advantage of transparency and ease of use. It is possible to fully explore their dynamic behavior space for all parameter combinations, including any emergent behavior and the borders between stability and instability. This can provide guidance for decision making but at a strategic level. Their disadvantage is the difficulty of estimating realistic parameters since they represent aggregate behavior that cannot directly be translated into local scale decisions.

Developing these dynamic models in a spatial context bears a lot of resemblance with the research in the PUB initiative of IAHS, however, this time framed in a dynamic way. The research also provides important opportunities to assist the spatial dimensions of water resource management from a practical perspective, including

- facilitating stakeholder participation,
- helping decision-makers through the generation and assessment of alternative futures, and
- learning from the experiences of other similar places, and move towards generalizations beyond individual case studies (Sivapalan and Blöschl, 2015).

The new IAHS initiative on change in hydrology and society, known as Panta Rhei (Montanari et al., 2013), is geared towards a better understanding of the changing dynamics of the water cycle related to changing human systems. Clearly, there are important synergies between the research on predictions in ungauged basins and Panta Rhei in addressing new exiting science questions from a spatio-temporal perspective in order to further the science of hydrology.

References

Blöschl, G., Sivapalan, M., Wagener, T., Viglione, A., and Savenije, H. H. G. (Eds.): Runoff Prediction in Ungauged Basins – Synthesis across Processes, Places and Scales, Cambridge University Press, Cambridge, UK, 465 pp., 2013.

Blöschl, G., Horváth, Z., Kiss, A., Komma, J., Nester, T., Perdigão, R. A. P., Viglione, A., and Waser, J.: Neue Methoden für das Hochwasserrisikomanagement (New methods for flood risk management), Österreichische Ingenieur- und Architekten-Zeitschrift (ÖIAZ), 160, 15–27, 2015.

Di Baldassarre, G., Viglione, A., Carr, G., Kuil, L., Yan, K., Brandimarte, L., and Blöschl, G.: Debates – Perspectives on socio-hydrology: Capturing feedbacks between physical and social processes, Water Resour. Res., 51, 4770–4781, doi:10.1002/2014WR016416, 2015.

Hrachowitz, M., Savenije, H. H. G., Blöschl, G., McDonnell, J. J., Sivapalan, M., Pomeroy, J. W., Arheimer, B., Blume, T., Clark, M. P., Ehret, U., Fenicia, F., Freer, J. E., Gelfan, A., Gupta, H. V., Hughes, D. A., Hut, R. W., Montanari, A., Pande, S., Tetzlaff, D., Troch, P. A., Uhlenbrook, S., Wagener, T., Winsemius, H. C., Woods, R. A., Zehe, E., and Cudennec, C.: A decade of Predictions in Ungauged Basins (PUB) – a review, Hydrol. Sci. J., 58, 1198–1255, doi:10.1080/02626667.2013.803183, 2013.

Merz, B., Aerts, J., Arnbjerg-Nielsen, K., Baldi, M., Becker, A., Bichet, A., Blöschl, G., Bouwer, L. M., Brauer, A., Cioffi, F., Delgado, J. M., Gocht, M., Guzzetti, F., Harrigan, S., Hirschboeck, K., Kilsby, C., Kron, W., Kwon, H.-H., Lall, U., Merz, R., Nissen, K., Salvatti, P., Swierczynski, T., Ulbrich, U., Viglione, A., Ward, P. J., Weiler, M., Wilhelm, B., and Nied, M.: Floods and climate: emerging perspectives for flood risk assessment and management, Nat. Hazards Earth Syst. Sci., 14, 1921–1942, doi:10.5194/nhess-14-1921-2014, 2014.

Montanari, A., Young, G., Savenije, H. H. G., Hughes, D., Wagener, T., Ren, L. L., Koutsoyiannis, D., Cudennec, C., Grimaldi, S., Blöschl, G., Sivapalan, M., Beven, K., Gupta, H., Arheimer, B., Huang, Y., Schumann, A., Post, D., Srinivasan, V., Boegh, E., Hubert, P., Harman, C., Thompson, S., Rogger, M., Hipsey, M., Toth, E., Viglione, A., Di Baldassarre, G., Schaefli, B., McMillan, H., Schymanski, S. J., Characklis, G., Yu, B., Pang, Z., and Belyaev, V.: Panta Rhei – Everything Flows": Change in hydrology and society – The IAHS Scientific Decade 2013–2022, Hydrol. Sci. J., 58, 1256–1275, doi:10.1080/02626667.2013.809088, 2013.

Sivapalan, M. and Blöschl, G.: Time scale interactions and the coevolution of humans and water, Water Resour. Res., 51, 6988–7022, doi:10.1002/2015WR017896, 2015.

Sivapalan, M., Takeuchi, K., Franks, S. W., Gupta, V. K., Karambiri, H., Lakshmi, V., Liang, X., McDonnell, J. J., Mendiondo, E. M., O'Connell, P. E., Oki, T., Pomeroy, J. W., Schertzer, D., Uhlenbrook, S., and Zehe, E.: IAHS Decade on Predictions in Ungauged Basins (PUB), 2003–2012: Shaping an exciting future for the hydrological sciences, Hydrol. Sci. J., 48, 857–880, 2003.

Takeuchi, K., Blöschl, G., Savenije, H. H. G., Schaake, J. C., Sivapalan, M., Viglione, A., Wagener, T., and Young, G.: Recommendations, Chapter 13, in: Runoff Prediction in Ungauged Basins – Synthesis across Processes, Places and Scales, edited by: Blöschl, G., Sivapalan, M., Wagener, T., Viglione, A., and Savenije, H., Cambridge University Press, Cambridge, UK, 384–387, 2013.

Proc. IAHS, 373, 61–67, 2016
proc-iahs.net/373/61/2016/
doi:10.5194/piahs-373-61-2016

Open Access
PIAHS

A regional look at the selection of a process-oriented model for flood peak/volume relationships

Ján Szolgay[1], Ladislav Gaál[1,2], Tomáš Bacigál[3], Silvia Kohnová[1], Kamila Hlavčová[1], Roman Výleta[1], and Günter Blöschl[4]

[1]Department of Land and Water Resources Management, Faculty of Civil Engineering, Slovak University of Technology, Bratislava, Slovakia
[2]MicroStep-MIS, spol. s r.o., Bratislava, Slovakia
[3]Department of Mathematics and Descriptive Geometry, Faculty of Civil Engineering, Slovak University of Technology, Bratislava, Slovakia
[4]Institute for Hydraulic and Water Resources Engineering, Vienna University of Technology, Vienna, Austria

Correspondence to: Ján Szolgay (jan.szolgay@stuba.sk)

Published: 12 May 2016

Abstract. Recent research on the bivariate flood peak/volume frequency analysis has mainly focused on the statistical aspects of the use of various copula models. The interplay of climatic and catchment processes in discriminating among these models has attracted less interest. In the paper we analyse the influence of climatic and hydrological controls on flood peak and volume relationships and their models, which are based on the concept of comparative hydrology in the catchments of a selected region in Austria. Independent flood events have been isolated and assigned to one of the three types of flood processes: synoptic floods, flash floods and snowmelt floods. First, empirical copulas are regionally compared in order to verify whether any flood processes are discernible in terms of the corresponding bivariate flood-peak relationships. Next the types of copulas, which are frequently used in hydrology are fitted, and their goodness-of-fit is examined in a regional scope. The spatial similarity of copulas and their rejection rate, depending on the flood type, region, and sample size are examined, too. In particular, the most remarkable difference is observed between flash floods and the other two types of flood. It is concluded that treating flood processes separately in such an analysis is beneficial, both hydrologically and statistically, since flood processes and the relationships associated with them are discernible both locally and regionally in the pilot region.

However, uncertainties inherent in the copula-based bivariate frequency analysis itself (caused, among others, also by the relatively small sample sizes for consistent copula model selection, upper tail dependence characterization and reliable predictions) may not be overcome in the scope of such a regional comparative analysis.

1 Introduction

Bivariate distributions of flood peaks and flood event volumes may be needed for solving a range of practical problems, including, e.g., the design of retention basins and identifying the extent and duration of flooding in flood hazard zones. For the statistical analysis of flood peaks and volumes, identical marginal distributions for both random variables were used in the past (e.g., Goel et al., 1998; Yue et al., 2002). Recently, the use of copula-based multivariate models has become widespread. These allow for separate studies of the marginal distributions of the component variables and the correlation/dependence structure between them. Numerous studies have been published on this topic (e.g., Shiau, 2003; De Michele et al., 2005; Chowdhary et al., 2011; Requena et al., 2013), including recommendations how to select appropriate copula models (e.g., Favre et al., 2004; Genest and Favre, 2007). Despite numerous studies, as mentioned by Chowdhary et al. (2011), the use of copula-based multivariate distributions for hydrological design still cannot be regarded as having been satisfactorily resolved. The selection of the types of bivariate distributions and the estimation of their parameters from observed peak-volume pairs are associated with far greater uncertainties when compared to univariate distributions, since observed flood records of the required length are rarely available. This poses a problem for reliable estimations of flood risks in bivariate design cases. It is being increasingly recognized that the problem cannot be approached from only a purely statistical perspective. The crucial step for predictions in the multivariate modelling of flood characteristics by copulas is the choice of the copula which best fits the data (Favre et al., 2004). In this respect Serinaldi and Kilsby (2013) highlighted the importance of studying the relationships between processes that generate the design variables and also the statistical techniques used to model them. In previous studies we have attempted to better understand the hydrological factors controlling flood hydrograph shapes, which also have implications for the dependence between flood peaks and volumes. In Gaál et al. (2012), we analyzed the ratio of both quantities (flood time-scales) based on the concept of comparative hydrology in a regional context in Austria. We have compared catchments with contrasting characteristics in order to understand roles of climate (through the type of precipitation generated), together with the attributes of the environment and flow generation processes (e.g., through antecedent soil moisture and soil characteristics) in a holistic way. In Gaál et al. (2015), our aim was to understand the causal hydrological factors controlling the strength of the relationship (quantified by Spearman's rank correlation coefficient) between flood peaks and volumes for the same dataset. These coefficients ranged from about 0.2 in high alpine catchments to about 0.8 in lowlands. The weak dependence in high alpine catchments was attributed to the characteristic mix of flood types in the region. The results also suggested that the factors controlling the strength of the dependence may be more related to the climate than catchment characteristics. In Szolgay et al. (2015), we aimed at analyzing the formal suitability of various copula-based bivariate relationships between flood peaks and flood volumes, with a particular focus on two basic flood generating seasons (summer and winter floods) with the goal of going a little beyond the statistics alone in the choice of the copula functions for the engineering applications. It was concluded that, for rainfall-fed floods, three extreme value copulas performed best in the pilot region (the Galambos, Gumbel-Hougaard and Hüsler-Reiss copulas) followed by a normal copula. The other copulas were not regarded as regionally acceptable. For winter floods the best performer was the Frank copula, followed by the normal and Plackett copulas and the three extreme value models. The Clayton and Joe copulas indicated an unacceptable performance for both seasons. In Szolgay et al. (2015), we also illustrated the importance of considering the influence of the length of data series through two simple simulation experiments. The preferences for the choice of copulas were still visible, but less evident. The results indicated that the acceptance of copula models can be conditioned on the flood types but the length of the series and possibly also the homogeneity of the flood types within a region may play an important role.

Table 1. The basic catchments characteristics used in this study. ALL SITES = all sites included (72 sites with no regional delineation), SUBREGION #1 = Innviertel + Hausruckviertel (24 sites), SUBREGION #2 = Traunviertel + Flysh (22 sites) and SUBREGION #3 = Mühlviertel + Waldviertel (26 sites). AREA – catchment area [km^2], MCE – mean catchment elevation [m a.s.l.], MRC – mean runoff coefficient of maximum annual flood events [–], MCS – mean catchment slope [–], PFA – percentage of forest area [–].

		AREA [km^2]	MCE [m a.s.l.]	MRC [–]	MCS [–]	PFA [–]
ALL SITES	maximum	444.3	888.0	0.85	0.33	1.00
	median	78.6	571.0	0.49	0.09	0.30
	minimum	10.6	342.0	0.19	0.02	0.00
SUBREGION #1	maximum	361.8	598.0	0.77	0.14	0.46
	median	67.4	450.5	0.60	0.10	0.13
	minimum	14.2	385.0	0.36	0.07	0.01
SUBREGION #2	maximum	444.3	839.0	0.74	0.33	1.00
	median	55.8	571.0	0.49	0.10	0.34
	minimum	12.0	342.0	0.20	0.02	0.00
SUBREGION #3	maximum	305.9	888.0	0.85	0.16	0.82
	median	120.6	707.5	0.42	0.09	0.48
	minimum	10.6	480.0	0.19	0.03	0.26

Here, these approaches are followed in a more differentiated regional look at the selection of a hydrological process-oriented copula model for flood peak/volume relationships. Specifically, we are again interested in the following questions:

- how similar are flood peak/volume relationships of different flood types in a climatically rather homogeneous region but in geologically different subregions,
- which factors may play a role in forming the similarity of copula models in the subregions and
- if recommendations could be formulated for engineering studies with respect to the suitability of some copula types for the given flood processes?

2 The pilot region

There are a wide variety of flood-generation mechanisms across Austria (e.g., Merz and Blöschl, 2003), which result in complex flood peak-volume relationships (Gaál et al., 2012, 2015; Szolgay et al., 2015). In order to decrease the complexity of runoff generation schemes in this analysis, we decided to reduce the climate variability through the selection of a geographically limited area, i.e., the Northern Lowlands region of Austria (Fig. 1), as a pilot region. The area is dominated by lowlands and hilly sites, with elevations ranging

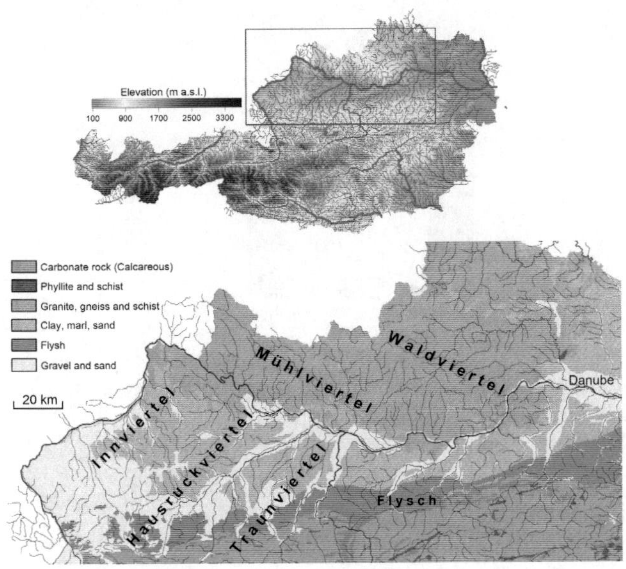

Figure 1. Location and geology of the pilot region in Austria and the subregions considered.

from about 400 to 1500 m a.s.l. The region is generally under the influence of air masses from the Atlantic. The mean annual precipitation in the target region shows a decreasing western-to-eastern gradient. Orographic enhancement is not significant; the annual rainfall amounts (from about 500 to 1500 mm) are significantly lower than in the Austrian Alps. The basic climatic and physiographic characteristics of the catchments are listed in Table 1.

Floods in the Northern Lowlands region may occur during the whole year. The winter floods are usually induced by snowmelt and rain-on-snow processes, when antecedent snowmelt saturates the soils, and temperature increases and/or relatively low rainfall intensities may then cause significant floods. Flash floods are mostly caused by convective events or cold fronts; synoptic floods depend on the particular circulation pattern, but westerly circulation prevails.

The data set used in this paper builds on the Austrian flood data described in detail in Szolgay et al. (2015) and the papers referenced therein. The 72 small and mid-sized catchments analyzed have areas in the range of 10.6 to 444.3 km^2 (median: 78.6 km^2), while the range of their mean elevations is from 342 to 888 m a.s.l. (median: 571 m a.s.l.). The time resolution of the runoff data was one hour. Three subregions were formed on the basis traditional geographical/geomorphological units based on characteristic geomorphology, geology and pedology (Fig. 1) using results of Gaál et al. (2012). The first subregion consists of the Weinviertel and Mühlviertel areas in the north, with a prevailing geology made of granite. In this region floods appear throughout the year. The spring floods are apparently associated with snow as indicated by the daily cycle of some of the events. In this subregion, the meteorological forcing appears to be due to a range of processes, including convective storms in summer as indicated by the slim shape of some of the summer hydrographs. The second subregion consists of the Innviertel and Hausruckviertel areas, where catchments have more winter floods than in the other subregions. The geology in this subregion is prevailingly formed of gravel and sand. In the third subregion, which consists of the Traunviertel and parts of the Flysh, the catchments have more summer floods and flash floods. For geology in this subregion marl and sandstone are characteristic.

3 Methodology

Instead of the traditional engineering approach which often deals with flood volumes associated with the annual maxima of flood events, the current analysis intends to include all the flood events in the region which can be hydrologically regarded as independent. According to our understanding, two subsequent flood events are independent, when they do not originate from the same synoptic situation. We assumed that after a seven-day period, on the average, a completely different atmospheric circulation situation occurs in Central Europe (Gaál et al., 2015; Szolgay et al., 2015). The flood type classification according to the genesis of events in the region was introduced by Merz and Blöschl (2003) and modified in Gaál et al. (2015). As a further modification to that, the 25 697 flood events from the 1976–2007 period were classified as synoptic floods (originally long or short rain-induced floods in Merz and Blöschl, 2003), flash floods, and snowmelt floods (originally rain-on-snow floods or snowmelt floods).

In this paper we are interested in the similarity of empirical copulas of flood peak-volume pairs, which we tested accordingly to the approach of Remillard and Scaillet (2009). It is based on the Cramér-von Mises type of distance measure:

$$S = \sum_i \left[C_{n1}\left(U_{1,i}, U_{2,i}\right) - C_{n2}\left(U_{1,i}, U_{2,i}\right)\right]^2 \qquad (1)$$

where C_{n1} and C_{n2} denote two empirical copulas. The probability distribution of S, given that the null hypothesis (H_0: the empirical copulas come from the same – unknown – bivariate distribution) holds, is unknown and needs to be bootstrapped. The result of the similarity testing is a p value, which is the percentage of how many simulations of the test statistic (under H_0) exceeded the estimator from the observations. Since the test statistic represents a distance, the lower values (a smaller departure from the null hypothesis statement) lead to larger p values, which however should not be used as a measure of dissimilarity.

As in Szolgay et al. (2015), nine frequently-used copulas were chosen to be fitted locally, specifically from the Archimedean class (Clayton, Frank, Gumbel-Hougaard and Joe copulas), the extreme-value class (Gumbel-Hougaard, Galambos, Hüsler-Reiss), the elliptical class (normal, Student t), and finally the Plackett copula. (The abbrevi-

Table 2. Number of identified independent flood events, stratified by regions and flood types. ALL SITES = all sites included (72 sites with no regional delineation), SUBREGION #1 = Innviertel + Hausruckviertel, SUBREGION #2 = Traunviertel + Flysh and SUBREGION #3 = Mühlviertel + Waldviertel.

	Number of catchments	All floods	Synoptic floods	Flash floods	Snowmelt floods
ALL SITES	72	25697	19093	1733	4871
SUBREGION #1	24	9396	7325	596	1475
SUBREGION #2	22	8788	6448	652	1688
SUBREGION #3	26	7513	5320	485	1708

ations of the copulas used throughout the paper are: cla = Clayton, fra = Frank, gal = Galambos, gum = Gumbel-Hougaard, hus = Hüsler-Reiss, joe = Joe, nor = Normal, pla = Plackett, tco = Student t copula). All the copulas are single parameter copulas with the exception of the Student t copula. In order to be consistent with the other copulas, we fixed the second parameter to effectively make it a single parameter copula. We chose a value of four for the second parameter to distinguish it from the normal copula.

As in Szolgay et al. (2015), the parameter θ of the copulas was estimated by maximizing the so-called pseudo-likelihood function:

$$L(\theta) = \sum\nolimits_i \log\left[c_\theta\left(U_{1,i}, U_{2,i}\right)\right], \quad (2)$$

which, in addition to the copula density c_θ, contains pseudo-observations $U_{j,i}$ ($i = 1, \ldots n$, $j = 1, 2$), i.e., a transformation of n real observations of random variable X_j, by means of a corresponding empirical distribution function (sometimes referred to as the "plotting position"). The goodness-of-fit was examined locally as well as analyzed in a regional scope. The goodness-of-fit of the parametric copulas under consideration was tested by a "blanket" test (Genest et al., 2009) with the Cramér-von Mises measure of distance:

$$S_n = \sum\nolimits_{i=1}^{n}\left[C_\theta\left(U_{1,i}, U_{2,i}\right) - C_n\left(U_{1,i}, U_{2,i}\right)\right]^2 \quad (3)$$

between the parametric copula C_θ and the empirical copula defined by:

$$C_n\left(u_1, u_2\right) = \frac{1}{n}\sum\nolimits_{i=1}^{n} 1\left(U_{1,i} \le u_1\right) 1\left(U_{2,i} \le u_2\right). \quad (4)$$

The probability distribution of the test statistic S_n, given that the null hypothesis (H_0: C_θ fits well) holds, is unknown and needs to be bootstrapped. Consequently, the p value is a percentage of how many simulations of the test statistic (under H_0) exceeded the estimator from the actual observations.

4 Results and discussion

The regional distribution of the flood types for each subregion is shown in Table 2. Locally, the percentage of the flood types ranges between 51 and 74 % in the case of synoptic floods and 8 and 46 and 2 and 19 % for the snowmelt and flash floods, respectively. In the selection there are no catchments with a missing flood type. The relatively smaller number of snowmelt floods is due to the modest elevations in the region. The fact that flash floods represent 6.7 % of all the events is not negligible, since in Gaál et al. (2015), the flash floods only represented 1.8 % based on the annual maxima flood events.

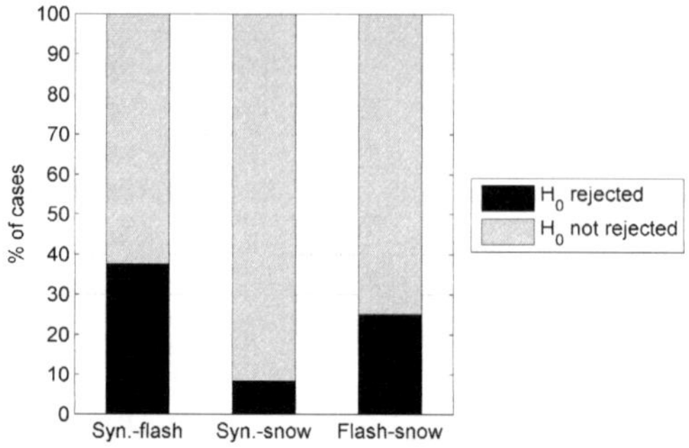

Figure 2. Results of the comparison of the empirical copulas locally: per cent ratio of the catchments where the equivalence of the pairs of empirical copulas was rejected or could not be rejected at the given level of significance (here $\alpha = 10\,\%$). The black color indicates the per cent ratio of the catchments where the null hypothesis about the equality of the empirical copulas for the given two flood processes was rejected.

First, a comparison of the empirical copulas for the different flood types locally was performed. In this analysis we were interested in whether different flood types, for the same catchment were distinguishable in terms of their empirical flood peak-volume copulas. The analysis was carried out for each catchment separately, and the flood samples of the process types were compared pairwise, i.e., synoptic floods vs. flash floods, synoptic floods vs. snowmelt floods, and flash floods vs. snowmelt floods. The results, which are shown in Fig. 2, suggest that synoptic and snowmelt floods could belong more often to the same unknown copula than is the case for the other combinations. This suggests that the synoptic and snowmelt floods are more similar to each other (in terms of their empirical copulas) than the other process pairs; or, in other words, flash floods tend to be more dissimilar from both the synoptic and snowmelt floods. This could be partly related to much stronger upper tail dependence of flash floods and their specific (similar) hydrograph shapes (Gaál et al., 2015). However, the relatively small number of events for such type of analysis in general and the differing number of events in the respective flood types in particular, may also play a role in the fact, that the analysis has not brought really conclusive results. These are objective factors, which cannot be overcome in the framework of comparative hydrology, when using only data available in practice.

Next, a comparison of the empirical copulas for each flood type regionally was performed where we were interested in whether different catchments with the same flood type are distinguishable in terms of their empirical peak-volume copulas (Fig. 3). It can be seen that the empirical copulas of synoptic floods are the least similar between the catchments. This seems to be surprising; a closer look into the subregions showed that this phenomenon is more pronounced in the southwest than in the north. The high ratio of rejections of the synoptic events across different pairs of sites is therefore likely to be related to the more complex temporal rainfall structure, the mix of long and short rain processes in the dataset, and the more complex geology resulting in a lower degree of similarity between the different events and sites which may become more evident, when the sample size increases (as it is in the case of synoptic floods when compared to the other two types). A more detailed analysis of the phenomenon is beyond the goals of the present study. In the case of flash floods and snowmelt, the difference between the process types is smaller; the analysis suggests that most catchment pairs, for a given flood type are not distinguishable in terms of their empirical peak-volume copulas.

Next, we attempted a more detailed flood typology and regional differentiation in fitting the copulas to the data than that in Szolgay et al. (2015). The goodness-of-fit test of the nine copula types at the 72 catchments for all the floods merged into one set; and the three flood types separately are shown in Fig. 4, stratified by subregions. The copula types (each column represents a copula type) are organized alphabetically, while the subregions are visualized by different color bars, which indicate $p \leq 0.05$, i.e., a rejection of the null hypothesis H_0. The number of catchments in the subregions is comparable, but the number of events analyzed was naturally different for each flood type.

In the case of analyzing all the floods together, we can see, that the three extreme value copulas (the Galambos, Gumbel-Hougaard and Hüsler-Reiss copulas) and the normal copula performed best in all the subregions (except in the southwest, where the extreme value models clearly outperform the normal). In 15 % of the catchments all nine models were rejected, the acceptance rate oscillates around 50 per cent in general. This relatively low rate (when compared to the process-wise analysis below) could be attributed to the mix of flood types in the merged data sample, but the effect of a relatively large sample size when compared to the other types also cannot be excluded (larger sample decreases the uncertainty even to an extent, where the models usually considered in practical applications are not suitable at all in some cases). The acceptance rate of these four copulas improved for the synoptic floods (still in 7 % of the catchment no one model was found suitable), but was highly variable across the subregions, which was not expected. This could be the result of the mix of short and long rain processes (Merz and Blöschl, 2003) and a smaller number of events in the subsamples; however, a detailed analysis is beyond the goals of this

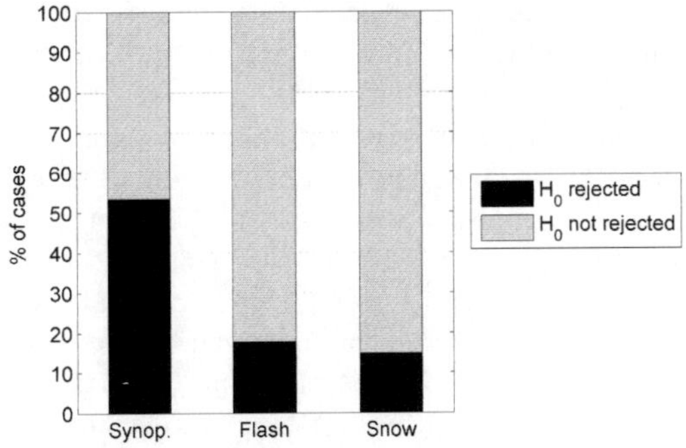

Figure 3. Results of the comparison of the empirical copulas regionally: per cent ratio of the cases where the equivalence of the empirical copulas was rejected or could not be rejected at the given level of significance (here $\alpha = 10\,\%$). Within the set of 72 sites, and for each flood process separately, altogether 2556 ($72 \times 71/2$) pairwise comparisons were carried out. The black color indicates the percent ratio of the cases where the null hypothesis about the equality of the empirical copulas for the given flood process was rejected.

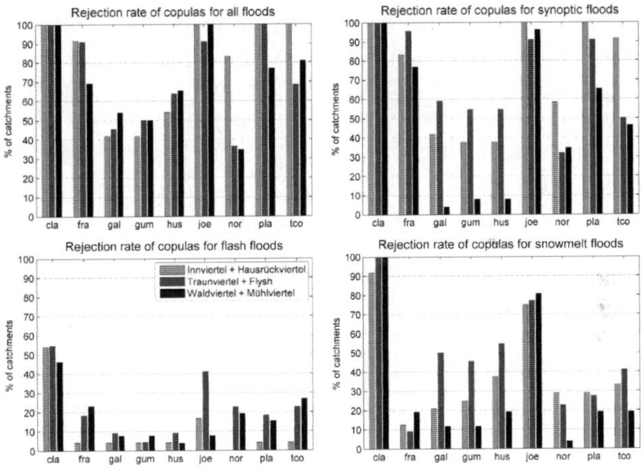

Figure 4. Results of the goodness-of-fit test of the selected nine copula types within the three subregions (indicated by colors), for the all the floods merged in a single data set (top left) and for flood types treated separately. The bars indicate per cent ratio of the catchments where the given copula type was rejected.

study. Interestingly, the extremal copulas exhibited a larger rejection rate in the Traunviertel and Flysh subregion, and the Student t copula can also be considered as applicable in practice in the region. The rejection/acceptance rate pattern in the case of snowmelt floods is clearly different and adds the Frank, Plackett and Student models into the basket of acceptable models (again in two catchments no suitable model was found). Nevertheless, the behavior of snowmelt floods within the subregion Traunviertel and Flysh is again different. An explanation could be found in the fact, that in this region, the underlying geology is more variable; the variability of the mean catchment elevation covers the other two regions (for two clearly disjunctive sets) and the time scales and their variability partially overlaps with the Waldviertel–Mühlviertel region (Gaál et al., 2012). However deepening such an analysis was beyond the scope of the research in this case. Further, one must also suspect, that the different regional patterns of the copula suitability in winter can also be related to the fact that the number of winter events is lower compared to the summer floods (though larger than in many copula studies based on annual maximum floods; Szolgay et al., 2015). The behavior of the flash floods (the smallest sample size) in the subregions is very similar to that observed in Szolgay et al. (2015) for the whole region and allows for a choice of almost all the models (despite the rather small variability of the flood wave shapes in the region; see, e.g., Gaál et al., 2015). In all the catchments at least one suitable model was found.

These results indicate that the acceptance of a particular copula model can be conditioned on the processes but that the size of the data samples and possibly also the homogeneity of the region with respect to the flood formation factors and flood types within the data set plays a role. However, uncertainties inherent in the bivariate frequency analysis methodology itself in real world applications (e.g., small samples for reliable model identification and upper tail dependence description) may not be overcome in the scope of such a regional comparative analysis using real data as conducted here.

5 Conclusions

Not much attention has been paid so far to directing a multivariate analysis of floods toward the selection of models for specific runoff generation processes. Here, this issue was addressed in a regional context by the differentiation of the flood types into three categories. Based on the results, it can be concluded that modeling dependence structure by treating flood processes separately in a regional context may prove beneficial with respect to narrowing the choice of acceptable models, since the suitability patterns of acceptable copula types are distinguishably different for the subregions/flood-types considered. This could help analysts to overcome some difficulties in the choice of the model caused by the inadequate length of a data series. On the other hand, it was also shown that a more detailed differentiation of the flood types and subregions opens in the selection of the model a greater degree of uncertainty than expected in Szolgay et al. (2015), which does not make the task easier for an analyst in practice. Despite that shortcoming, given that usually more than one statistically suitable dependence model exists, a regional analysis and an uncertainty analysis of the design values in the engineering studies resulting from the choice of a model can be recommended, especially for important water resources projects.

Our results support Favre et al. (2004) and Serinaldi and Kilsby (2013), who emphasized that further work is needed to choose the best copulas capable of reproducing the dependence structure of multivariate hydrological variables. But, as shown in a comparative hydrology framework above, the choice of the copula model that best fits the observed data and is regionally acceptable in term of flood typology, is not a trivial issue, even if more than statistical aspects are taken into consideration, since the lack of sufficient data makes the analysis difficult (if not even impossible). As a general recommendation resulting from this study, it is advisable to select models from the extreme value class of copulas (in the given region).

Note that even the adoption of generally accepted and widely used copula models may not lead to a successful bivariate fitting. Uncertainties inherent in the copula-based bivariate frequency analysis itself (caused, among others, also by the relatively small samples sizes for consistent copula model selection, upper tail dependence characterization and reliable predictions) may not be overcome in the scope of such a regional comparative analysis.

Based on this comparative study and results of other more advanced studies (e.g., Serinaldi, 2013, 2015) it can be concluded, if reliable predictions will be required for an important engineering application, the benefits of regional bivariate frequency analysis methods could be further explored (e.g., Ben Aissisa et al., 2015) or the potential of the combination of rainfall generators, rainfall runoff models, analysis of historical floods and advanced statistics considering uncertainty might be utilized.

Acknowledgements. We would like to thank the Austrian Academy of Sciences (International Strategy for Disaster Reduction Programme, IWHRE2008, 2008–2013), for its financial support. This research was also supported by the Slovak Research and Development Agency under Contract no. APVV 0496-10, by the Slovak Grant Agency VEGA under Project no. 1/0776/13 and by the IMPALA project (FP7-PEOPLE-2011-IEF-301953) of the Marie Curie Intra European Fellowship. This publication was further supported by Competence Center for SMART Technologies for Electronics and Informatics Systems and Services, ITMS 26240220072, funded by the Research & Development Operational Programme from the ERDF.

The authors would also like to thank Francesco Serinaldi for a very inspiring open discussion on the topic during a review of their other paper.

References

Ben Aissia, M. A., Chebana, F., Ouarda, T. B. M. J., Bruneau, P., and Barbet, M.: Bivariate index-flood model: case study in Québec, Canada, Hydrolog. Sci. J., 60, 247–268, doi:10.1080/02626667.2013.875177, 2015.

Chowdhary, H., Escobar, L. A., and Singh, V. P.: Identification of suitable copulas for bivariate frequency analysis of flood peak and flood volume data, Hydrol. Res., 42, 193–216, doi:10.2166/nh.2011.065, 2011.

De Michele, C., Salvadori, G., Canossi, M., Petaccia, A., and Rosso, R.: Bivariate statistical approach to check adequacy of dam spillway, J. Hydrol. Eng., 10, 50–57, 2005.

Favre, A.-C., El Adlouni, S., Perreault, L. Thiémonge, N., and Bobeé, B.: Multivariate hydrological frequency analysis using copulas, Water Resour. Res., 40, W01101, doi:10.1029/2003WR002456, 2004.

Gaál, L., Szolgay, J., Kohnová, S., Parajka, J., Merz, R., Viglione, A., and Blöschl, G.: Flood timescales: Understanding the interplay of climate and catchment processes through comparative hydrology, Water Resour. Res., 48, W04511, doi:10.1029/2011WR011509, 2012.

Gaál, L., Szolgay, J., Kohnová, S., Hlavčová, K., Parajka, J., Viglione, A., Merz, R., and Blöschl, G.: Dependence between flood peaks and volumes – A case study on climate and hydrological controls, Hydrolog. Sci. J., 60, 968–984, doi:10.1080/02626667.2014.951361, 2015.

Genest, C. and Favre, A.-C.: Everything you always wanted to know about copula modeling but were afraid to ask, J. Hydrol. Eng., 12, 347–368, 2007.

Genest, C., Rémillard, B., and Beaudoin, D.: Goodness-of-fit tests for copulas: A review and a power study, Ins. Math. Econ., 44, 199–213, doi:10.1016/j.insmatheco.2007.10.005, 2009.

Goel, N. K., Seth, S. M., and Chandra, S.: Multivariate modeling of flood flows, ASCE, J. Hydraul. Eng., 124, 146–155, 1998.

Merz, R. and Blöschl, G.: A process typology of regional floods, Water Resour. Res., 39, 1340, doi:10.1029/2002WR001952, 2003.

Requena, A. I., Mediero, L., and Garrote, L.: A bivariate return period based on copulas for hydrologic dam design: accounting for reservoir routing in risk estimation, Hydrol. Earth Syst. Sci., 17, 3023–3038, doi:10.5194/hess-17-3023-2013, 2013.

Remillard, B. and Scaillet, O.: Testing for equality between two copulas, J. Multivariate Anal., 100, 377–386, doi:10.1016/j.jmva.2008.05.004, 2009.

Shiau, J. T.: Return period of bivariate distributed extreme hydrological events, Stoch. Env. Res. Risk A., 17, 42–57, doi:10.1007/s00477-003-0125-9, 2003.

Serinaldi, F.: An uncertain journey around the tails of multivariate hydrological distributions, Water Resour. Res., 49, 6527–6547, doi:10.1002/wrcr.20531, 2013.

Serinaldi, F.: Can we tell more than we can know? The limits of bivariate drought analysis in the United States, Stoch. Env. Res. Risk A., 1–14, doi:10.1007/s00477-015-1124-3, 2015.

Serinaldi, F. and Kilsby, C. G.: The intrinsic dependence structure of peak, volume, duration, and average intensity of hyetographs and hydrographs, Water Resour. Res., 49, 3423–3442, doi:10.1002/wrcr.20221, 2013.

Szolgay, J., Gaál, L., Kohnová, S., Hlavcová, K., Výleta, R., Bacigál, T., and Blöschl, G.: A process-based analysis of the suitability of copula types for peak-volume flood relationships, Proc. IAHS, 370, 183–188, doi:10.5194/piahs-370-183-2015, 2015.

Yue, S., Ouarda, T. B. M. J., Bobeé, B., Legendre, P., and Bruneau, P.: Approach for describing statistical properties of flood hydrograph, J. Hydrol. Eng., 7, 147–53, 2002.

Proc. IAHS, 373, 69–72, 2016
proc-iahs.net/373/69/2016/
doi:10.5194/piahs-373-69-2016

Added-value from a multi-criteria selection of donor catchments in the prediction of continuous streamflow series at ungauged pollution control-sites

Gilles Drogue[1], **Wiem Ben Khediri**[1], **and Céline Conan**[2]

[1]Laboratoire LOTERR, Université de Lorraine, Metz, 57045 CEDEX 1, France
[2]Agence de l'Eau Rhin-Meuse, Moulins Lès Metz, 57161 BP 30019, France

Correspondence to: Gilles Drogue (gilles.drogue@univ-lorraine.fr)

Published: 12 May 2016

Abstract. We explore the potential of a multi-criteria selection of donor catchments in the prediction of continuous streamflow series by the spatial proximity method. Three criteria have been used: (1) spatial proximity; (2) physical similarity; (3) stream gauging network topology. An extensive assessment of our spatial proximity method variant is made on a 149 catchment-data set located in the Rhine-Meuse catchment. The competitiveness of the method is evaluated against spatial interpolation of catchment model parameters with ordinary kriging. We found that the spatial proximity approach is more efficient than ordinary kriging. When distance to upstream/downstream stream gauge stations is considered as a second order criterion in the selection of donor catchments, an unprecedented level of efficiency is reached for nested catchments. Nevertheless, the spatial proximity method does not take advantage from physical similarity between donor catchments and receiver catchments because catchments that are the most hydrologically similar to each catchment poorly match with the catchments that are the most physically similar to each catchment.

1 Introduction

Because quantitative hydrological information contribute to the understanding of short and medium-term fluctuations in pollutants concentrations recorded in river flows (e.g. Burt et al., 2010), there is a strong demand to reconstruct continuous flow time series at ungauged pollution-control sites for short (i.e. up to 15 days) and medium (i.e. up to 180 days) lead times, before and after the river water sampling has occurred. The best way to handle this problem would be to set up a coordinated monitoring network in order to get continuous streamflow series at the point where water quality data are collected. Unfortunately, different brakes like financial costs, logistics and most often diverging interest between water flow managers prevent the design of such a coordinated data-acquisition network. Therefore, rainfall-runoff modelling strategies can be viewed as surrogate models for reconstructing and simulating continuous flow time series at ungauged pollution control-sites. In the French part of the Rhine-Meuse catchment, we made different attempts to produce daily streamflow series through regionalisation of catchment model parameters (Drogue and Plasse, 2014; Plasse et al., 2014). In the worst case, where no hydrological information is available at the point of interest, we came to the conclusion that, in our temperate non Mediterranean area, the spatial proximity method is the best approach to solve the regionalisation problem of hydrograph prediction provided for the hydrological network is sufficiently dense (i.e. 1 station per 250 km^2). But our results have also shown that there is still a considerable room for progress in reducing the prediction error in such hydrological information at ungauged catchments. For that reason and also because there is an important part of nested catchments in our catchment-data set, we test a variant of the spatial proximity approach where the donor catchments are not only selected according to their spatial proximity but also according to their physical similarity and gauging network connectivity from the receiver catchment.

2 Setting the scene: study area and datasets

2.1 Study area

The investigated territory corresponds to the French part of the Rhine-Meuse catchment (North-Eastern part of France). The presence of the Vosges Mountains induces climatic gradients among highest of France. Considering the weak influence of snow on the hydrological regime of the upstream mountain rivers, the snow component is not taken into account in the regionalisation study reported in this paper.

2.2 Datasets

We split the entire period of observations (1990–2002) into two periods: from 1 January 1990 to 31 December 1995 and from 1 January 1996 to 31 December 2002. Warm up periods of one year has been used in both cases. For the efficiency estimation of the catchment model regionalisation we used the second sub-period as a confirmatory period.

2.2.1 Catchment-data set

We made an extensive assessment of our regionalisation scheme on a dense stream gauging network (approximately 1 gauging station for 250 km^2) comprising 149 reliable stations gauging non-regulated rivers and providing daily streamflow values validated on the target period 1990–2002. More than half of these stations are also dedicated to water quality monitoring. The drainage areas lie between 5 and 11 500 km^2. The catchment set includes 40 % of nested catchments (i.e. 60 % of adjacent catchments). The territory covered by our catchment sample is approximately 38 000 km^2. This accounts for 15 % of the total drainage area of the Rhine-Meuse catchment ($\approx$ 200 000 km^2 at the outlet).

2.2.2 Climate forcings and rainfall-runoff model

In addition to streamflow data, we also collected daily precipitation and potential evapo-transpiration from the SAFRAN gridded climatology data. The daily lumped GR4J rainfall–runoff model (Perrin et al., 2003) has been calibrated over the training period (1990–1995) by using the Broyden–Fletcher–Goldfarb–Shanno algorithm (hill climbing optimization technique; see Byrd et al., 1995). Parameters have been optimized with the Nash–Sutcliffe coefficient computed on the square root of the daily streamflow series (NS_{sqrtQ}). This power transformation makes possible to define a parameter set representative of the catchment behaviour on all the streamflow range. The repetition of the split-sample test for all the catchment-data set allows setting up a regional library containing 149 vectors of four optimal parameters.

3 How strengthening the spatial proximity approach?

In a first simulation experiment, we optimized the performance of the spatial proximity approach allowing the output averaging option and four geographic neighbours (Plasse et al., 2014). In this basic version of the spatial proximity approach, we filter out poorly modelled donor catchments, i.e. the ones having a NS_{sqrtQ} below 0.7 in calibration mode. In this study we go one step further by selecting geographic neighbours according to a regional composite rank mixing elementary ranks related to spatial proximity, physical similarity and stream gauging network topology. The experiment design is described in Fig. 1.

As for a truly ungauged catchment, streamflow hydrograph is unkown, we apply a method that introduces hydrological catchment behaviour in the assessment of catchment physical similarity (Oudin et al., 2010). We compute two catchment classifications (a hydrological one and a physical one) where similarity is defined as an Euclidean distance in the catchment property space: for the hydrological classification we use seven hydrological signatures calculated on the 1990–2002 period (runoff coefficient, lag time, Base Flow Index, slopes of the Flow Duration Curve for the high flow range, the low flow range and the medium flow range, rising limb density); for the physical classification we use 70 catchment attributes related to climate, geology, land cover, hydrology and morphology and a weighted version of the Euclidean distance for estimating the catchment similarity. Then, we sought physical attributes allowing the best matching between both classifications according to the Adjusted Rand Index – ARI – (Hubert and Arabie, 1985). We also incorporate the hydraulic connectivity of nested catchments in the composite rank calculation (Fig. 1). We proceed in a very simple way: for a target point having upstream/downstream neighbour(s), we assign a topological rank to neighbour(s) according to the stream network distance between the target point and its neighbours. The closest is the gauging station the smallest is the rank. For adjacent stations, we add one to the r_{topo} value of the furthest upstream/downstream neighbour(s) rank value. For an adjacent receiver catchment, the topological rank is set to 0. We examine the efficiency of our regionalisation method by jakknife cross validation (Fig. 1). As the catchment-data set is quite large, a $[-1; 1]$ bounded version of the NS_{sqrtQ} criterion (called C2M) is calculated for efficiency estimation (Mathevet et al., 2006). For sake of robustness evaluation, the GR4J model parameters are also regionalised by ordinary kriging (OK). The variogram properties of GR4J model parameters are summarized in Table 1. As the X2 parameter has no spatial autocorrelation, the median value is used.

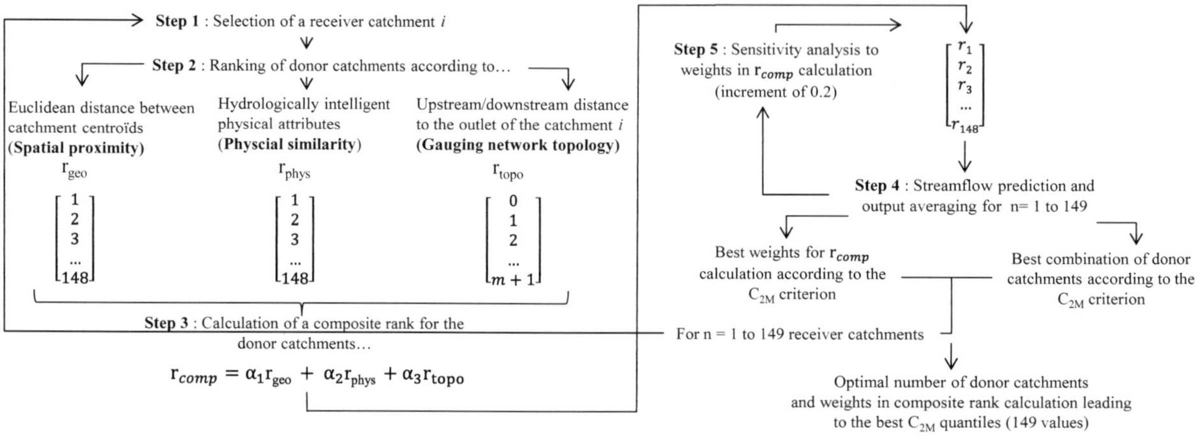

Figure 1. Flowchart of the regionalisation scheme applied to our catchment-data set.

Table 1. Variogram models used to interpolate the X1, X3 and X4 GR4J model parameters in the study area.

model parameter	Definition	Variogram model
X1 (mm)	maximal capacity of the production reservoir	spherical; range: 100 km; sill: 250 000 mm^2
X3 (mm)	capacity of the non linear routing reservoir	spherical; range: 37 km; sill: 3100 mm^2
X4 (day)	unit hydrograph time base	spherical; range: 35 km; sill: 0.25 day^2

Table 2. Physical attributes maximizing the Adjusted Rand Index (ARI) computed between the classification based on physical attributes and the classification based on hydrological signatures.

Iterations	Physical attributes	Number of clusters	Weights in Euclidean distance metric	Adjusted Rand Index (–)
1	Catchment perimeter	13	1	0.333
2	Catchment perimeter (CP) + proportion of Strahler stream order 5 (SSO5)	13	CP: 0.5 SSO5: 0.5	0.545

4 Results and discussion

4.1 Which physical attribute(s) of catchment could be used as proxie(s) for hydrological similarity?

The catchment classification based on hydrological signatures leads to nine clusters. The two physical attributes optimizing the ARI index are the catchment perimeter and the proportion of Strahler stream order 5 (Table 2). Nevertheless, the intermediate value of ARI (0.545) shows that physical catchment characteristics are poor proxies for describing hydrological behaviour similarity patterns. In order to orient future researches, a focus should be made on hydrological signatures that discriminate the most hydrologically similar catchments and physically similar catchments (Oudin et al., 2010).

4.2 Overall performance of the multi-criteria spatial proximity approach

Looking at Fig. 2 we can see that the spatial proximity method produces less prediction error than OK over the confirmatory period. An improvement of model hydrograph prediction at ungauged sites could be obtained when using the multi-criteria selection approach with four donor catchments, especially for well regionalised catchments (Fig. 2). Optimal weights in composite rank leading to that result are respectively $\alpha_1 = 0.8$ for r_{geo} (spatial proximity), $\alpha_2 = 0.1$ for r_{phys} (physical similarity) and $\alpha_3 = 0.1$ for r_{topo} (distance between upstream/downstream gauging stations). It means that for the considered monitoring stream gauge network, physical similarity and network topology have a second order effect in the selection of pertinent donor catchments in comparison to Euclidean distance between catchment centroids. Therefore, as a first guess and with the specificity of our study area,

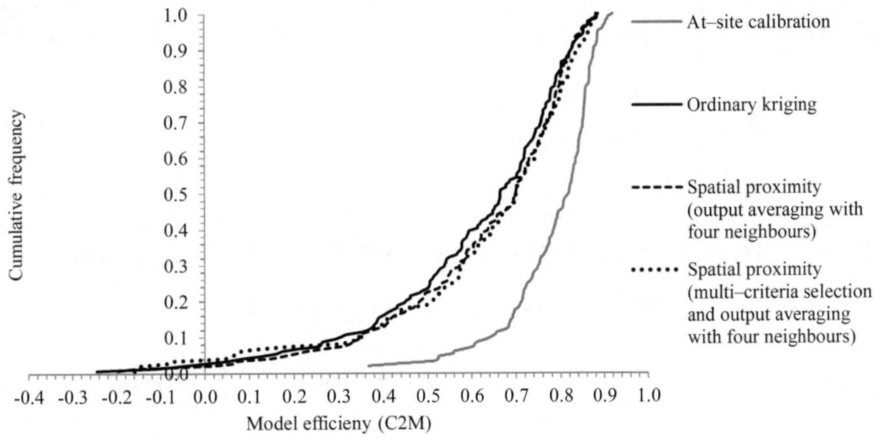

Figure 2. Cumulative distribution functions (CDFs) of GR4J model efficiencies using model parameters regionalised by three methods. CDF for at-site calibration efficiencies is also shown. Results are given in validation mode.

the simple spatial proximity method can be regarded as good enough (see also Parajka et al., 2005).

5 Conclusions

We tested a multi-criteria variant of the spatial proximity approach for daily hydrograph prediction at ungauged sites. Three criteria were used to choose the neighbours of a target catchment: spatial proximity, physical similarity (conditioned by catchment hydrological similarity patterns) and distances between upstream/downstream neighbours for nested catchments. No added value comes up from using physical similarity in the selection process of donor catchments. Poorly regionalised catchments do not take advantage from the multi-criteria approach. For well modelled nested catchments, prediction error of hydrograph in ungauged conditions could be reduced by prioritizing upstream/downstream neighbours among the closest donor catchments. In light of these results, hydrograph prediction over ungauged catchments through catchment model regionalisation definitely appears as a learning process.

Acknowledgements. We would like to thank Météo France for providing the SAFRAN reanalysis data, the SCHAPI for providing access to the HYDRO streamflow archive, the German Federal Waterways and Shipping Administration as well as the German Federal Institute of Hydrology (BfG) for providing the German streamflow data, and the Service public de Wallonie, Direction générale opérationnelle "Mobilité et Voies hydrauliques", Direction de la Gestion hydrologique intégrée, Service d'Etudes Hydrologiques (SETHY) for making the Belgian streamflow data freely available. We also would like to acknowledge the Rhine-Meuse French Water Agency for its financial support (Grant no. 15C54112) and the two anonymous reviewers for their thorough review.

References

Burt, T. P., Howden, N. J. K., Worrall, F., and Whelan, M. J.: Long-term monitoring of river water nitrate: how much data do we need?, J. Environ. Monitor., 12, 71–79, doi:10.1039/b913003a, 2010.

Byrd, R. H., Lu, P., Nocedal, J., and Zhu, C.: A limited memory algorithm for bound constrained optimization, SIAM J. Sci. Comput., 16, 1190–1208, doi:10.1137/0916069, 1995.

Drogue, G. and Plasse, J.: How can a few streamflow measurements help to predict daily hydrographs at almost ungauged sites?, Hydrolog. Sci. J., 59, 2126–2142, doi:10.1080/02626667.2013.865031, 2014.

Hubert, L. and Arabie, P.: Comparing partitions, J. Classif., 2, 193–218, 1985.

Mathevet, T., Michel, C., Andreassian, V., and Perrin, C.: A bounded version of the Nash-Sutcliffe criterion for better model assessment on large sets of basins, IAHS Red Books Series, 307, 211–219, 2006.

Oudin, L., Kay, A., Andreassian, V., and Perrin, C.: Are seemingly physically similar catchments truly hydrologically similar? Water Resour. Res., 46, W11558, doi:10.1029/2009WR008887, 2010.

Parajka, J., Merz, R., and Blöschl, G.: A comparison of regionalisation methods for catchment model parameters, Hydrol. Earth Syst. Sci., 9, 157–171, doi:10.5194/hess-9-157-2005, 2005.

Perrin, C., Michel, C., and Andreassian, V.: Improvement of a parsimonious model for streamflow simulation, J. Hydrol., 279, 275–289, doi:10.1016/S0022-1694(03)00225-7, 2003.

Plasse, J., Drogue, G., François, D., Battaglia, P., and Conan, C.: Using point flow measurements for guided reconstruction of daily streamflow time series through continuous flow simulation: a regional case study on the Rhine-Meuse district, La Houille Blanche, 1, 45–52, doi:10.1051/lhb/2014007, 2014.

Proc. IAHS, 373, 73–80, 2016
proc-iahs.net/373/73/2016/
doi:10.5194/piahs-373-73-2016

Open Access
PIAHS

Spatially-smooth regionalization of flow duration curves in non-pristine basins

Daniele Ganora, Francesco Laio, Alessandro Masoero, and Pierluigi Claps

Politecnico di Torino, Department of Environment, Land and Infrastructure Engineering, Torino, Italy

Correspondence to: Daniele Ganora (daniele.ganora@polito.it)

Published: 12 May 2016

Abstract. The flow duration curve (FDC) is a fundamental signature of the hydrological cycle to support water management strategies. Despite many studies on this topic, its estimation in ungauged basins is still a relevant issue as the FDC is controlled by different types of processes at different time-space scales, thus resulting quite sensitive to the specific case study.

In this work, a regional spatially-smooth procedure to evaluate the annual FDC in ungauged basins is proposed, based on the estimation of the L-moments (mean, L-CV and L-skewness) through regression models valid for the whole case study area. In this approach, homogeneous regions are no longer required and the L-moments are allowed to continuously vary along the river network, thus providing a final FDC smoothly evolving for different locations on the river. Regressions are based on a set of topographic, climatic, land use and vegetation descriptors at the basin scale. Moreover, the model ensures that the mean annual runoff is preserved at the river confluences, i.e. the sum of annual flows of the upstream reaches is equal to the predicted annual downstream flow.

The proposed model is adapted to incorporate different "sub-models" to account for local information within the regional framework, where man-induced alterations are known, as common in non-pristine catchments. In particular, we propose a module to consider the impact of existing/designed water withdrawals on the L-moments of the FDC.

The procedure has been applied to a dataset of daily observation of about 120 gauged basins on the upper Po river basin in North-Western Italy.

1 Introduction

Flow duration curves (FDC) are widely used tools to represent water availability in a river basin and are thus considered for many water resources planning and management purposes, like concessions for water uses and the planning of new hydropower plants. The FDC represents the percentage of time a certain value of discharge (usually at the daily scale) is equaled or exceeded in a river section over a specified a period of time. A FDC computed for a single year is called "annual"; if the observations of multiple years are merged together, the FDC is usually referred to as "period-of-record" or "total". The empirical FDC can be easily built up by plotting the sorted the observations versus their frequency of non-exceedance computed with the Weibull plotting position, although the FDC is more frequently represented with respect to the exceedance frequency, thus resulting as a decreasing function. The evaluation of the FDC in ungauged basins is still a major issue in hydrological modeling, despite a large body of literature available on this topic, as recently reviewed by Castellarin et al. (2013).

In this work, we develop a regional model for prediction of the FDC in ungauged basins, developed and applied in the upper Po river basin (in North-Western Italy, an area mainly characterized by alpine and piedmont environments). The statistical framework developed is able to take advantage of local information about some types of anthropic effects. The aim of the work is to provide a regional tool to estimate the mean annual FDC in a generic watershed based on morphometric and climatic descriptors.

2 Methods

The regional spatially-smooth (SS) statistical estimation method proposed in this paper is based on a framework developed by Laio et al. (2011) in the context of regional flood frequency analysis. Analogies between the procedures are obtained by representing the mean annual FDC by means of its L-moments, i.e. distribution-free statistics which describe the mean, the variability and the skewness of the empirical FDC curve. L-moments are linear combinations of order statistics and are widely used in statistical hydrology; here we will refer to as μ for the mean value, while the dimensionless coefficients τ (L-CV) and τ_3 (L-skewness) respectively describe the variability and the skewness of the curve (see e.g., Hosking and Wallis, 1997, and Grimaldi et al., 2011, for more details on L-moments).

In the SS method the L-moments are the variables to be regionalized, i.e. the variables that will be predicted in ungauged basins. This is not so common in the literature as most of the models use as regionalization variables directly the FDC quantiles or the parameters of an analytical function fitted to the FDC. Then, the complete FDC can be calculated by fitting a probability distribution function to the set of predicted regional L-moments. In this way, the errors introduced by the preliminary choice of a distribution to fit the empirical curve do not enter in the regionalization procedure. The issue of the choice of a suitable distribution to represent the estimated FDC represents the last step in the regionalization process and will be discussed in the results section.

The present approach is referred to as spatially-smooth because it does not require the delineation of homogeneous regions, i.e. groups of basins sharing the same statistical characteristics (and thus the same L-moments τ and τ_3 within the region). Rather, L-moments are allowed to continuously vary in the space, so the FDC can vary accordingly. Following Wagener et al. (2013), the SS approach belong to the "mapping function" paradigms in which the set of relationship between the L-moments and the descriptors (and not the L-moments themselves) remain valid for the whole area of interest.

Ultimately, this approach can be considered an extension of the index-flow approach, as the dimensionless FDC is not constant in a region but is allowed to change site by site.

To build the regional model, the sample L-moments of the mean annual FDC must be computed at each available gauging station; estimation in ungauged sites is then obtained by building multiple linear regression models based on some basins descriptors (i.e. morphologic, climatic, geological, etc characteristics of the catchments), also known as descriptors. For a generic variable y to be regionalized, two different model structures have been considered:

$$y = a_0 + a_1 \cdot x_1 + \ldots + a_\mathrm{d} \cdot x_\mathrm{d} + \varepsilon \quad (1)$$

$$y = a_0 \cdot x_1^{a_1} \cdot \ldots \cdot x_\mathrm{d}^{a_\mathrm{d}} \cdot \varepsilon \quad (2)$$

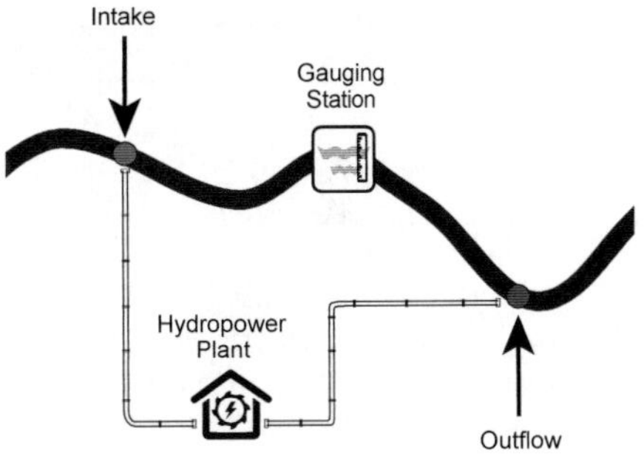

Figure 1. Example of gauging station bypassed by a water flow.

where x_d represents the generic basin descriptor, a_d are the coefficients to be estimated, and ε is the residual error to be minimized. Equation (1) can be solved by the Least Squares method (LS), while Eq. (2) can be still solved within the LS framework after the log-transformation of both side of the equation. After the selection of a set of suitable regression models, the prediction of μ, τ and τ_3 in a generic ungauged basin can be performed.

To provide reliable results, observed data should be natural flow observations, i.e. with no alterations due to upstream water uses. However, many gauging stations are located on rivers affected by man-induced alterations: in such cases, the observations should be previously processed to provide "naturalized" values. Where the actual amount of daily derived flow is known, corrections are straightforward, but this information is seldom available. This is the case (see Fig. 1) of run-on-river power plants where the gauging station is located between the intake and the outflow of the system and no information are available about the actual flow derived by the plant.

To overcome this problem, and exploiting also the data relative to such stations, a new methodology, developed by Ganora et al. (2013), has been used to obtain the L-moments of the "natural". In essence, the L-moments computed on the measurements affected by changes (i.e., downstream the intake) are converted to a set of naturalized L-moments subsequently used in the analysis. The method only requires the maximum discharge that can be withdrawn (said ΔQ) and the affected (observed) L-moments and does not need to assess the complete time series of the natural flow. Note that ΔQ is a known characteristic of the plants, available, in our case, from a regional database of water infrastructures.

The correction method uses the parameter:

$$K = \frac{\Delta Q}{\mu^\mathrm{D}}, \quad (3)$$

where μ^D is the mean observed discharge (i.e. downstream the intake; superscript D) to correct the L-moments by means of the equations

Table 1. Observed and corrected L-moments for gauging stations affected by relevant water intakes.

Basin	K	μ obs.	μ corr.	τ obs.	τ corr.	τ_3 obs.	τ_3 corr.
Devero at Baceno	6.069	1.73	4.51	0.385	0.218	0.374	0.161
Germanasca at Perrero	0.637	3.14	4.6	0.708	0.528	0.627	0.465
Isorno at Pontetto	4	0.75	1.93	0.485	0.285	0.677	0.315
Po at Castiglione Torinese	2.016	54.57	117.5	0.585	0.371	0.607	0.33
S.Bernardino at S.Bernardino Santino	0.414	4.35	5.82	0.728	0.576	0.67	0.55
Stura di Demonte at Vinadio	3.053	2.62	5.45	0.482	0.291	0.536	0.258
Stura di Viù at Germagnano	0.734	4.77	7.26	0.596	0.435	0.568	0.406
Toce at Domodossola	3.391	12.21	29.34	0.375	0.224	0.469	0.226
Varaita at Torrette	8.824	1.36	3.2	0.383	0.212	0.541	0.221

$$\begin{cases} \mu^{\mathrm{U}} = \mu^{\mathrm{D}} \cdot \Psi[K] \\ \tau^{\mathrm{U}} = \tau^{\mathrm{D}} \cdot \dfrac{0.5}{L_2^*[K]} \\ \tau_3^{\mathrm{U}} = \tau_3^{\mathrm{D}} \cdot \dfrac{1}{3} \dfrac{L_2^*[K]}{L_3^*[K]} \end{cases} . \tag{4}$$

In Eq. (4) the superscript D indicates the downstream (observed) value, while U is the upstream (naturalized) statistic. Functions $\Psi[K]$, $L_2^*[K]$ and $L_3^*[K]$ give the correction factor for each L-moment and are reported in the Appendix A. Details about the correction method are discussed in Ganora et al. (2013).

3 Data availability

The available hydrological dataset refers to North-West of Italy (see Fig. 2) and includes 129 gauging stations with a total of 1438 station-year of daily observations, that have been selected after different quality controls. Only years with no more than 3 missing values have been included in the dataset. The available records have a mean length of 11 years, with actual durations ranging between 2 and 52 years. Many stations are quite recent and their data do not completely overlap with longer records, however they allow a larger spatial coverage of the model. More details about the dataset can be found in Ganora et al. (2013).

For each station, the annual FDCs have been computed based on the daily discharge values; the $N = 365$ values have been sorted and associated to the non-exceedance frequency $F_i = \frac{i}{N+1}$, with $i = 1 \ldots N$ being the index of the sorted sample (leap years curves have been resampled on the frequency domain to obtain 365 values). Finally, the mean annual FDC at each station has been computed by taking the average of each annual FDC quantile (i.e., the average frequency-by-frequency of all the annual values).

The database encompasses a number of basins affected by different water withdrawals for different uses and with different intake systems. In many cases, for our purposes, such anthropic effects can be neglected (e.g., negligible intake with

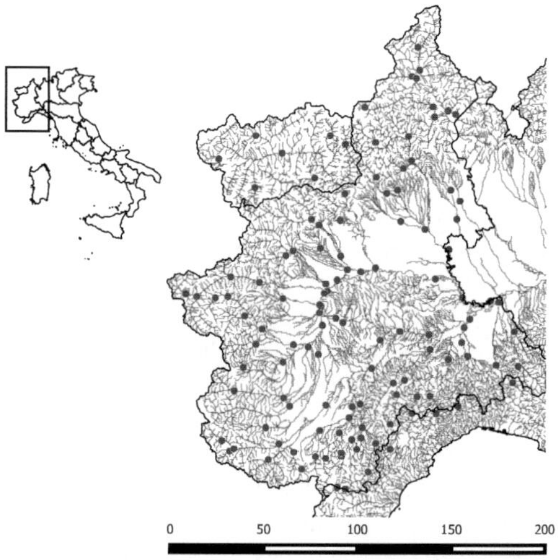

Figure 2. Location of gauging stations used in the analysis.

respect to the average flow; high-elevation reservoir affecting only the upper part of the river and with negligible effect on the long-term statistics). In other cases alterations can be corrected with measurements of real intake (e.g. large irrigation canals). However, other cases require a correction of the observed streamflow statistics to provide reliable "natural" values to be used in the regional analysis (with the method described in the previous section). This is the case of 9 gauging stations located downstream the intake of run-of-the-river hydropower plants and upstream their water return. For these intakes no withdrawal data were available, as no gauging recording system was installed. Table 1 reports the main characteristics of the above mentioned stations with man-induced alterations, as well as the corrected ("naturalized") L-moments.

4 Spatially-Smooth regional estimation

Sample L-moments, i.e. μ, τ and τ_3, have been computed from the available average FDCs with the necessary corrections to obtain "naturalized" values in the 9 cases described above. To support the implementation of the regional model, a number of catchment descriptors (about 120), including morphologic and climatic characteristics of the basin, soil and vegetation type, land use, etc, have been computed for each basin. The most relevant descriptors have been also reported in an "Atlas of basin characteristics" (Gallo et al., 2013). The descriptors act as independent variable in the regression models and some of them are expected to provide a statistical dependence with each L-moments.

Concerning the mean annual flow, to implement the regression model we considered the transformed variable

$$h = (\mu \cdot 31\,536)/A, \tag{5}$$

which represents the flow in mm when μ in in $m^3\,s^{-1}$ and A the catchment area in km^2. Using h in the additive model structure (Eq. 1) and considering only descriptors that represents area-averaged values (e.g. mean basin elevation, spatial average of annual rainfall, etc.), the model guarantees the congruence of the mean discharge along the stream, i.e. the mean annual flow downstream of a confluence is equal to the sum of the two mean annual flow values computed upstream. This is a notable feature of the method that allows to keep the estimates of mean annual flow congruent at the confluences.

In order to select the most appropriate regional model, we performed all the possible regressions combining 1 to 4 descriptors as independent variables to estimate h; each regression was tested for significance with the t Student test (5 % level of significance) and for multicollinearity with the VIF test (limit value equal to 5, see e.g. Montgomery et al., 2001). Models passing the tests were finally sorted according to their adjusted coefficient of determination and the best-performing ones were further evaluated by visual check of diagnostic error plots. Regressions have been solved using the Weighted Least Square method (e.g. Montgomery et al., 2001) considering the inverse of the square root of record length as weighting coefficients.

Concerning the L-moments τ and τ_3, an analogous procedure has been performed but without requiring any congruence at the confluences, and thus without restricting the set of possible descriptors. In this case, both the additive and the multiplicative model structures have been considered, thus considering a larger number of possible models (about 10^6 models have been checked for each L-moment and each model structure).

The final models, applicable to the whole area of interests are:

$$h = -7.3605 \times 10^2 + 1.2527 \times \mathrm{MAP} + 3.2569 \times 10^{-1} \times z_m + 5.2674 \times \mathrm{fourier_{B1}} - 6.7185 \times \mathrm{clc_2} \tag{6}$$

$$\tau = -2.896 \times 10^{-1} - 2.688 \times 10^{-3} \times \mathrm{clc_3} + 9.643 \times 10^{-5} \times a_{75} + 1.688 \times 10^{-4} \times \mathrm{MAP} + 2.941 \times 10 \times c_{\mathrm{int}}, \tag{7}$$

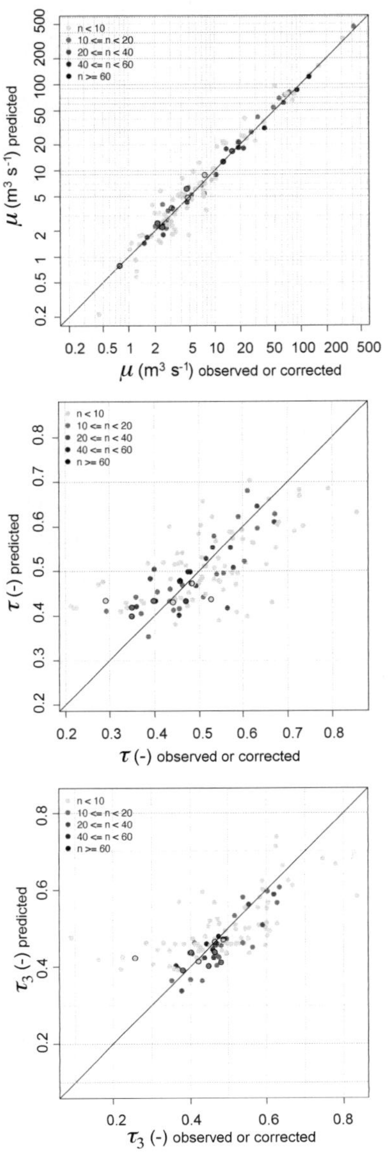

Figure 3. Natural values of the L-moments (observed or corrected) compared with regional predictions. Colors represent record lengths; circled points are the corrected values.

$$\tau_3 = 4.755 \times z_{\max}^{-0.2702} \times \mathrm{SD(IDF}_a)^{0.06869} \times \mathrm{CV_{rp}^{0.2106}},$$

where MAP is the mean annual precipitation, z_m and z_{max} are the mean and maximum basin elevation, $\mathrm{fourier_{B1}}$ and $\mathrm{CV_{rp}}$ are rainfall regime parameters, $\mathrm{clc_2}$ and $\mathrm{clc_3}$ are land use parameters, a_{75} is the 75th percentile of the hypsographic curve, c_{int} and $\mathrm{SD(IDF}_a)$ are extreme-rainfall statistics. All the parameters are computed at the basin scale; details are reported in Gallo et al. (2013).

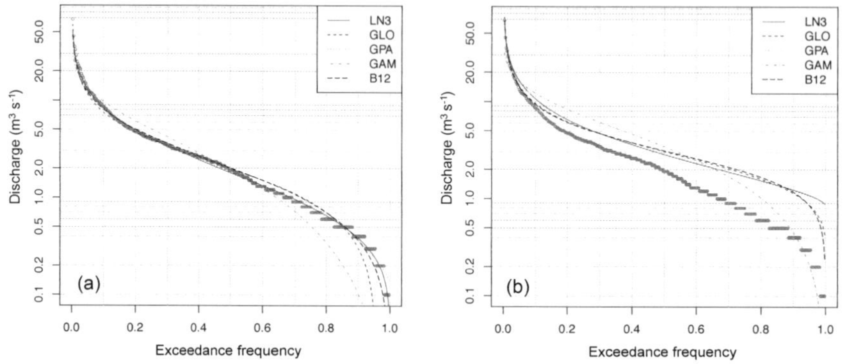

Figure 4. Empirical (points) and analytical (lines) flow duration curve for the Agogna at Momo river: **(a)** fitting on sample L-moments; **(b)** fitting on regional L-moments. Probability distributions are: log-Normal (LN3), Generalized Logistic (GLO), Generalized Pareto (GPA), Gamma (GAM) and Burr XII (B12).

Performances of the selected regression models can be examined in Fig. 3 where predictions in cross-validation are reported. The mean annual discharge prediction presents little scatter, while L-CV and L-skewness tend to be more dispersed. Worse performances in higher-order L-moments prediction were in fact expected, but it is worth nothing that the points with larger errors are in general characterized by shorter time series and thus by larger sample uncertainty. Further diagnostics on residuals (not reported) have been performed to check possible heteroskedasticity and residual correlation.

5 Selection of FDC analytical form

The regional statistical framework presented is related to the estimation of the L-moments, similarly to Laio et al. (2011). The L-moments can be used to fit different probability distribution functions to represent the FDC and the choice of the analytical form of the FDC can be done as the last step of the procedure. To suggest a suitable distribution to represent the FDC in ungauged basins, different distributions have been investigated, all with 3 parameters: the log-Normal (LN3) the Generalized Logistic (GLO), the Generalized Pareto (GPA) and the Gamma (GAM) (details can be found in Hosking and Wallis, 1997), and the Burr XII (also called B12) described, for example, in Ganora and Laio (2015).

A first series of test of fitting was performed with the more common distributions (LN3, GLO, GPA, GAM). The main drawback observed was that in many cases negative streamflow values in the lower tail were obtained. This occurs when the distributions have a variable lower bound that can allow (depending on the parameter set) negative predictions for high exceedance frequencies. Further constraints to the parameters of the distribution can be added to keep the result consistent, but such controls seem hardly applicable in a regional context, for two reasons: first, the adjustments in general require numerical optimization that become difficult to be implemented in large-scale models; second, it is in general preferable to avoid complex (and likely unstable) estimation methods when the model will be used by practitioners that may not have confidence with numerical techniques.

To overcome the problem of negative quantiles without optimizing the distribution's parameters, we introduce the use of the Burr probability distribution, well known in different scientific communities but rarely used in the hydrological field (Ganora and Laio, 2015). In particular, the proposed function is the Burr XII distribution, originally proposed by Burr (1942) as a two-parameter function, and later extended in the three-parameter form (cumulative frequency function):

$$
\begin{aligned}
F(x) &= 1-\left[1-k\left(\frac{x}{\lambda}\right)^{c}\right]^{\frac{1}{k}} \qquad k \neq 0 \\
&= 1-\exp\left[-\left(\frac{x}{\lambda}\right)^{c}\right] \qquad k = 0,
\end{aligned}
\tag{8}
$$

where k and c are the shape parameters and λ is the scale parameter. The parameter k can be any real number ($-\infty < k < \infty$), while $c > -k$ and $\lambda > 0$. The domain is always positive, being $0 \leq x \leq \infty$ for $k \leq 0$ and $0 \leq x \leq \lambda k^{-1/c}$ otherwise; for this kind of application we considered only non-positive values of k in order to use only unbounded (from above) curves. The condition $k = 0$ is the limiting case for which $F(x)$ becomes the two-parameter Weibull distribution. Another limiting case appears for $k \to -\infty$ (with $c \to \infty$ simultaneously) and lead to the Pareto distribution:

$$
\begin{aligned}
F_P(x) &= 1-\left(\frac{x_0}{x}\right)^{\gamma} \qquad x > x_0 \\
&= 0 \qquad \text{otherwise},
\end{aligned}
\tag{9}
$$

where $-c/k = \gamma$ and $x_0 = \lambda(-k)^{-1/c}$.

The Burr function results quite "flexible", and thus suitable to fit many different shapes of empirical data, thanks to the wide range of skewness and kurtosis values it covers; nevertheless it has only three parameters, thus avoiding overfitting problems. On the other hand, despite the simple analytical form of the Burr XII, the estimation of the two shape parameters is not straightforward as it requires the joint inversion of two nonlinear equations (not shown here). For details about the estimation of the parameters with the method of L-moments as well for other characteristics of the distribution the reader is referred to Ganora and Laio (2015) and references therein.

To proceed with the model selection, a comparison between the different distributions is in order. This comparison has been performed by fitting the analytical curves to 365 quantiles corresponding to the non-exceedance probabilities $F_i = i/366$ with $i = 1 \ldots 365$. For each distribution, the difference between the empirical and the estimated quantiles has been calculated (on both untransformed and log-transformed discharge values). The RMSE provided a error statistic for each considered station.

Results show that LN3 and GPA have generally good fitting performances; GLO and B12 have slightly larger errors, but comparable with LN3 and GPA, while the GAM does not provide reliable results. However, the LN3 provides negative values in 21 % of the stations, the GLO in 51 % and the GPA in 11 %. The average duration (in days per year) affected by negative quantiles is 10, 14 and 19 days for the LN3, GLO and GPA respectively (the average is computed considering only the stations with at least one negative value). In the context of the data set analysed here the Burr distribution results the best choice, as it provides adequate fitting capabilities without the need of further optimization to avoid negative or inconsistent quantile predictions. An example of fitting is reported in Fig. 4a where the studied analytical functions have been superimposed on the empirical FDC.

Panel b of Fig. 4 shows the same record, but compared to the analytical FDCs computed from the regional L-moments. An overestimation of the mid-right part of the FDC is evident, while the left side of the curve seems properly recognized. In this case, although the shift could be due to an inaccurate estimation of the L-moments, it can also be caused by the presence of unknown (or underestimated) water intakes upstream the gauging station. Under this perspective, the regional model can be also used as a diagnostic tool to identify basins in which an improvement of the knowledge on the actual water use is recommended. This kind of analysis is currently under development.

6 Conclusions

A regional model for the estimation of the mean annual flow duration curve (FDC) has been presented in the paper. The model is made of three regression-based equations to estimate the L-moments (i.e., the mean, the L-CV and the L-skewness) of the FDC at any ungauged basin, which can be subsequently used to fit the whole curve. In particular, concerning the mean annual flow, the river network structure is explicitly accounted for: at any confluence, the mean annual flow in the downstream section is the sum of the mean values in the two upstream river reaches, thus ensuring the congruence of the prediction.

Moreover, the method allows the use in the calibration phase of records affected by water derivations, even if the actual derived flow is not known or hardly retrievable. This is possible thanks to a parsimonious corrections method that provide "naturalized" L-moments considering only simple "geometrical" information about the water intake and does not require the reconstruction of the whole natural time series.

Finally, among different analytical representations of the FDC, the three-parameter Burr XII distribution has been proved to be the most suitable for the case study. In fact, while it provides good fitting performances, it does not allow the estimation of negative discharge values, which frequently appear by applying other analytical forms. This is advantageous as it is not necessary to optimize the parameters of a distribution to force the predicted quantiles to positive values only, thus making the model suitable for large-scale unsupervised applications.

Due to its flexibility, the present modeling framework is easily adaptable to further extensions like, for instance, the inclusion of observations from new gauging stations (by combining local and regional L-moments), while its use as a diagnostic tools to investigate the actual effect of existing water derivations is already in progress.

Appendix A

The correction method developed by Ganora et al. (2013) is based on the hypothesis that the upstream FDC L-moments can be obtained by correcting the observed ones through Eq. (4) in a parsimonious way, i.e. using just simple information about the intake system. This is necessary when scarce information are available about the plant, while more sophisticated models could be adopted where the available information is more detailed.

The method assumes the intake flow is exactly corresponding to the actual available flow below the threshold ΔQ. In this work we also consider the plant is always in exercise, even though the original method allows to account for a number of days of stop during the year. Correction coefficients are obtained by studying the effect of water withdrawal on a virtual FDC defined by an exponential distribution; in this way the problem can be treated analytically and solutions are based o a single parameter.

Applicability of correction factors has been tested with real data (i.e. also with non-exponential FDC) showing that μ and τ can be very efficiently reconstructed while, as expected, corrections for τ_3 provide larger errors, but still useful for the scope of the analysis.

To allow the implementation of Eq. (4), from Ganora et al. (2013), we define:

$d^* = \text{Exp}[W[K]]$ where $W[K]$ is the Lambert W-function, i.e. the inverse of equation $K = We^W$;

$$\psi[K] = K/W[K];$$

$$L_2 = \psi[K] \cdot S_2[d^*] + k \cdot \left(d^{*2} - d^*\right);$$

$$L_3 = \psi[K] \cdot S_3[d^*] + k \cdot (-2d^{*3} + 3d^{*2} - d^*);$$

$$S_2[d^*] = \frac{1}{2} d^* \left(2 - d^* + 2(d^* - 1) \cdot \text{Ln}[d^*]\right);$$

$$S_3[d^*] = \frac{1}{6} d^* \left(6 + d^*(4d^* - 9) - 6(d^* - 1)(2d^* - 1) \cdot \text{Ln}[d^*]\right).$$

Acknowledgements. Funding from the Italian Ministry of Education, Universities and Research (FIRB-RBFR12BA3Y project), and from the European Union – Horizon 2020 ERC-2014-CoG ("CWASI" project, ref. 647473) are acknowledged.

References

Burr, I. W.: Cumulative frequency functions, Ann. Math. Stat., 13, 215–232, 1942.

Castellarin, A., Botter, G., Hughes, D. A., Liu, S., Ouarda, T. B. M. J., Parajka, J., Post, D. A., Sivapalan, M., Spence, C., Viglione, A., and Vogel, R. M.: Prediction of flow duration curves in ungauged basins, in: Runoff Prediction in Ungauged Basins: Synthesis across Processes, Places and Scales, edited by: Bloeschl, G., Sivapalan, M., Wagener, T., Viglione, A., and Savenije, H., Cam bridge University Press, 2013.

Gallo, E., Ganora, D., Laio, F., Masoero, A., and Claps, P.: Atlante dei bacini imbriferi piemontesi ISBN:978-88-96046-06-7, available at: http://www.idrologia.polito.it/web2/open-data/Renerfor/atlante_bacini_piemontesi_LR.pdf (last access: 4 April 2016), 2013 (in Italian).

Ganora, D. and Laio, F.: Hydrological Applications of the Burr Distribution: Practical Method for Parameter Estimation, J. Hydrol. Eng., 20, 11, doi:10.1061/(ASCE)HE.1943-5584.0001203, 2015.

Ganora, D., Gallo, E., Laio, F., Masoero, A., and Claps, P.: Analisi idrologiche e valutazioni del potenziale idroelettrico dei bacini piemontesi ISBN:978-88-96046-07-4, available at: http://www.idrologia.polito.it/web2/open-data/Renerfor/analisi_idrologiche_LR.pdf (last access: 4 April 2016), 2013 (in Italian).

Grimaldi, S., Kao, S. C., Castellarin, A., Papalexiou, S. M., Viglione, A., Laio, F., Aksoy, H., and Gedikli, A.: Statistical hydrology, in: Treatise on Water Science, Vol. 2: The Science of Hydrology, edited by: Wilderer, P., Chap. 2.18, 479–517, Elsevier, doi:10.1016/B978-0-444-53199-5.00046-4, 2011.

Hosking, J. and Wallis, J.: Regional Frequency Analysis: An Approach Based on L-Moments, Cambridge University Press, 1997.

Laio, F., Ganora, D., Claps, P., and Galeati, G.: Spatially smooth regional estimation of the flood frequency curve (with uncertainty), J. Hydrol., 408, 67–77, doi:10.1016/j.jhydrol.2011.07.022, 2011.

Montgomery, D., Peck, E., and Vining, G.: Introduction to linear regression analysis, in: Regional flood frequency, edited by: Ouarda, T., Girard, C., Cavadias, G., and Bobee, B., Wiley Series Probability and Statistics, 3rd Edn., Wiley, New York, 2001.

Wagener, T., Bloeschl, G., Goodrich, D. C., Gupta, H. V., Sivapalan, M., Tachilawa, Y., Troch, P. A., and Weiler, M.: A synthesis framework for runoff prediction in ungauged basins, in: Runoff Prediction in Ungauged Basins: Synthesis across Processes, Places and Scales, edited by: Bloeschl, G., Sivapalan, M., Wagener, T., Viglione, A., and Savenije, H., Cambridge University Press, 2013.

Proc. IAHS, 373, 81–85, 2016
proc-iahs.net/373/81/2016/
doi:10.5194/piahs-373-81-2016

Open Access
PIAHS

The value of weather radar data for the estimation of design storms – an analysis for the Hannover region

Uwe Haberlandt and Christian Berndt

Institute of Water Resources Management, Hydrology and Agricultural Hydraulic Engineering, Leibniz University of Hannover, Hannover, Germany

Correspondence to: Uwe Haberlandt (haberlandt@iww.uni-hannover.de)

Published: 12 May 2016

Abstract. Pure radar rainfall, station rainfall and radar-station merging products are analysed regarding extreme rainfall frequencies with durations from 5 min to 6 h and return periods from 1 year to 30 years. Partial duration series of the extremes are derived from the data and probability distributions are fitted. The performance of the design rainfall estimates is assessed based on cross validations for observed station points, which are used as reference. For design rainfall estimation using the pure radar data, the pixel value at the station location is taken; for the merging products, spatial interpolation methods are applied. The results show, that pure radar data are not suitable for the estimation of extremes. They usually lead to an overestimation compared to the observations, which is opposite to the usual behaviour of the radar rainfall. The merging products between radar and station data on the other hand lead usually to an underestimation. They can only outperform the station observations for longer durations. The main problem for a good estimation of extremes seems to be the poor radar data quality.

1 Introduction

Design storms are required for the planning and evaluation of hydraulic structures and flood risk management in urban and rural catchments. The design storms are derived from frequency analyses of annual maximum rainfall or rainfall above a threshold for specific durations. The storms are usually condensed for different durations and frequencies to intensity-duration-frequency (IDF) curves or depth-duration-frequency curves (DDF) for a certain location. In order to obtain reliable estimation of design rainfall, long-term precipitation observations in high temporal resolution are required. Especially short duration observations are often only available with poor spatial density, which demands regionalisation. There have been different studies about regionalisation of DDF curves over the last years (Durrans and Kirby, 2004; Johnson et al., 2016; Madsen et al., 2002). Also scaling methods have been applied to derive IDF curves for short durations from better available daily observations (Yu et al., 2004).

An alternative would be to use weather radar for the estimation of design rainfall, which is available in a high spatial and temporal resolution, or at least to use it as an additional information for regionalisation. Meanwhile the observation length of many operational radar instruments extend over a time period of 10 years, which suggests to analyse their benefits for estimating design rainfall. Rainfall derived from radar data is usually biased and needs some kind of correction. This can be done by adjusting the radar data (Krajewski and Smith, 2002) or by merging radar data and observations (Berndt et al., 2014). So far, only a few investigations have been carried out utilising radar rainfall for extreme value analyses. Marra and Morin (2015) used a 23-year radar record to estimate IDF curves for different climatic zones in Israel. They found a general overestimation of radar based rainfall extremes compared to the gauge data, but with 70 % of the cases within the uncertainty bounds of the rain gauge derived IDF's. Eldardiry et al. (2015) analysed the contribution of different factors to the uncertainty in the estimation of design storms. They employed a 13-year data set from the NEXRAD radar network for the Louisiana region in the USA and found that radar data underestimate the observed gauge based IDF curves due to the conditional bias of the radar product. They also found that a regional estimation of the IDF curves, e.g. using the index flood method, reduces the uncertainty significantly compared to the at site estimations. Overeem et al. (2009) used a 11-year radar data set for extreme value analyses in the Netherlands. They found that the radar data are suitable for the estimation of DDF curves if regional frequency analyses is applied. However, the uncertainty for the estimation of storms with longer durations becomes large due to the short sample and stronger spatial correlation of events.

In the current study, different regionalisation methods are compared to estimate DDF curves from interpolated rainfall products with and without utilising radar information. This is supposed to provide insights about the real benefit of radar data for at site estimation of DDF curves compared to using gauge based rainfall data only.

2 Methodology

2.1 Radar data pre-processing

The radar data pre-processing was performed according to Berndt et al. (2014). In the following, those steps are briefly summarized: (1) raw radar reflectivities at 5 min resolution were transformed into rainfall intensities using a standard $Z - R$ relationship; (2) a simple statistical clutter correction method was applied; (3) the data were interpolated on a 1 km × 1 km grid; (4) a space-time filter was applied on this grid for smoothing; (5) outliers were removed considering the cumulative distribution function of standard errors between rain gauge and radar data. In addition, different approaches for radar data adjustment to gauge data are employed, which are described together with the interpolation methods below.

2.2 Rainfall estimation for unobserved locations

Continuous 5 min point precipitation time series are estimated for a set of locations for which observed rainfall is available, however without using the observations at the target location in the estimation procedure; i.e. a cross- validation is performed. Thus, a real validation of the estimation method is possible assuming that the observations are error free. The following methods are used for the estimation/interpolation of rainfall data sets:

1. REF – This represents the observed reference rainfall time series, which is taken without modification.
2. NN – A nearest neighbour interpolation using recording rainfall stations is carried out.
3. OK – Ordinary kriging is applied using the m closest surrounding recording rainfall stations.
4. Radar – Pre-processed radar data as described above are extracted from the nearest 1 km × 1 km pixel and taken without further adjustment.
5. RadarADJ – Radar data are adjusted with daily rainfall using the denser network of non-recording stations.
6. CM – Conditional merging interpolation (Sinclair and Pegram, 2005) is applied using data from recording stations and radar data without adjustment (Radar).

   CMADJ – Conditional merging interpolation is applied using data from recording stations and radar with adjustment (RadarADJ).

For performance assessment the relative bias

$$r\text{bias} = \frac{1}{S \cdot R} \sum_{i=1}^{S} \sum_{j=1}^{R} \left(\frac{\hat{y}_{i,j} - y_{i,j}}{y_{i,j}} \right) \quad (1)$$

and the relative root mean squared error

$$r\text{rmse} = \sqrt{\frac{1}{S \cdot R} \sum_{i=1}^{S} \sum_{j=1}^{R} \left(\frac{\hat{y}_{i,j} - y_{i,j}}{y_{i,j}} \right)^2} \quad (2)$$

are used, were $\hat{y}$ and y are the estimated and observed design storm quantiles, respectively, and S and R are the number of stations and return periods, respectively. The calculations are carried out separately for different storm durations D.

2.3 Extreme value analyses

Basis for the extreme value analyses are the 5 min time series obtained from the interpolation methods listed above. The extreme value analysis is carried out according to the German standards for design storm estimation (DWA-M-531, 2012). Partial duration series (PDS) are built with a sample size of about e times the number of years for durations with $D = 5$, 10, 15, 30, 60, 120, 240, 360, 720, 1440 min. A minimum dry spell duration of Min[4 h, D] is applied to guarantee independence of the storms. The exponential probability distribution is fitted to the PDS for each duration. Finally, the parameters of the distributions are smoothed over the durations to allow a consistent estimation of DDF curves without jumps (see DWA-M-531, 2012).

3 Study area and data

The study area is the radar range with a radius of 128 km for the weather radar at the Hannover airport (see Fig. 1). This region has elevations from the sea level in the northern mostly flat part up to 1141 m a.s.l. in the Harz Mountains in the South. The average annual precipitation varies between 500 and 1700 mm yr^{-1}. Radar data were available for the period from 2000 to 2012 (13 years). The radar data preprocessing was carried out as explained in Sect. 2.1. Eight recording reference stations have been selected for which a validation of the DDF curves was carried out. In addition, 46 recording stations and 512 non-recording stations were

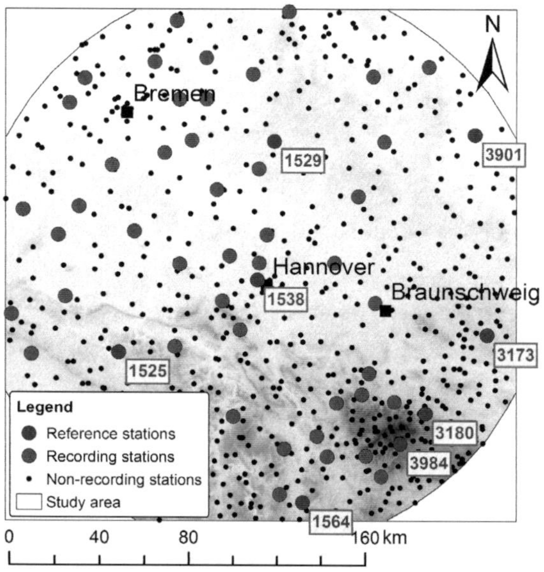

Figure 1. Study area with topography, radar circle and rainfall stations.

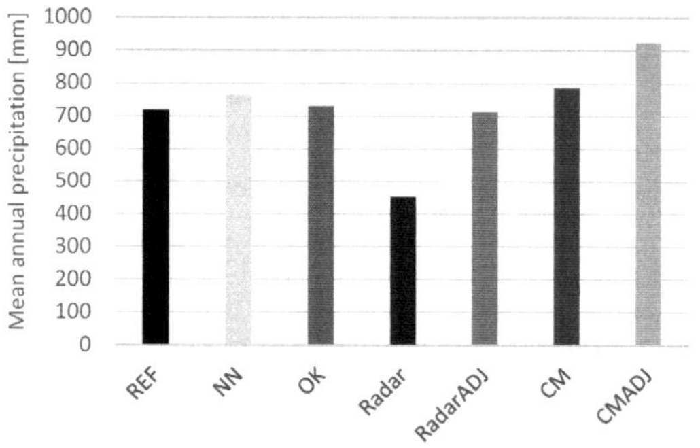

Figure 2. Mean annual rainfall from 5 min interpolated rainfall series for the period 2000 to 2012 averaged over the 8 reference stations.

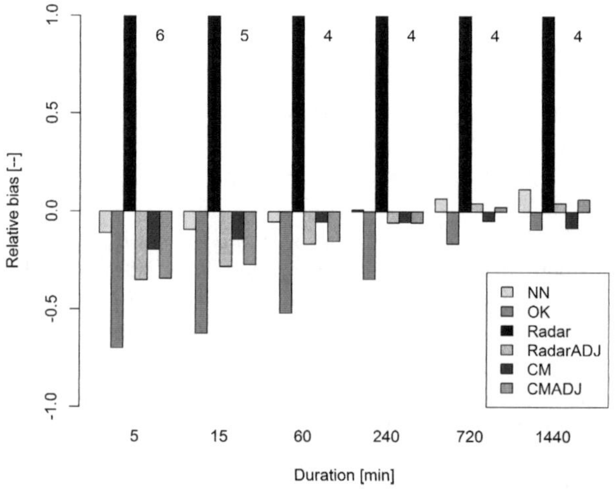

Figure 3. Relative bias averaged over the 8 reference stations and 8 return periods ($T = 1, 2, 3, 5, 10, 20, 25, 33$ years).

available for the interpolations within the study area, however with highly varying temporal coverage. For all stations, the same 13 years period as for the radar observations was used.

4 Results

As a starting point, the estimation of mean annual precipitation using the different interpolation methods is evaluated. Figure 2 shows a comparison of the mean annual precipitation for the 13 years averaged over the 8 reference stations using the different methods. The two approaches, which do not use radar data, NN and OK, provide interpolated time series almost without bias. Employing pure radar for the estimation leads to a significant underestimation, which is typical for comparisons between radar rainfall and station values. When radar data are adjusted with rainfall from daily stations, the bias in the annual values is removed. The interpolations using conditional merging with radar data (CM) and with adjusted radar data (CMADJ), respectively, lead to slight overestimations here.

In terms of the extremes, the results are different from the mean values. Figure 3 shows the relative bias for the design rainfall estimation using the different interpolation methods averaged over the 8 reference stations and 8 return periods $T = 1$, 2, 3, 5, 10, 20, 25 and 33 years. The relative bias using the pure radar data (Radar) reveals a huge overestimation of the extremes. Comparing the selected events from pure radar data and station data for the same locations shows only little temporal overlap. This indicates that there might be still considerable errors and outliers in the radar data, which do not represent real rainfall. Ordinary kriging (OK), which does not include radar data, shows the largest negative bias. This is likely due to the strong smoothing behaviour of this method. The smallest bias is obtained for the simple nearest neighbour (NN) interpolation. Also acceptable is the bias for the conditional merging (CM) technique without radar data adjustment. The methods with daily adjusted radar rainfall, RadarADJ and CMADJ, still express significant biases for short durations.

Figure 4 shows a comparison of the relative root mean squared errors obtained from the different methods, again averaged of the 8 reference stations and 8 return periods. Using pure radar data (Radar) produces the largest error followed by OK, which uses only the recording stations. The overall smallest error is found when applying the simple nearest neighbour technique (NN) for interpolation. The second smallest error is obtained when conditional merging (CM) is used, which can outperform the NN approach only for longer durations. The methods with daily adjustments, RadarADJ

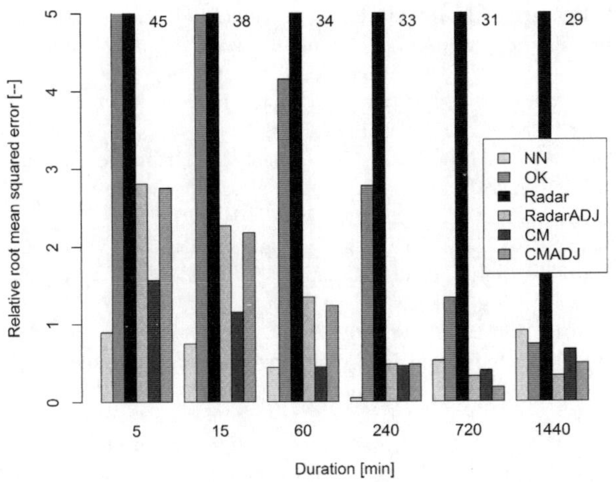

Figure 4. Relative rout mean squared error averaged over the 8 reference stations and 8 return periods ($T = 1, 2, 3, 5, 10, 20, 25, 33$ years).

and CMADJ, have still considerable high errors for short rainfall durations.

In Fig. 5 exemplarily estimated DDF curves for the rainfall station Hannover are presented for return periods $T = 1$, 5, 10 and 20 years. The pure radar DDF is only visible for $T = 1$ years and outside the x axes range for the other return periods. This shows again the large overestimation if pure radar data are used to estimate the extremes. The errors for the estimation of the DDF curves increase with increasing return period. The DDF curves based on OK show again a significant underestimation. The best methods for the station Hannover is CMADJ. The best method varies between the stations with an average performance as indicated in Fig. 4.

5 Summary and conclusions

This study has investigated the benefit of radar data for the estimation of design rainfall. Different interpolation methods were applied on 5 min time series from recording rain stations, radar data and merging products. The interpolated data sets were used for extreme value analyses and the estimation performance was assessed based on observations. The results and conclusions from this analysis can be summarized as follows:

- Using pure radar data leads to large overestimation of DDF curves. This is probably due to measurement errors for the weather radar, which could not be detected and corrected in pre-processing.

- The nearest neighbour approach gives overall the best results. This is partly due to the non-smoothing character of this method and due to the sufficiently dense network of recording rainfall stations.

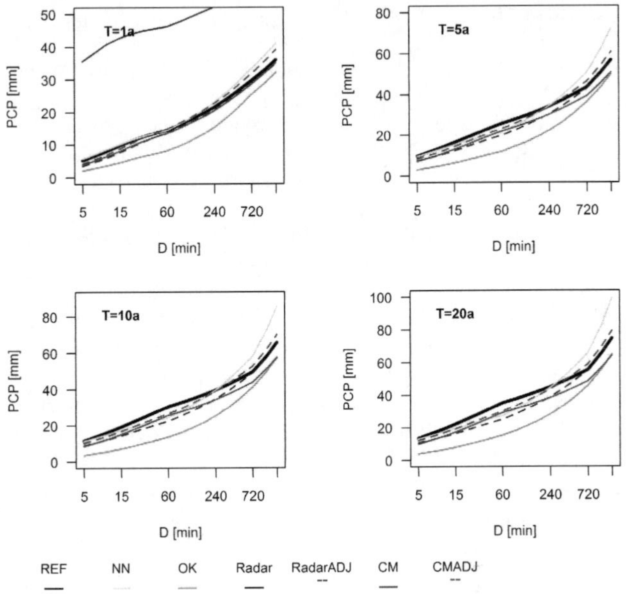

Figure 5. Depth duration frequency curves for the station Hannover (id 1538 in Fig. 1) for four selected return periods $T = 1, 2, 10, 20$ years.

- The radar-gauge merging methods reduce the error considerably but are best only for long durations. However, they all provide a negative bias, which can be explained by the smoothing effect of the interpolation methods.

- Ordinary kriging leads to the strongest underestimation of the design storms due to these smoothing effects. This approach cannot be recommended when extreme value analysis is required.

Overall, the results indicate that using radar data in this manner will not benefit the estimation of design rainfall, although this might be different for other study regions and radar instruments. Anyway, for improvements it is necessary to better correct the radar data for errors. A regional frequency analysis of extreme rainfall using methods similar like the index flood method might lead to better results. In addition, a procedure, which first estimates extreme value distributions locally and then interpolates the parameters in space, might be better for avoiding negative biases.

Acknowledgements. We thank the German Weather Service (DWD) for providing the precipitation and radar data.

References

Berndt, C., Rabiei, E., and Haberlandt, U.: Geostatistical merging of rain gauge and radar data for high temporal resolutions and various station density scenarios, J. Hydrol., 508, 88–101, 2014.

Durrans, S. R. and Kirby, J. T.: Regionalization of extreme precipitation estimates for the Alabama rainfall atlas, J. Hydrol., 295, 101–107, 2004.

DWA-M-531: Merkblatt: Starkregen in Abhängigkeit von Wiederkehrzeit und Dauer, Deutsche Vereinigung für Wasserwirtschaft, Abwasser und Abfall e. V., Hennef, 2012.

Eldardiry, H., Habib, E., and Zhang, Y.: On the use of radar-based quantitative precipitation estimates for precipitation frequency analysis, Part 2, J. Hydrol., 531, 441–453, 2015.

Johnson, F., Hutchinson, M. F., The, C., Beesley, C., and Green, J.: Topographic relationships for design rainfalls over Australia, J. Hydrol., 533, 439–451, 2016.

Krajewski, W. F. and Smith, J. A.: Radar Hydrology: rainfall estimation, Adv. Water Resour., 25, 1387–1394, 2002.

Madsen, H., Mikkelsen, P. S., Rosbjerg, D., and Harremoës, P.: Regional estimation of rainfall intensity-duration-frequency curves using generalized least squares regression of partial duration series statistics, Water Resour. Res., 38, 1239, doi:10.1029/2001WR001125, 2002.

Marra, F. and Morin, E.: Use of radar QPE for the derivation of Intensity–Duration–Frequency curves in a range of climatic regimes, Part 2, J. Hydrol., 531, 427–440, 2015.

Overeem, A., Buishand, T. A., and Holleman, I.: Extreme rainfall analysis and estimation of depth-duration-frequency curves using weather radar, Water Resour. Res., 45, W10424, doi:10.1029/2009wr007869, 2009.

Sinclair, S. and Pegram, G.: Combining radar and rain gauge rainfall estimates using conditional merging, Atmos. Sci. Lett., 6, 19–22, 2005.

Yu, P.-S., Yang, T.-C., and Lin, C.-S.: Regional rainfall intensity formulas based on scaling property of rainfall, J. Hydrol., 295, 108–123, 2004.

Proc. IAHS, 373, 87–94, 2016
proc-iahs.net/373/87/2016/
doi:10.5194/piahs-373-87-2016

Spatial variability of the parameters of a semi-distributed hydrological model

Alban de Lavenne, Guillaume Thirel, Vazken Andréassian, Charles Perrin, and Maria-Helena Ramos

Irstea, Hydrosystems and Bioprocesses Research Unit (HBAN), 1, rue Pierre-Gilles de Gennes, CS 10030, 92761 Antony Cedex, France

Correspondence to: Alban de Lavenne (alban.de-lavenne@irstea.fr)

Published: 12 May 2016

Abstract. Ideally, semi-distributed hydrologic models should provide better streamflow simulations than lumped models, along with spatially-relevant water resources management solutions. However, the spatial distribution of model parameters raises issues related to the calibration strategy and to the identifiability of the parameters. To analyse these issues, we propose to base the evaluation of a semi-distributed model not only on its performance at streamflow gauging stations, but also on the spatial and temporal pattern of the optimised value of its parameters. We implemented calibration over 21 rolling periods and 64 catchments, and we analysed how well each parameter is identified in time and space. Performance and parameter identifiability are analysed comparatively to the calibration of the lumped version of the same model. We show that the semi-distributed model faces more difficulties to identify stable optimal parameter sets. The main difficulty lies in the identification of the parameters responsible for the closure of the water balance (i.e. for the particular model investigated, the intercatchment groundwater flow parameter).

1 Introduction

1.1 What hydrological good sense suggests

Developing modelling tools that help to understand the spatial distribution of water resources is a key issue for better management. The dynamics of streamflow depends on (i) the spatial variability of precipitation (which, a priori, should be better handled by a semi-distributed hydrological model), (ii) the heterogeneity of catchment behavior (which can be dealt explicitly with by spatially-variable model parameters), and, increasingly, (iii) localized human regulations (for instance, water reservoirs). Since calibration is generally based on discharge measurements at the outlet of the catchment only, and gauging stations are not available everywhere, semi-distributed hydrological models are often difficult to parameterize. As argued by Pokhrel and Gupta (2011), difficulties are due to the smoothing effect of catchments and to the dispersive effect of flow routing combined with numerical issues and measurement uncertainty. The authors state that the impact of spatial variability could become "virtually non-detectable by conventional performance measures by the time the water reaches the catchment outlet".

This raises the need to better understand how well parameters are identified in a semi-distributed model compared to a lumped model. The variability of catchment model parameters calibrated over different periods ("time variability") is one way of approaching this question. Indeed, as reminded by Merz et al. (2011), parameters of a rainfall–runoff model are supposed to represent stable catchment conditions, while the time-varying conditions are supposed to be triggered by the time-series of meteorological inputs. Thus, optimised parameter values should not be overly sensitive to changes of climatic conditions, and one would expect a semi-distributed model to be more stable than a lumped one (because the parameters of the lumped model would have to account implicitly for changing spatial precipitation patterns).

1.2 What the literature says

However, literature provides many examples showing that this assumption is hardly satisfied. Merz et al. (2011) raise two main difficulties. First, problems in model structure and data measurements tend to be compensated by calibration. For instance, Wagener et al. (2003) identify inconsistencies in the structure of a rainfall–runoff model by highlighting instabilities of the optimal values of the parameters between periods with and without rainfall events. In contrast, Juston et al. (2009) found that data subset of their input data (from daily to quaterly sampling intervals) can provide very similar constraints on model calibration and parameter identification.

Secondly, conditions of the catchment itself may change over time, which consequently, and understandably, shifts optimal model parameters, and justifies rigourous evaluations of the model robustness (Thirel et al., 2015). For instance, changes of land use might directly impact optimal parameter sets (Andréassian et al., 2003; Brown et al., 2005; Verstegen et al., 2016). Trends on parameters can also be related to changes in climatic conditions (Merz et al., 2006; Merz and Blöschl, 2009). On a study based on 273 Austrian catchments, Merz et al. (2011) attribute a doubling of the parameter that controls runoff generation in their model to be related to hydrological changes (such as higher evapotranspiration and drier catchment conditions) rather than calibration artifacts. For about one third of 17 African catchments, Niel et al. (2003) find their model parameters to be unstable, but they did not identify any climate-related reason. Wilby (2005) analyses this question in the context of climate change impact assessment for the River Thames. The author finds model parameters to be highly sensitive to training periods, and recommends the quantitification of those large uncertainties due to parameter instability, identifiability and non-uniqueness. Similarly, Brigode et al. (2013) found that the uncertainty due to the climate characteristics of the calibration period is larger than the uncertainty in the estimation of parameters that is often quantified on the basis of Bayesian inference. They attribute this finding to a lack of robustness of the model and recommend more efforts to be put into this aspect.

1.3 Scope of the paper

This paper investigates the procedure of parameter identifiability in a semi-distributed model by comparing model calibration schemes and results with a lumped model on which it is based. From this comparison, we address two main questions: (1) Does spatial distribution of parameters interfere with parameters identifiability? Indeed, one could hope that applying parameters to a more geographically-limited area tends to facilitate their identification. (2) What are the parameters that are the most variable in the lumped and in the semi-distributed models? In this way, we aim to diagnose which components of the model are the least robust, in the sense that their parameterisation is difficult to transpose in time and space.

2 Material and methods

2.1 Study area and hydro-meteorological data

The model is implemented in Eastern France, close to the border with Germany, over 64 sub-catchments of French tributaries of the River Rhine (Fig. 1), namely the rivers Moselle, Sarre and the smaller tributaries of the Rhine located in the Vosges massif. Catchments size vary from 27 to 770 km^2 and this represents a total area of about 4340 km^2. Climate is predominantly oceanic with continental influence. Annual precipitation (P) varies from about 700 mm in the lowland to about 1600 mm in the Vosges massif. Average daily temperature (T) and potential evapotranspiration (PE) in the catchment vary from 7 to 10 °C and 540 to 690 mm, respectively.

The two hydrological models implemented (lumped and semi-distributed) require daily time series of P and PE as input data. We used climate data from the SAFRAN meteorological reanalysis of Météo-France (Vidal et al., 2010), which is provided on a square grid of 8 km × 8 km. Discharge data were extracted from the French Hydro database (Leleu et al., 2014) at the daily time step. They were used to perform the calibration and the evaluation of the models. The study period is 1971–2000.

2.2 The GR5J lumped and GRSD semi-distributed rainfall–runoff models

The GRSD semi-distributed rainfall–runoff model was developed by Lerat (2009) and Lobligeois (2014). It is based on the GR5J lumped model (Fig. 2) proposed by Le Moine (2008), which has five free parameters to calibrate (Table 1). The main components of the model are two stores: a production store (with maximum capacity X1) and a routing store (capacity X3), which is filled by the output of a unit hydrograph (of time base X4). Two other parameters, X2 and X5, are used to quantify the intercatchment groundwater flows (IGF). In order to account for snow accumulation and melt, the model is combined with a degree-day snow module (Valéry et al., 2014), which contains two additional parameters (C_{TG} and K_f). However, in our study, these parameters were not calibrated and fixed at their default values, respectively set at 0.2 and 4.5 mm °C^{-1}, as proposed by previous studies in France (Valéry et al., 2014).

The semi-distributed model is applied on sub-catchments. The delineation of sub-catchments is performed only at gauging stations, which means that discharge measurements are available for every hydrological units of the model. The lumped GR5J model is applied on hydrological units composed of upstream catchments (headwater catchments) or intermediate sub-catchments (drained area between downstream and upstream stations). In that way, each hydrological unit receives its own meteorological inputs (P and PE) and uses a distinct parameter set (see Sect. 2.3.2).

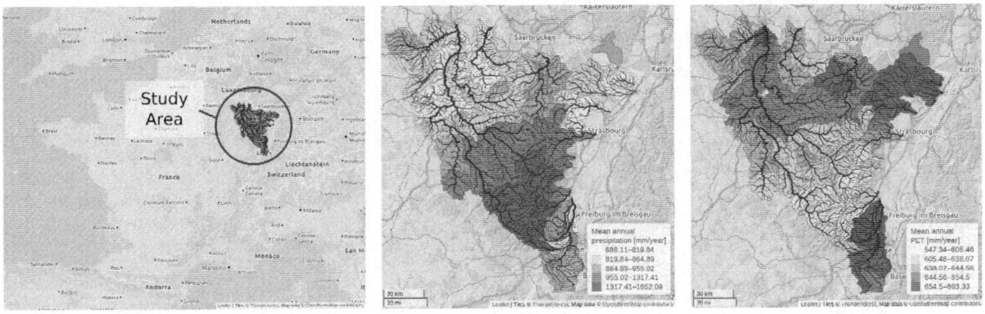

Figure 1. Location of the study area (left), and average annual precipitation (centre) and annual potential evapotranspiration (right) for each sub-catchment (climate data are estimated from the 1971–2000 SAFRAN database).

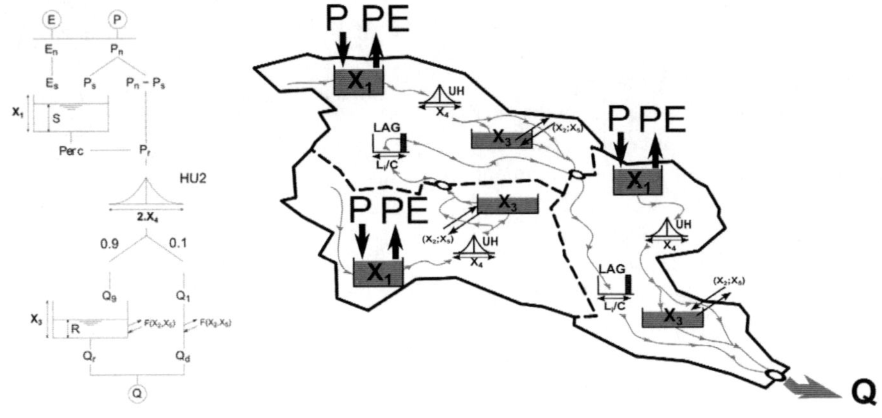

(a) *GR5J* (b) *GRSD*

Figure 2. Schematic representation of the GR5J and GRSD semi-distributed model (from Lobligeois et al., 2014).

Table 1. List of the parameters for the semi-distributed conceptual rainfall–runoff model GRSD.

Model parameter	GR5J	GRSD	Description
X1	Free	Free	Production store capacity [mm]
X2	Free	Free	Groundwater exchange coefficient [$\mathrm{mm\,d^{-1}}$]
X3	Free	Free	Routing store capacity [mm]
X4	Free	Free	Time base of the unit hydrograph [d]
X5	Free	Free	Threshold for groundwater exchange [–]
C	–	Free	Average streamflow velocity [$\mathrm{m\,s^{-1}}$]
C_{TG}	Fixed	Fixed	Ponderation coefficient of the snow thermic state [–]
K_{f}	Fixed	Fixed	Degree-day factor [$\mathrm{mm\,^{\circ}C^{-1}}$]

The outflow of each GR5J model is finally routed to its downstream catchment using a linear lag propagation model (Bentura and Michel, 1997). Previous studies have shown that this propagation model gives a satisfactory level of efficiency compared to more sophisticated channel routing methods (Lobligeois et al., 2014). This routing functionality implies an additional free parameter (compared to the lumped model GR5J) that needs to be calibrated on each hydrological unit: the average flow velocity C.

2.3 Methodology

2.3.1 Goodness of fit criteria

In order to quantify the agreement between simulations (S) and observations (O), we used the Kling–Gupta Efficiency (KGE) (Gupta et al., 2009), which is based on a decomposition of the Nash–Sutcliffe efficiency (Nash and Sutcliffe, 1970). Moreover, in order to evaluate performances on high and low flows, we used an objective function $\mathrm{KGE_m}$ composed of two criteria (Eq. 1): a KGE applied on discharge values to emphasize high flows and a KGE applied on inverse discharge values to emphasize low flows. Both criteria are applied on the selected discharge time series. Similarly to the KGE criteria, KGE_m varies between $-\infty$ and 1, which is its optimal value.

$$KGE_m(S,O) = 0.5 \cdot \left(KGE(S,O) + KGE\left(\frac{1}{S+\epsilon}, \frac{1}{O+\epsilon}\right)\right), \quad (1)$$

where O and S are the observed and simulated discharges. In order to face numerical problems in case of zero discharge when using the inverse transformation, an ϵ constant is used and set to 1 % of the mean value of O (Pushpalatha et al., 2012).

2.3.2 Calibration strategy of the semi-distributed model

Following Lerat et al. (2012), we performed a multi-site calibration of the GRSD semi-distributed model. Streamflow data at interior points are used to calibrate the model at one outlet. Each intermediate catchment is allowed to have a different parameter set. This is done sequentially, from upstream to downstream points: once the upstream catchment is calibrated, its parameters remain fixed during the calibration of the downstream intermediate catchment.

Sequential calibration is a common strategy for semi-distributed models (see e.g., Andersen et al., 2001; Moussa et al., 2007). It needs as much calibration runs as there are interior points. However, it only uses successive single objective functions, rather than using multi-response objective function to optimize every interior points simultaneously.

2.3.3 Rolling calibration periods

Similarly to Coron et al. (2012), we calibrated the parameters using 10-year long consecutive periods between 1971 and 2000, and used the rest of the time series (20 years) for validation (Fig. 3). This is equivalent to 21 split-sample tests (Klemeš, 1986) performed every year.

Following the work of Merz and Blöschl (2009) and Merz et al. (2011), this enables to provide 21 parameter sets for each of the 64 catchments in order to analyse the temporal and spatial changes of the calibrated parameters. This testing strategy is applied to both models, the lumped model and the semi-distributed model. Parameter variability can thus be compared between the two modelling strategies.

3 Results and discussion

3.1 Performance of the streamflow simulations

The comparison of the goodness-of-fit between the lumped GR5J model and the semi-distributed GRSD model shows slightly better results of the lumped model during calibration and identical results during validation (Fig. 4). One would expect higher performances of the semi-distributed model because it accounts for spatial heterogeneities and nonlinearities that can influence the response of the system. How-

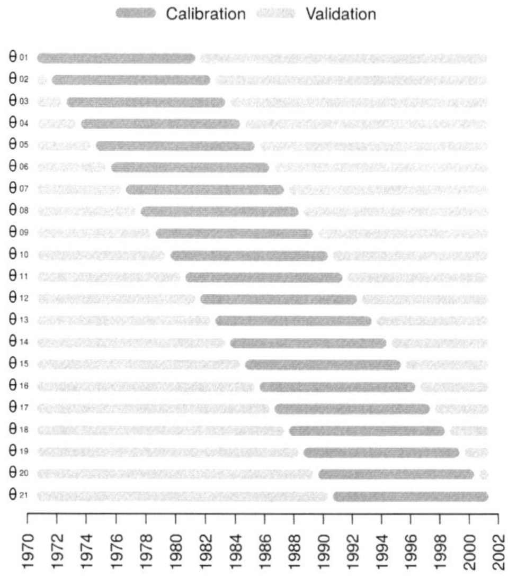

Figure 3. Illustration of the rolling calibration period methodology: 21 parameter sets θ can be identified for each catchment.

ever, literature provides numerous examples of similar results (e.g., Reed et al., 2004), where lumped models perform better. To explain such behaviour, calibration strategies, which are not as well defined for semi-distributed models as for lumped models, are often pointed out (Pokhrel and Gupta, 2011).

As expected, performance on upstream catchments are similar between the lumped and the semi-distributed catchments. Indeed, for those catchments, models are strictly identical (Sect. 2.2). Minor differences can be explained by calibration artifacts and by the fact that both models are implemented in two different modelling environments.

We did not detect any significant performance trends in time. Calibration performances are rather stable, whereas validation performances are subject to more fluctuations. These are similar between the lumped and the semi-distributed models. Results illustrate that both models are potentially able to produce stable efficiency (KGE_m) all along 1971–2000 period, but each calibration period does not provide the same robustness (as observed by validation).

3.2 Temporal trends and variability of parameters

Here, we compare the temporal trends and the variability of parameters, and differences between the two models. The distribution of parameter values according to the 21 calibration periods is given in Fig. 5.

As expected from the structure of the models (Sect. 2.2), upstream catchments have similar parameter sets for both models, whereas different optimum parameter values are obtained for downstream catchments. The main differences concern the parameter X1 (capacity of production store), which is higher for the lumped model than for the semi-

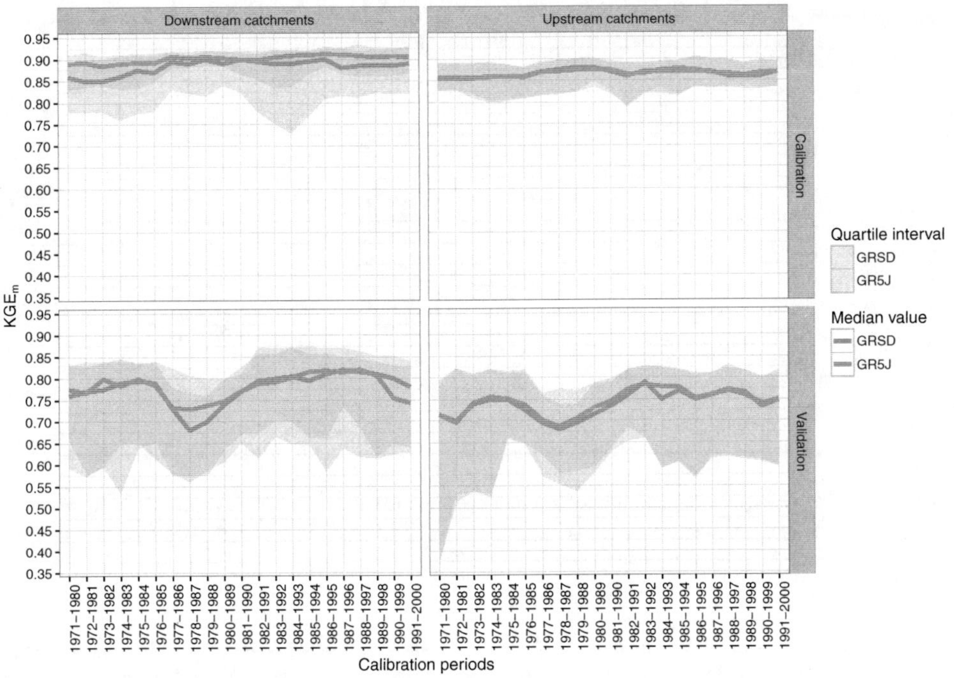

Figure 4. Quantile values of the goodness-of-fit (KGE_m) during calibration and validation over the 64 catchments for the lumped model (GR5J) and the semi-distributed model (GRSD) along the 21 calibration periods. Upstream catchments are all headwater hydrological units, and downstream catchments are all the others.

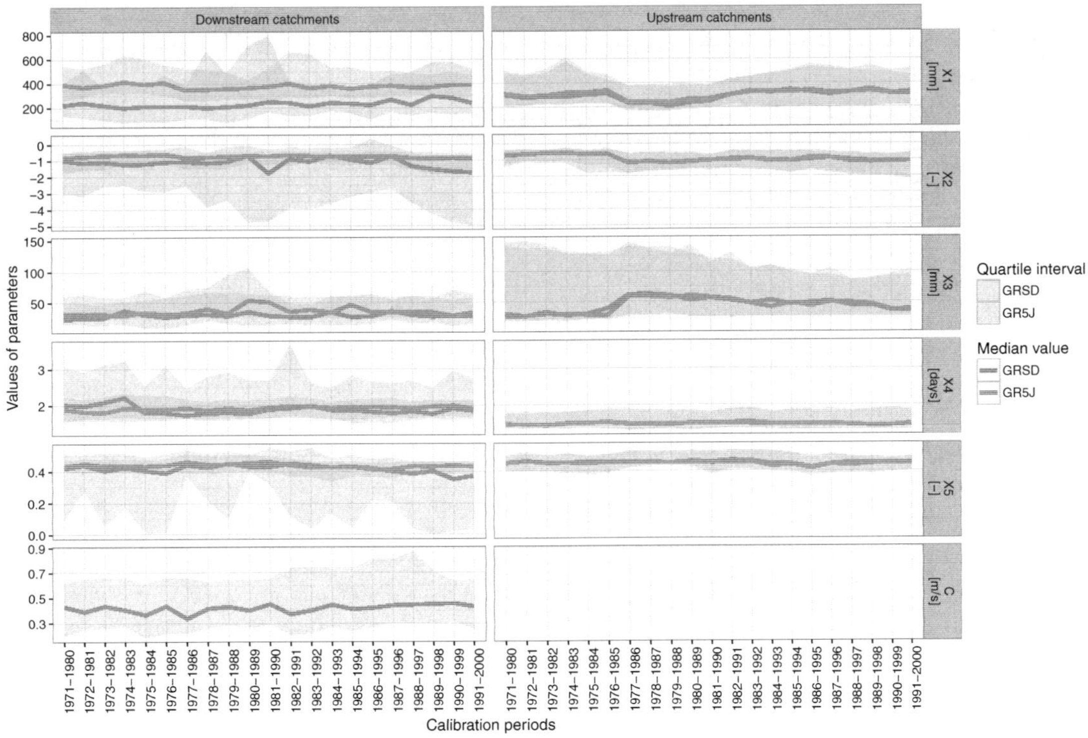

Figure 5. Quantile values of optimised parameter values according to the calibration periods for the 64 catchments. Upstream catchments are all headwater hydrological units, and downstream catchments are all the others.

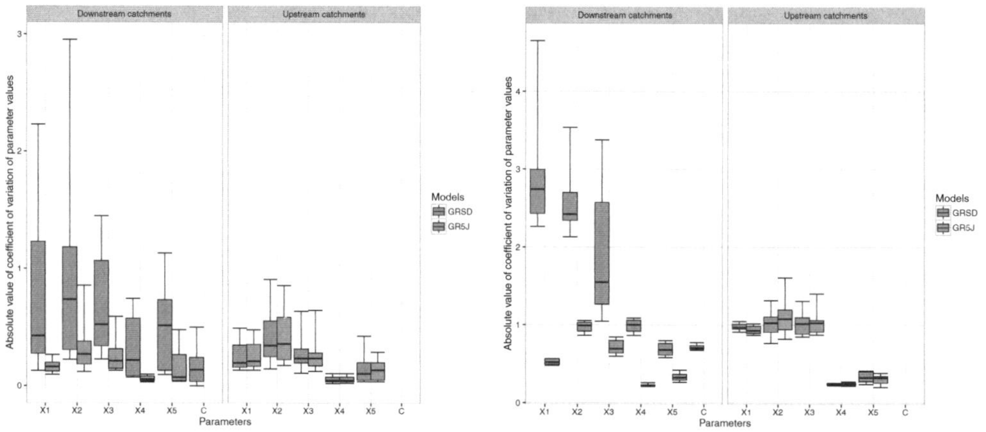

(a) *Temporal variability* (b) *Spatial variability*

Figure 6. Variability of parameters values among calibration periods within one catchment (64 catchments summarized by boxplots, **a**) and variability of parameters values among catchments within one calibration period (21 calibration periods summarized by boxplots, **b**). Boxplot limits describe the 10th, 25th, 50th, 75th and 90th quantiles.

distributed model. This means a more dynamic response of the downstream hydrological units of the GRSD model and lower evapotranspiration losses. The smoothing of the hydrographs may, in fact, be achieved by the succession of responses of the sub-catchments, from upstream to downstream.

The smaller production store in GRSD appears to be compensated by X2 and X5. Indeed, those parameters aim to quantify intercatchment groundwater flows (IGF), which is the amount of water that daily gets out/in of the catchment to fill/empty the routing store and the direct flow component. The X2 parameter quantifies IGF according to a linear relation with the routing store rate ($\frac{S_{\text{rout}}(t)}{\text{X3}}$), whereas X5 allows changing the sign of IGF during the year (Eq. 2). In GRSD, where we observe greater negative values of X2 and smaller values of X5 (Fig. 5), IGF may be higher to compensate lower evapotranspiration losses in the production store.

$$\text{IGF}(t) = \text{X2} \cdot \left(\frac{S_{\text{rout}}(t)}{\text{X3}} - \text{X5} \right) \quad (2)$$

By looking at parameter values according to the calibration periods, a relative stability of the median value appears among catchments. Only the parameters X2 and X5 in GRSD tend to slowly decrease. These trends are not observed in the lumped model, whose parameters appear more stable for downstream catchments.

From the relative stability in Fig. 5, one could conclude about overall relative robustness of the calibration. However, this does not evaluate parameter stability for each catchment. Therefore, for each hydrological unit of both models, we calculated the coefficient of variation of the parameter values using the 21 parameter sets (Fig. 6).

First, results clearly show a higher temporal variability of the parameters of the semi-distributed model, comparatively to the variability observed with the lumped model (Fig. 6). It appears that limiting the geographical extent of the area on which the parameter set is applied does not facilitate its identification. One reason for this can come from the calibration strategy, in which one sub-catchment receives simulated outflows from upstream catchments. This upstream volume of water can already represent most of the observed downstream hydrograph. Therefore, parameters applied just on a downstream intermediate sub-catchment might have a minor impact, with problems of sensitivity. Thus calibration may converge to more unstable values, which brings only small improvements to downstream simulations.

Second, it is shown that the most important temporal instability of parameter values is related to the parameter X2, followed by X3 and X5 parameters. All three parameters are used to quantify IGF (Eq. 2). If parameter X2 is showing the most important temporal instability for both models, it is not the case of parameter X5 with the lumped model, where it appears to be one of the least variable parameter (just after parameter X4). This result highlights the problems encountered in quantifying IGF. It seems to be the least identifiable parameter in the lumped model (with high variability of parameter X2). The problem of IGF parameter identifiability is even exacerbated with the semi-distributed model. This might be due to the high inter-dependency of parameters in the formulation of IGF (Eq. 2) in the models. Future improvements of the GRSD semi-distributed model should focus on this issue.

The most stable parameter appears to be X4 (time base of unit hydrograph), which is consistent with previous works (Lobligeois, 2014), where it has been shown that this parameter can easily be related to catchment physical characteristics, such as catchment size.

3.3 Spatial variability of parameters

We also analysed the spatial variability of the parameter values, considering variability between parameters and between models. To this end, we estimated the coefficient of variation of the parameter values among catchments (one performed by calibration period). The aim is to quantify how much parameters can be different between catchments.

Similarly to the temporal variability of parameters, spatial variability appears to be higher with the semi-distributed model than with the lumped model. However, contrary to the temporal variability, spatial variability is more expected here, as one of the objectives of a semi-distributed model is precisely to consider those spatial heterogeneities of the hydrological response. However, we noticed again that parameters X1 and X2 are the most variable parameters for the semi-distributed model. Therefore they appear among the most variable parameters for both analyses, the time variability and the space variability analyses. These two parameters control water balance. Similarly to the temporal variability, they are expected to be highly variable in space in order to get along with sequential observations at each downstream station during calibration.

Spatial variabilities are not constant over time (as observed by the boxplot widths on Fig. 6). Particularly, the spatial variability of parameter X3 (capacity of routing store) is stable for the lumped model, whereas it appears to be very dependent on the calibration period for the semi-distributed model. For instance, it has the highest spatial variability during the "1984–1994" calibration period, and is among the lowest variability just 3 years after (not shown). Once again, a robustness problem of GRSD is identified and needs to be addressed in further investigations.

4 Conclusion

In this paper, we compared the spatio-temporal variability of the parameters of a semi-distributed model (GRSD) and a lumped model (GR5J) on which it is based. We applied a rolling calibration strategy over 21 periods and 64 French catchments.

A classical evaluation of discharge simulations using goodness-of-fit criteria was applied to the outputs of both models. It illustrates a slightly better performance of the lumped model during calibration, and similar performance of the models during validation. However, further investigation on parameter identifiability highlighted much higher temporal variabilities of the semi-distributed model. This study also showed that it is more difficult to identify catchment's specific parameter sets with the semi-distributed model than with the lumped model.

The methodology applied also enabled to identify the more unstable parameters. Results showed that the parameters related to the quantification of intercatchment groundwater flows (IGF) are the most unstable. We conclude that further modelling efforts should focus on the model structure in order to better quantify IGF.

This work also emphasizes the fact that the calibration strategy and the evaluation approach of a semi-distributed model should not focus only on goodness-of-fit performance, but also on parameter identifiability, especially if the model aims to be used to explore future scenarios in a changing world. Such an approach would also facilitate the application of the model at ungauged locations, since parameters that depict high variability in time and space might be more difficult to regionalize.

Data availability

The streamflow data used in this study are freely available at: http://www.hydro.eaufrance.fr. The climatic data are produced by Métèo-France. Information can be found at: http://publitheque.meteo.fr.

Acknowledgements. The first author was partially funded by the French Waterboard "Agence de l'Eau Rhin-Meuse" within the MOSARH21 project (project #15C92002). Météo-France and SCHAPI are thanked for making climatic and hydrological data available for this study. The authors thank the anonymous reviewer for his comments on the manuscript.

References

Andersen, J., Refsgaard, J. C., and Jensen, K. H.: Distributed hydrological modelling of the Senegal River Basin – model construction and validation, J. Hydrol., 247, 200–214, doi:10.1016/S0022-1694(01)00384-5, 2001.

Andréassian, V., Parent, E., and Michel, C.: A distribution-free test to detect gradual changes in watershed behavior, Water Resour. Res., 39, 1252, doi:10.1029/2003WR002081, 2003.

Bentura, P. L. and Michel, C.: Flood routing in a wide channel with a quadratic lag-and-route method, Hydrolog. Sci. J., 42, 169–189, doi:10.1080/02626669709492018, 1997.

Brigode, P., Oudin, L., and Perrin, C.: Hydrological model parameter instability: A source of additional uncertainty in estimating the hydrological impacts of climate change?, J. Hydrol., 476, 410–425, doi:10.1016/j.jhydrol.2012.11.012, 2013.

Brown, A. E., Zhang, L., McMahon, T. A., Western, A. W., and Vertessy, R. A.: A review of paired catchment studies for determining changes in water yield resulting from alterations in vegetation, J. Hydrol., 310, 28–61, doi:10.1016/j.jhydrol.2004.12.010, 2005.

Coron, L., Andréassian, V., Perrin, C., Lerat, J., Vaze, J., Bourqui, M., and Hendrickx, F.: Crash testing hydrological models in contrasted climate conditions: An experiment on 216 Australian catchments, Water Resour. Res., 48, W05552, doi:10.1029/2011WR011721, 2012.

Gupta, H. V., Kling, H., Yilmaz, K. K., and Martinez, G. F.: Decomposition of the mean squared error and NSE performance criteria: Implications for improving hydrological modelling, J. Hydrol., 377, 80–91, doi:10.1016/j.jhydrol.2009.08.003, 2009.

HYDRO: http://www.hydro.eaufrance.fr, last access: May 2016.

Juston, J., Seibert, J., and Johansson, P.-O.: Temporal sampling strategies and uncertainty in calibrating a conceptual hydrological model for a small boreal catchment, Hydrol. Process., 23, 3093–3109, doi:10.1002/hyp.7421, 2009.

Klemeš, V.: Operational testing of hydrological simulation models, Hydrolog. Sci. J., 31, 13–24, doi:10.1080/02626668609491024, 1986.

Leleu, I., Tonnelier, I., Puechberty, R., Gouin, P., Viquendi, I., Cobos, L., Foray, A., Baillon, M., and Ndima, P.-O.: La refonte du système d'information national pour la gestion et la mise à disposition des données hydrométriques, La Houille Blanche, 25–32, doi:10.1051/lhb/2014004, 2014.

Le Moine, N.: Le bassin versant de surface vu par le souterrain: une voie d'amélioration des performances et du réalisme des modèles pluie-débit?, PhD thesis, Université Pierre et Marie Curie, Paris, France, 348 pp., 2008.

Lerat, J.: Quels apports hydrologiques pour les modèles hydrauliques? Vers un modèle intégré de simulation des crues, PhD thesis, University of Pierre et Marie Curie, Paris, France, 390 pp., 2009.

Lerat, J., Andréassian, V., Perrin, C., Vaze, J., Perraud, J. M., Ribstein, P., and Loumagne, C.: Do internal flow measurements improve the calibration of rainfall-runoff models?, Water Resour. Res., 48, W02511, doi:10.1029/2010WR010179, 2012.

Lobligeois, F.: Mieux connaître la distribution spatiale des pluies améliore-t-il la modélisation des crues?, Diagnostic sur 181 bassins versants français, PhD thesis, AgroParisTech, Paris, France, 310 pp., 2014.

Lobligeois, F., Andréassian, V., Perrin, C., Tabary, P., and Loumagne, C.: When does higher spatial resolution rainfall information improve streamflow simulation? An evaluation using 3620 flood events, Hydrol. Earth Syst. Sci., 18, 575–594, doi:10.5194/hess-18-575-2014, 2014.

Merz, R. and Blöschl, G.: A regional analysis of event runoff coefficients with respect to climate and catchment characteristics in Austria, Water Resour. Res., 45, W01405, doi:10.1029/2008WR007163, 2009.

Merz, R., Blöschl, G., and Parajka, J.: Spatio-temporal variability of event runoff coefficients, J. Hydrol., 331, 591–604, doi:10.1016/j.jhydrol.2006.06.008, 2006.

Merz, R., Parajka, J., and Blöschl, G.: Time stability of catchment model parameters: Implications for climate impact analyses, Water Resour. Res., 47, W02531, doi:10.1029/2010WR009505, 2011.

Météo-France: http://publitheque.meteo.fr, last access: May 2016.

Moussa, R., Chahinian, N., and Bocquillon, C.: Distributed hydrological modelling of a Mediterranean mountainous catchment – Model construction and multi-site validation, J. Hydrol., 337, 35–51, doi:10.1016/j.jhydrol.2007.01.028, 2007.

Nash, J. and Sutcliffe, J.: River flow forecasting through conceptual models part I – A discussion of principles, J. Hydrol., 10, 282–290, doi:10.1016/0022-1694(70)90255-6, 1970.

Niel, H., Paturel, J.-E., and Servat, E.: Study of parameter stability of a lumped hydrologic model in a context of climatic variability, J. Hydrol., 278, 213–230, doi:10.1016/S0022-1694(03)00158-6, 2003.

Pokhrel, P. and Gupta, H. V.: On the ability to infer spatial catchment variability using streamflow hydrographs, Water Resour. Res., 47, W08534, doi:10.1029/2010WR009873, 2011.

Pushpalatha, R., Perrin, C., Le Moine, N., and Andréassian, V.: A review of efficiency criteria suitable for evaluating low-flow simulations, J. Hydrol., 517, 1176–1187, doi:10.1016/j.jhydrol.2011.11.055, 2012.

Reed, S., Koren, V., Smith, M., Zhang, Z., Moreda, F., Seo, D.-J., and DMIP Participants: Overall distributed model intercomparison project results, J. Hydrol., 298, 27–60, doi:10.1016/j.jhydrol.2004.03.031, 2004.

Thirel, G., Andréassian, V., and Perrin, C.: On the need to test hydrological models under changing conditions, Hydrolog. Sci. J., 60, 1165–1173, doi:10.1080/02626667.2015.1050027, 2015.

Valéry, A., Andréassian, V., and Perrin, C.: 'As simple as possible but not simpler': What is useful in a temperature-based snow-accounting routine? Part 2 – Sensitivity analysis of the Cemaneige snow accounting routine on 380 catchments, J. Hydrol., 517, 1176–1187, doi:10.1016/j.jhydrol.2014.04.058, 2014.

Verstegen, J. A., Karssenberg, D., van der Hilst, F., and Faaij, A. P.: Detecting systemic change in a land use system by Bayesian data assimilation, Environ. Modell. Softw., 75, 424–438, doi:10.1016/j.envsoft.2015.02.013, 2016.

Vidal, J.-P., Martin, E., Franchistéguy, L., Baillon, M., and Soubeyroux, J.-M.: A 50-year high-resolution atmospheric reanalysis over France with the Safran system, Int. J. Climatol., 30, 1627–1644, doi:10.1002/joc.2003, 2010.

Wagener, T., McIntyre, N., Lees, M. J., Wheater, H. S., and Gupta, H. V.: Towards reduced uncertainty in conceptual rainfall-runoff modelling: dynamic identifiability analysis, Hydrol. Process., 17, 455–476, doi:10.1002/hyp.1135, 2003.

Wilby, R. L.: Uncertainty in water resource model parameters used for climate change impact assessment, Hydrol. Process., 19, 3201–3219, doi:10.1002/hyp.5819, 2005.

Proc. IAHS, 373, 95–100, 2016
proc-iahs.net/373/95/2016/
doi:10.5194/piahs-373-95-2016

Climate, orography and scale controls on flood frequency in Triveneto (Italy)

Simone Persiano[1], Attilio Castellarin[1], Jose Luis Salinas[2], Alessio Domeneghetti[1], and Armando Brath[1]

[1]Department DICAM, School of Civil Engineering, University of Bologna, Bologna, Italy
[2]Institute of Hydraulic Engineering and Water Resources Management, Vienna University of Technology, Vienna, Austria

Correspondence to: Simone Persiano (simone.persiano2@unibo.it)

Published: 12 May 2016

Abstract. The growing concern about the possible effects of climate change on flood frequency regime is leading Authorities to review previously proposed reference procedures for design-flood estimation, such as national flood frequency models. Our study focuses on Triveneto, a broad geographical region in North-eastern Italy. A reference procedure for design flood estimation in Triveneto is available from the Italian NCR research project "VA.PI.", which considered Triveneto as a single homogeneous region and developed a regional model using annual maximum series (AMS) of peak discharges that were collected up to the 1980s by the former Italian Hydrometeorological Service. We consider a very detailed AMS database that we recently compiled for 76 catchments located in Triveneto. All 76 study catchments are characterized in terms of several geomorphologic and climatic descriptors. The objective of our study is threefold: (1) to inspect climatic and scale controls on flood frequency regime; (2) to verify the possible presence of changes in flood frequency regime by looking at changes in time of regional L-moments of annual maximum floods; (3) to develop an updated reference procedure for design flood estimation in Triveneto by using a focused-pooling approach (i.e. Region of Influence, RoI). Our study leads to the following conclusions: (1) climatic and scale controls on flood frequency regime in Triveneto are similar to the controls that were recently found in Europe; (2) a single year characterized by extreme floods can have a remarkable influence on regional flood frequency models and analyses for detecting possible changes in flood frequency regime; (3) no significant change was detected in the flood frequency regime, yet an update of the existing reference procedure for design flood estimation is highly recommended and we propose the RoI approach for properly representing climate and scale controls on flood frequency in Triveneto, which cannot be regarded as a single homogeneous region.

1 Introduction

One of the most common tasks in hydrology is to produce an accurate estimation of the design flood at ungauged or sparsely gauged river cross-sections (see e.g. Salinas et al., 2014, and references therein). This task is often addressed by means of regional flood frequency analysis by collecting flood data from gauged basins which are supposed to be hydrologically similar to the target (ungauged or sparsely gauged) basin (see Hosking and Wallis, 1993). Our study focuses on Triveneto, a broad mountainous geographical area in North-eastern Italy consisting of the administrative districts of Trentino-Alto Adige, Veneto and Friuli-Venezia Giulia; the study area counts numerous dams that routinely undergo hydrologic and hydraulic risk assessments. A reference procedure for design flood estimation in Triveneto is available from the Italian NCR research project "VA.PI.", which developed an index-flood regional model based upon annual maximum series (AMS) of peak discharges that were collected up to the 1980s. We consider a very detailed AMS database that we recently compiled for 76 catchments located in Triveneto. Our annual flood sequences include historical data together with more recent data, and the data sets spans from 1913 to 2012.

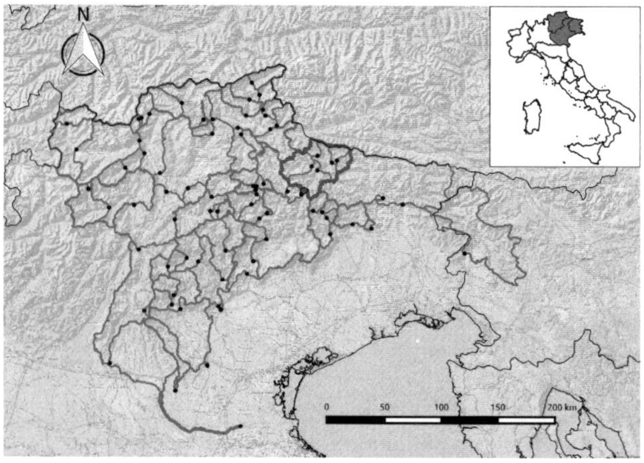

Figure 1. Location of the 76 Triveneto gauging stations (black points) considered in the present study. Grey lines represent catchments boundaries. Pieve di Cadore, an example target basin, is highlighted in red.

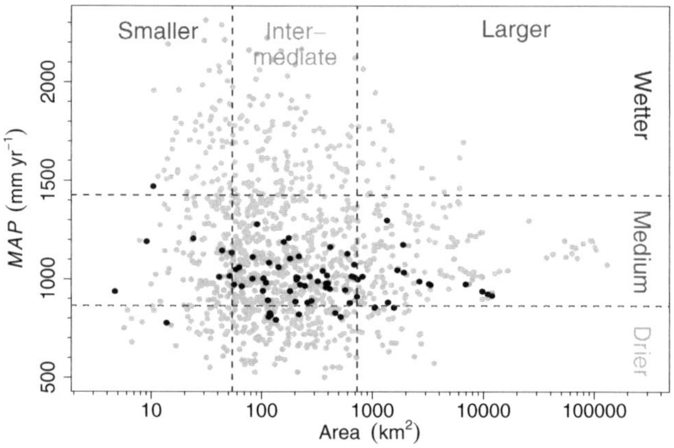

Figure 2. Catchment characteristics of the Triveneto data set (black points). Grey points represent the Austrian, Slovakian and Italian data sets considered in Salinas et al. (2014). Black-dashed lines identify the subdivision into the six subsets based on the 20 and 80 % quantiles of the catchments descriptor values for the European data sets.

The objective of our study is threefold: (1) to check whether climatic and scale controls on flood frequency regime in Triveneto are similar to the controls that were recently found in Europe; (2) to verify the possible presence of changes in flood frequency regime by looking at changes in time of regional L-moments of annual maximum floods (see Hosking and Wallis, 1993, for an introduction to L-moments); (3) to develop an updated reference procedure for design flood estimation in Triveneto by using a focused-pooling approach (i.e. Region of Influence, RoI; see Burn, 1990).

2 Area and MAP control on regional flood frequency distribution

The selection of the most suitable regional parent distribution is a key aspect in any regional flood frequency analysis. The scientific literature recommends using the L-moment ratio diagrams for addressing this task (see e.g. Hosking and Wallis, 1993). Because of the unavailability or high uncertainty of sample statistics for ungauged or poorly gauged regions, many studies have focused on the relationships between sample L-moments and catchment descriptors. In particular, a recent study by Salinas et al. (2014) on a data set of annual maximum series (AMS) of peak flow from a total of 813 catchments (in Austria, Italy and Slovakia) shows the great importance of incorporating into regional models information on mean annual precipitation (MAP) and basin area as surrogates of climate and scale controls. The first main aim of our study is to check whether climatic and scale controls on flood frequency regime in Triveneto are similar to the controls that were recently found in Europe by Salinas et al. (2014).

The study region consists of 76 catchments (see Fig. 1) with minimum, mean and maximum length of observed AMS equal to 5, 31 and 87 years, in this order. To reduce the effects of sampling variability when estimating higher order L-moments (see e.g. Viglione, 2010), the information from each site was weighted proportionally to the site record length (Hosking and Wallis, 1993) during all regionalization phases. For comparison purposes, the data set was divided into the same six subsets identified in Salinas et al. (2014): smaller (area $< 55\,\mathrm{km}^2$), intermediate ($55\,\mathrm{km}^2 <$ area $< 730\,\mathrm{km}^2$) and larger (area $> 730\,\mathrm{km}^2$) catchments, drier (MAP $< 860\,\mathrm{mm\,yr}^{-1}$), medium ($860\,\mathrm{mm\,yr}^{-1} <$ MAP $< 1420\,\mathrm{mm\,yr}^{-1}$) and wetter (MAP $> 1420\,\mathrm{mm\,yr}^{-1}$) catchments. It is evident that the majority of the Triveneto catchments (black points in Fig. 2) belong to the central subset. So, in order to verify the impact of catchment size and MAP on the L-moment ratios for the Triveneto regional parent distribution, we considered only the two subsets associated with intermediate site and medium MAP. For each one of them, the two L-moment ratio diagrams L-Cv–L-Cs and L-Ck–L-Cs were plotted, including the theoretical lines for the most common two- and three-parameter distributions, respectively (see Fig. 3). Similarly to Salinas et al. (2014), the points drawn in the diagrams represent the record length weighted moving average (WMA) values of the corresponding sample L-moments and their colour intensity is proportional to the mean value of the descriptor not used for the stratification. WMA was computed for a window including the 35 most similar catchments in terms of the evaluated descriptor (i.e. area or MAP).

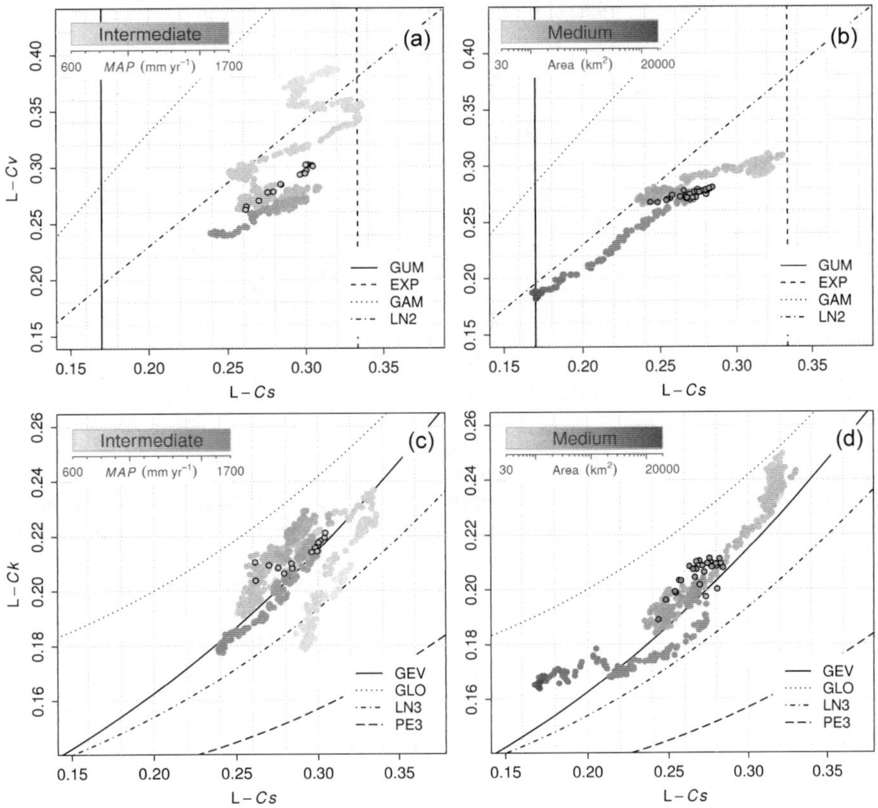

Figure 3. L-moment ratio diagrams for the subsets defined by intermediate area **(a, c)** and medium MAP **(b, d)**. Each point represents a record length weighted moving average (WMA) of L-Cv **(a, b)** or L-Ck **(c, d)** against the corresponding values of L-Cs and the colour intensity is proportional to the catchment descriptor of interest (area or MAP); WMAs are computed for either 35 Triveneto basins (black-border) or 70 European basins (no-border) having similar area or MAP value.

Figure 3 clearly shows the significant superimposition of Triveneto WMAs to WMAs obtained by Salinas et al. (2014). Figure 3a and b confirm that sample L-Cv and L-Cs are poorly described by the most common two-parameter distributions. On the other hand, Fig. 3c shows how medium-sized (intermediate) catchments associate with high to medium MAP are well described by the Generalized Extrem Value (GEV) distribution, and Fig. 3d clearly reports the same behaviour for medium MAP catchments. The strong control of area and MAP on regional L-moments (i.e. WMAs) is also confirmed by the intensity of gradation, which increases with decreasing L-moment ratios values for all the considered cases: average L-moment ratio values are higher for smaller than larger catchments and for drier than wetter ones.

3 Changes in flood frequency

The routinely hydrological and hydraulic risk assessments of dams located in Triveneto are related to the growing concern about the possible effects of climate change on the region flood-frequency regime. A second main objective of this study is therefore to verify the possible presence of changes in flood frequency regime by looking at the variability of regional L-moments of annual maximum floods in time. In fact, according to Beniston (2012), Alps are a particularly sensitive context for climate change, and Castellarin and Pistocchi (2012) detected noticeable variations in the frequency regime of AMF in the last five decades in the Swiss Alps, with significant and abrupt changes of the mean flood, with change-point years identified between 1942 and 1977. Changes in the flood frequency regime for alpine Swiss catchments are also highlighted in Allamano et al. (2009b) and Schmocker-Fackel and Naef (2010), from the 1960s and the 1970s, respectively.

In this context, we decided to check the presence of changes in the frequency regime of AMS in Triveneto with reference to 1965, by dividing the total AMS sample into two sub-samples: PRE1965 (all years before 1965, for a total of 1217 station-years of data) and POST1965 (all years from 1965, included; 1148 station-years of data). It is important to highlight that POST1965 sub-sample includes 1966, which was characterized by exceptionally heavy floods in Triveneto, and Italy in general. Year 1966 is well-known for the 4–5 November floods in Florence, but the entire autumn

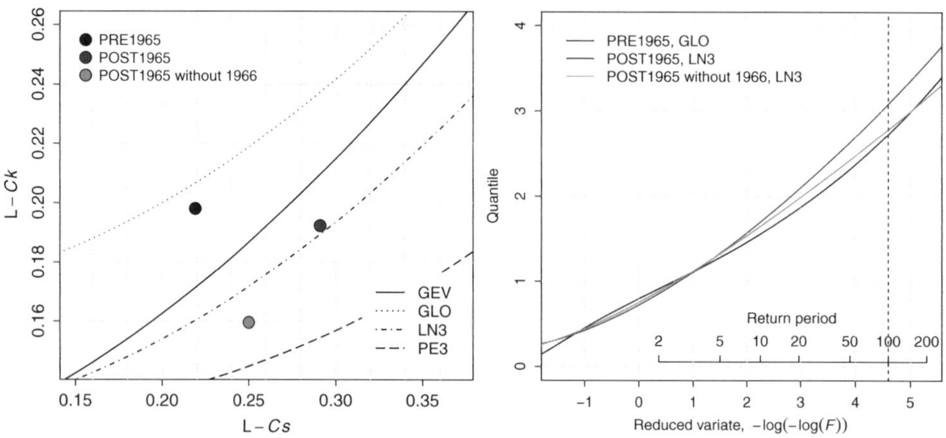

Figure 4. Left: L-moment ratio diagram for three-parameter distributions. Blue, red and green points represent the record length weighted average of L-Cv and L-Cs in Triveneto for PRE1965, POST1965 and POST1965 without 1966 sub-samples, in this order. Right: dimensionless growth-curves associated with the above mentioned three sub-samples. The choice of the corresponding three-parameter distribution was made by applying Hosking and Wallis test (see Hosking and Wallis, 1993).

of 1966 was particularly difficult for North-East Italian regions: Triveneto counted 87 deaths and about 42 000 displaced people (data from the Italian NCR). Hence, we decided to test the importance of this year in the flood frequency regime, by analysing a further sub-sample, obtained by removing AMF of 1966 from the POST1965 sub-sample (1117 station-years of data).

For each of the considered three sub-samples we plotted the record length weighted average of L-Cv and L-Cs in the corresponding diagram and estimated the associated regional three-parameter distribution (see Fig. 4). The assessment of the best fitting statistical models was done through a visual inspection (observing the distance between the averaged sample L-moment ratios and the theoretical lines), and confirmed by the Hosking and Wallis goodness-of-the-fit test (see Hosking and Wallis, 1993). As reported in Fig. 4, left, the differences shown by the three sub-samples are remarkable: whereas POST1965 sub-sample lays near the generalized logistic (GLO) theoretical curve, POST1965 and POST1965 without 1966 are associated with a three-parameter log-normal (LN3) distribution. Observing the shape of the dimensionless growth curve (see Fig. 4, right) differences are clear: for the 100-year return period, POST1965 shows a higher quantile than the other two sub-samples. In particular, it is important to highlight the high similarity between PRE1965 and POST1965 without 1966. This is emphasised by observing that the important differences observed between POST1965 and POST1965 without 1966 are due to 31 station-years of data only (the difference between 1148 and 1117 station-years of data). This behaviour shows how a single exceptional year characterized by extreme floods can heavily influence the observed flood frequency regime and therefore the results of the statistical analysis performed to detect changes.

4 Region of Influence approach

It is important to highlight that the dimensionless growth-curves in Fig. 4 do not refer to a homogeneous regional sample. In fact, although Triveneto was considered as a single homogeneous region by the Italian NCR research project "VA.PI." (by using AMS collected up to the 1980s, see Villi and Bacchi, 2001), the significant update done in our study showed that the hypothesis of homogeneity does not hold for Triveneto. The Hosking and Wallis test (see Hosking and Wallis, 1993) indicates for the study region a high heterogeneity degree ($H_1 > 8$, when the threshold value of H_1 for acceptably homogeneous regions is 1) for the overall Triveneto sample, and also for the three sub-samples shown in Fig. 4 (H_1 is approximately 4 and 9 for PRE1965 and the two POST1965 sub-samples, respectively).

The third and last main objective of our study is to update the Triveneto reference procedure for flood frequency estimation. Therefore we tested and applied the so called Region of Influence (RoI) approach (see Burn, 1990, and Zrinji and Burn, 1996), which replaces the idea of homogeneous regions consisting of contiguous and geographically identifiable regions with the most general idea of homogeneous groups of basins with similar hydrological behaviour, which may or may not be geographically close to each other. The RoI approach delineates homogeneous pooling-groups for a given target site referring to a minimum amount of information in terms of climatic and geomorphological descriptors of the basins that strongly influence the flood frequency regime at regional scale. In our study we consider the following descriptors: area and MAP (whose strong influence on the flood frequency regime has been confirmed in this study), mean and minimum elevation and latitude and longitude of catchment centroid. In particular, it is very important to con-

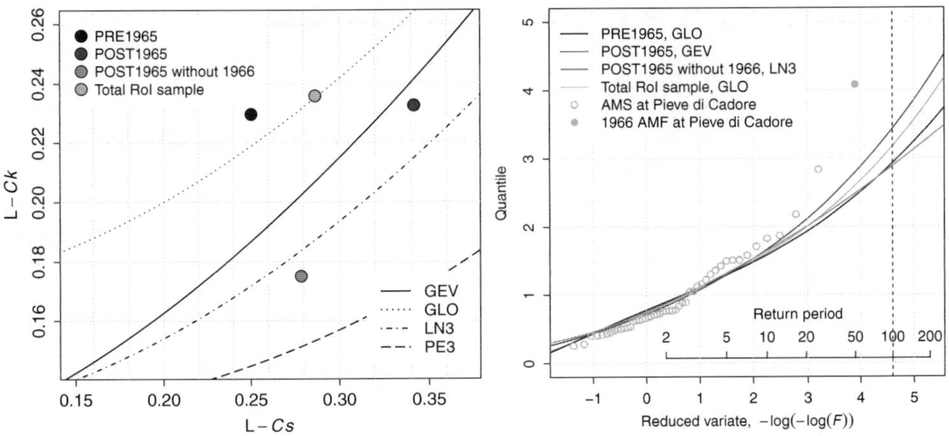

Figure 5. Left: L-moment ratio diagram for three-parameter distributions. Blue, red, green and grey points represent the record length weighted average of L-Cv and L-Cs for PRE1965, POST1965 and POST1965 without 1966 RoI sub-samples and total RoI sample, in this order. Right: dimensionless growth-curves associated with the above mentioned four sub-samples. The choice of the corresponding three-parameter distribution was made by applying Hosking and Wallis test (see Hosking and Wallis, 1993). Grey points indicate the empirical growth-curve for Pieve di Cadore (see Fig. 1, the full coloured point is the 1966 annual maximum flood).

sider catchment elevation as a descriptor of the orographic effect on the flood frequency regime (see e.g. Allamano et al., 2009a). The Euclidean distance in the 6-dimensional descriptor space was used as measure of hydrological dissimilarity (see e.g. Zrinji and Burn, 1996), where the same weight has been assigned to each of the mentioned descriptors.

The RoI approach arranges catchments according to their dissimilarity with the target site, pooling together only the most similar ones. The number of catchments to include in the RoI is determined by considering the return period of the design flood T: according to the "$5T$ rule" proposed by Jakob et al. (1999), a number of station-years of annual maxima $N \geq 5T$ is required to get a reliable estimation of the T-year flood (T-quantile). Therefore the RoI sample consists of a pooling-group of catchments that are the most hydrologically similar to the target site and whose overall station-years of data sums up to $5T$ (i.e. 500 station-years of data for $T = 100$ years).

We report herein an application example for the Pieve di Cadore dam, for which a 49 AMS (from 1951 to 1999) has been compiled from inflow series to the artificial reservoir. The RoI sample consists of 16 catchments, for a total of 555 observations, which enable one to get a reliable prediction of the 100-year quantile according to the "$5T$ rule". We can observe the significant advantage of the RoI approach in terms of homogeneity: the heterogeneity associated with the RoI sample ($H_1 \simeq 1.48$) is much lower than that of the total Triveneto sample. Therefore the grey point in Fig. 5, left, and the corresponding RoI dimensionless growth-curve (grey line in in Fig. 5, right) are characteristics of a regional parent distribution that is suitable for predicting dimensionless flood quantiles up to $T = 100$ years for Pieve di Cadore. Similarly to what was done for Triveneto, we checked the presence of changes in the frequency regime by splitting the total RoI sample into the three sub-samples PRE1965 (343 years of data), POST1965 (212 years of data) and POST1965 without 1966 (208 years of data). Figure 5 shows the remarkable differences between the three sub-samples: their shapes are associated to GLO, GEV (generalized extreme value) and LN3 distributions, respectively. In particular, the superimposition of PRE1965 and POST1965 without 1966, and the significant difference between the two POST1965 sub-samples, highlight again the great influence of a single exceptional year (1966, in our study) on the estimation of the flood frequency regime. This difference is even more striking in this case since it is due to 4 station-years of data.

5 Conclusions

Our study confirmed the value of including climatic and physiographic information in statistical regionalization. Catchment descriptors as mean annual precipitation (MAP), basin area, mean and minimum elevation and catchment location can be used as surrogates of climate and scale controls. In particular, our study shows that climatic and scale controls on flood frequency regime in Triveneto are similar to the controls that were recently found in Europe by Salinas et al. (2014): the regional flood frequency regime of medium-sized catchments associated with high to medium MAP is well described by the GEV distribution and average L-moment ratio values are higher for smaller than larger catchments and for drier than wetter ones.

By splitting the Triveneto sample into two distinct sub-samples (i.e. before and after 1965) and then removing the 1966 year from the second one, we observe very similar conditions in terms of flood frequency regime, highlighting the remarkable influence that a single exceptional year (1966) can have in the representation of flood frequency regime.

Finally, our update of Triveneto AMS data set falsified the main assumption of the current reference procedure for design-flood estimation in Triveneto (see Villi and Bacchi, 2001), showing that the study region is not homogeneous in terms of flood frequency regime. Therefore, we developed an updated reference procedure by using the Region of Influence (RoI) approach. The RoI approach delineates homogeneous pooling-groups of sites for any given target site referring to selected climatic and geomorphological descriptors that are particularly relevant for describing regional flood frequency (e.g. the ones listed above). We observe that the regional sample obtained through the RoI approach for Pieve di Cadore (used as an example) is characterized by a homogeneity degree which is much higher than the Triveneto one, and shows no differences between past and present flood frequency regime, confirming again the great importance of the flood information collected in the exceptional year 1966 for correctly representing the site flood frequency distribution.

Acknowledgements. The study is part of the research activities carried out by the working group: Anthropogenic and Climatic Controls on WateR AvailabilitY (ACCuRAcY) of Panta Rhei – Everything Flows Change in Hydrology and Society (IAHS Scientific Decade 2013–2022).

References

Allamano, P., Claps, P., and Laio, F.: An analytical model of the effects of catchment elevation on the flood frequency distribution, Water Resour. Res., 45, W01402, doi:10.1029/2007WR006658, 2009a.

Allamano, P., Claps, P., and Laio, F.: Global warming increases flood risk in mountainous areas, Geophys. Res. Lett., 36, L24404, doi:10.1029/2009GL041395, 2009b.

Beniston, M.: Impacts of climatic change on water and associated economic activities in the Swiss Alps, J. Hydrology, 412–413, 291–296, doi:10.1016/j.jhydrol.2010.06.046, 2012.

Burn, D. H.: Evaluation of regional flood frequency analysis with a region of influence approach, Water Resour. Res., 26, 2257, doi:10.1029/WR026i010p02257, 1990.

Castellarin, A. and Pistocchi, A.: An analysis of change in alpine annual maximum discharges: implications for the selection of design discharges, Hydrol. Process., 26, 1517–1526, doi:10.1002/hyp.8249, 2012.

Hosking, J. R. M. and Wallis, J. R.: Some statistics useful in regional frequency analysis, Water Resour. Res., 29, 271–281, doi:10.1029/92WR01980, 1993.

Jakob, D., Reed, D. W., and Robson, A. J.: Choosing a pooling-group, in: Flood Estimation Handbook, Institute of Hydrology, Wallingford, UK, Volume 3, 1999.

Salinas, J. L., Castellarin, A., Kohnová, S., and Kjeldsen, T. R.: Regional parent flood frequency distributions in Europe – Part 2: Climate and scale controls, Hydrol. Earth Syst. Sci., 18, 4391–4401, doi:10.5194/hess-18-4391-2014, 2014.

Schmocker-Fackel, P. and Naef, F.: Changes in flood frequencies in Switzerland since 1500, Hydrol. Earth Syst. Sci., 14, 1581–1594, doi:10.5194/hess-14-1581-2010, 2010.

Viglione, A.: Confidence intervals for the coefficient of L-variation in hydrological applications, Hydrol. Earth Syst. Sci., 14, 2229–2242, doi:10.5194/hess-14-2229-2010, 2010.

Villi, V. and Bacchi, B.: Valutazione delle piene nel Triveneto, CNR-GNDCI, Padova-Brescia, Italia, 2001.

Zrinji, Z. and Burn, D. H.: Regional Flood Frequency with Hierarchical Region of Influence, J. Water Res. Pl.-ASCE, 122, 245–252, 1996.

Proc. IAHS, 373, 101–107, 2016
proc-iahs.net/373/101/2016/
doi:10.5194/piahs-373-101-2016

Natural streamflow simulation for two largest river basins in Poland: a baseline for identification of flow alterations

Mikołaj Piniewski[1,2]

[1]Warsaw University of Life Sciences, Warsaw, Poland
[2]Potsdam Institute of Climate Impact Research, Potsdam, Germany

Correspondence to: Mikołaj Piniewski (m.piniewski@levis.sggw.pl)

Published: 12 May 2016

Abstract. The objective of this study was to apply a previously developed large-scale and high-resolution SWAT model of the Vistula and the Odra basins, calibrated with the focus of natural flow simulation, in order to assess the impact of three different dam reservoirs on streamflow using the Indicators of Hydrologic Alteration (IHA). A tailored spatial calibration approach was designed, in which calibration was focused on a large set of relatively small non-nested sub-catchments with semi-natural flow regime. These were classified into calibration clusters based on the flow statistics similarity. After performing calibration and validation that gave overall positive results, the calibrated parameter values were transferred to the remaining part of the basins using an approach based on hydrological similarity of donor and target catchments. The calibrated model was applied in three case studies with the purpose of assessing the effect of dam reservoirs (Włocławek, Siemianówka and Czorsztyn Reservoirs) on streamflow alteration. Both the assessment based on gauged streamflow (Before-After design) and the one based on simulated natural streamflow showed large alterations in selected flow statistics related to magnitude, duration, high and low flow pulses and rate of change. Some benefits of using a large-scale and high-resolution hydrological model for the assessment of streamflow alteration include: (1) providing an alternative or complementary approach to the classical Before-After designs, (2) isolating the climate variability effect from the dam (or any other source of alteration) effect, (3) providing a practical tool that can be applied at a range of spatial scales over large area such as a country, in a uniform way. Thus, presented approach can be applied for designing more natural flow regimes, which is crucial for river and floodplain ecosystem restoration in the context of the European Union's policy on environmental flows.

1 Introduction

The Vistula and the Odra basins (VOB), whose total area in Poland, Germany, Czech Republic, Slovakia, Ukraine and Belarus amounts to 313 000 km^2 are among the five largest river basins in the European Union. To date, the VOB has not been modelled in a comparable manner as the remaining three basins from the EU's top five: the Danube (Pagliero et al., 2014), the Rhine or the Elbe (Huang et al., 2015). The VOB is covered by larger-scale applications, e.g. for the Baltic Sea Basin Donnelly et al. (2014) or entire Europe Abbaspour et al. (2015), but they offer much coarser resolution than it would be desired to address locally specific water resource problems. Such problems can be more accurately addressed by a large-scale and high-resolution model that is tailored for this area.

The overwhelming majority of large river basins in the world are currently to some extent impacted by human pressure that usually causes departures of discharge from natural to altered conditions (Richter et al., 1997). The largest impact on streamflow regime is usually attributed to dam reservoirs, although the irrigation systems, inter-basin transfers, point source discharges and withdrawals can also have a considerable effect. The degree of flow alteration in the VOB is moderate compared to western Europe or other worlds regions (cf. Milliman et al., 2008) and it is possible to identify both a subset of near-pristine benchmark catchments and heavily-regulated catchments.

Assessment of anthropogenic streamflow alteration and identification of baseline streamflow conditions and their range of variability across different scales, is fundamental to water resources management, understanding subsequent ecological effects, and designing environmental flows (Poff et al., 2010). Flow gauge-based approaches to assessing streamflow alteration are popular, but their large-scale applicability is limited due to a low density of gauging stations or missing data records. An alternative approach is to apply a hydrological model that can prove useful in diverse ways. For example, a model that does not account for water management but is tailored for simulation of natural flows can serve as a tool providing a rapid estimate of the baseline hydrological conditions, which is the first step in a popular framework for assessing environmental flows called ELOHA (Ecological Limits of Hydrologic Alteration) framework (Poff et al., 2010).

The main objective of this study is to apply a previously developed large-scale and high-resolution SWAT model of the VOB, calibrated with the focus of natural flow simulation, in order to assess the impact of three different dam reservoirs on streamflow using the Indicators of Hydrologic Alteration (IHA).

2 Natural streamflow simulation with SWAT

A detailed description of the model setup, calibration protocol and model evaluation has been published elsewhere (Piniewski et al., 2016). Here we provide a short summary essential for better understanding of the core part related to the assessment of flow alteration.

2.1 Model inputs and calibration approach

SWAT is a process-based, semi-distributed, continuous-time hydrological model that simulates the movement of water, sediment and nutrients on a catchment scale with a daily time step (Arnold et al., 1998).

The whole area of the VOB was divided into 2633 sub-basins and 21 311 Hydrological Response Units. The 1951–2013 5 km resolution daily climate data were obtained from the CHASE-PL Forcing Data (CPLFD-GDPT5), the gridded daily temperature and precipitation dataset (Berezowski et al., 2015, 2016). The SUFI-2 algorithm (Abbaspour et al., 2004) within the SWAT-CUP software package was used for model calibration. We used Kling-Gupta Efficiency (KGE) as the objective function (Gupta et al., 2009). We selected 1991–2000 as the model calibration period and 2001–2010 as the validation period.

A spatial calibration approach accompanied by a parameter regionalisation scheme that leads to simulation of natural streamflow in the whole model domain was designed. The first step of the calibration protocol was selection of small, non-nested catchments located in the VOB, with relatively undisturbed flow regimes, the so-called "benchmark" catchments. 80 catchments fulfilling the specified criteria were left. In the next step benchmark catchments were grouped into clusters of similar flow regime properties. The rationale was to derive a smaller number of homogeneous calibration areas, each of which would have its own parameter set. We have followed the statistical approach applied in previous works on natural flow regime classifications (Mackay et al., 2014) in order to do the cluster analysis. Eight flow regime clusters showed in Fig. 1 were distinguished.

The total area of 80 benchmark catchments constitutes only 25 % of the VOB. Therefore a regionalisation strategy was necessary to transfer calibrated parameter values across the entire VOB. We used the hydrological distance approach (He et al., 2011), in which hydrological similarity was evaluated based on a set of climatic-physiographic properties of the donor and target catchments. This approach was evaluated by validating the model at two most downstream gauging stations on the two investigated rivers.

2.2 Model performance evaluation

The model performance at 80 benchmark gauges as well as at two most downstream basin outlets is illustrated in Fig. 2. Median KGE across all benchmark catchments was equal to 0.7 and 0.63 for calibration and validation period, respectively. Even though there are individual catchments with rather poor results (KGE below 0.4), the majority of KGE values falls into a satisfactory range. Figure 2b and c also shows that the parameter regionalisation approach worked very well: calibration results at two most downstream gauges are significantly better than the average results for the benchmark catchments.

The calibrated SWAT model of the VOB was then used to run a simulation for the whole period of availability of CPLFD data. The resulting output dataset CPL-NH (CHASE-PL – Natural Hydrology), consisting of monthly sub-basin water balance components and daily (natural) streamflow records for 2633 sub-basins and reaches, for the time period 1954–2013 is stored in an open online research data archive 3TU.Datacentrum (Piniewski et al., 2015). For further analysis, we have extracted output from this dataset.

3 Assessment of flow alteration caused by dams

In order to illustrate the effect of dam reservoirs on downstream flow alteration we selected three reservoirs situated in different parts of Poland (Fig. 1) and having different characteristics (Table 1). Selected reservoirs differ with respect to upstream catchment area, dam height, construction year, capacity and dominating functions. All of them were associated with the nearest flow gauge situated downstream of a dam and with respective model outlet for which simulated daily flows were available. The Nature Conservancy's IHA program (Mathews and Richter, 2007) was applied to calculate 33 flow statistics based on pre-dam and post-dam ob-

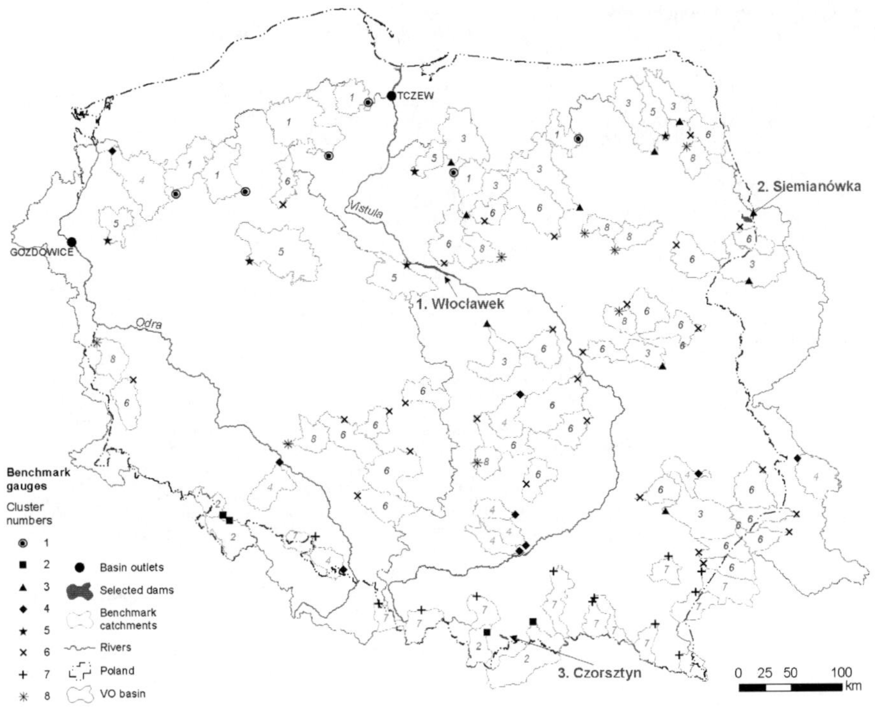

Figure 1. Division of benchmark catchments into 8 clusters and location of three studied dam reservoirs (modified after Piniewski et al., 2016).

served and modelled streamflow data. A number of comparisons were made, each of which served a different purpose:

1. Comparison of pre-impact and post-impact flow statistics using gauged flows provides an estimate of the magnitude of flow alteration due to dam effect. This can be referred to as the Before-After (BA) design (Peñas et al., 2016).
2. Comparison of pre-impact and post-impact flow statistics using simulated flows provides an estimate of the flow alteration due to climate effect.
3. Comparison of pre-impact gauged and simulated flow statistics is a way of model validation.
4. Comparison of post-impact gauged and simulated flow statistics is an alternative to the BA design. It can be referred to as the Control-Impact (CI) design (Peñas et al., 2016), where modelled flows serve as a control.

The IHA statistics refer to different aspects of flow regime (magnitude, duration, timing, high and low flow pulses, rate of change). For illustrative purposes three example statistics selected for each reservoir are discussed below.

3.1 Włocławek Reservoir

Constructed in 1970, Włocławek Reservoir with a surface area of 70 km^2 is the largest reservoir in Poland. It also has the largest upstream catchment area of 171 000 km^2. Since the ratio of capacity to inflow is relatively low, reservoir operation does not strongly affect such parameters as seasonal flows, extreme flows or timing. However, the parameters related to the rate of change (rise rate, fall rate and number of flow reversals) are affected, as shown in Fig. 3a–c. The analysis based on gauged flows demonstrates that all these parameters have at least doubled after dam construction. At the same time there is no significant change in the same parameters assessed using modelled flows, which shows that the natural climatic variability did not affect them. Comparison of pre-dam observed and simulated data shows on the other hand that the SWAT-based rate of change parameters, particularly fall rate and number of reversals, are underestimated. Thus, the effect of Włocławek dam assessed following the CI design is presumably over-estimated.

3.2 Siemianówka Reservoir

Constructed in 1992, Siemianówka Reservoir with a surface area of 32.5 km^2 is the third largest reservoir in Poland. It also has relatively small upstream catchment area of 1100 km^2. This is a lowland catchment situated mainly in Belarus. Due to a relatively large ratio of capacity to inflow, a long list of flow parameters are affected by this dam. Three of them showing significant impacts were selected for illustrative purposes: mean December flow, 1-day maximum flow

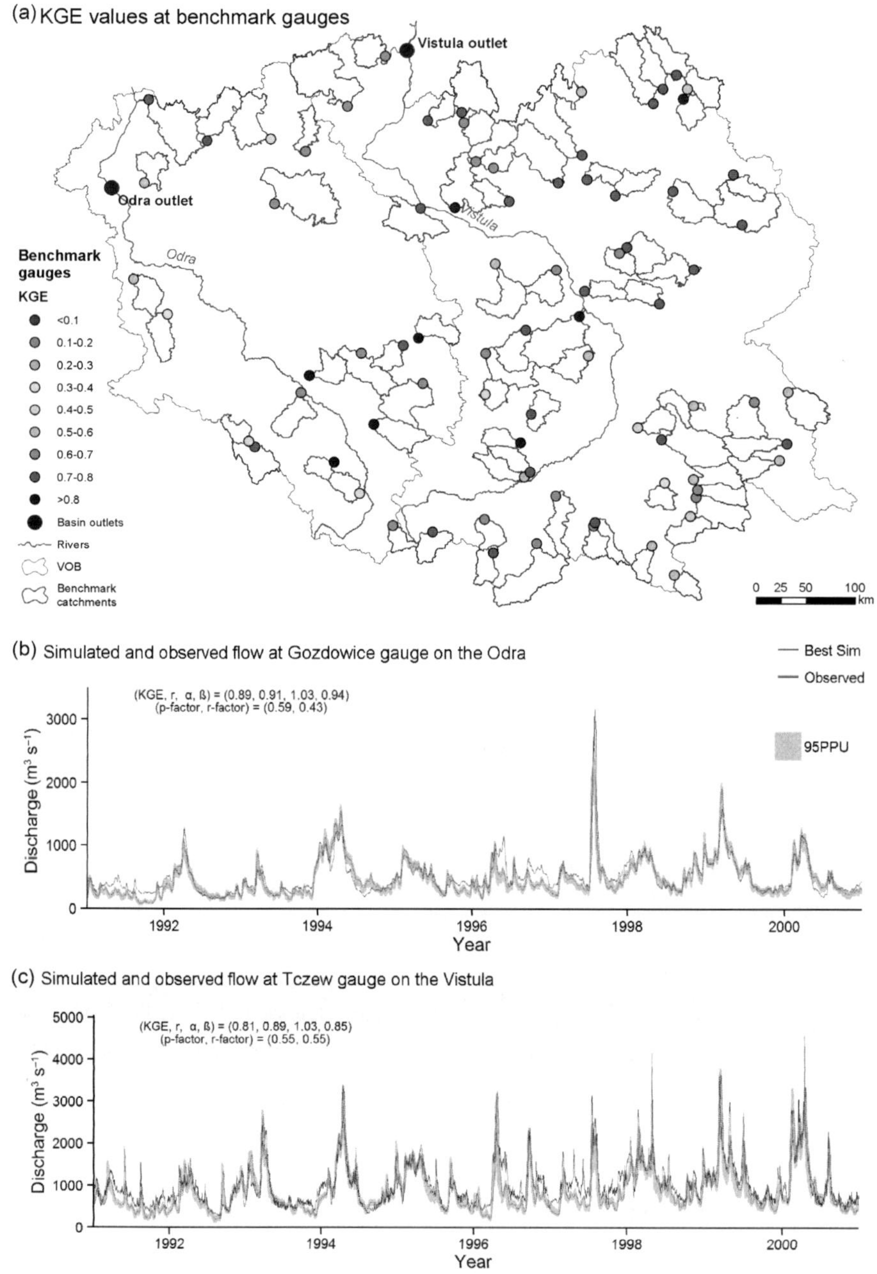

Figure 2. Calibration results: KGE values at benchmark gauges **(a)**, simulated and observed flows at Gozdowice gauge on the Odra **(b)** and at Tczew gauge on the Vistula **(c)** (modified after Piniewski et al., 2016).

and high pulse number (Fig. 3d–f). The analysis following the BA design demonstrates that post-dam period was not homogeneous, i.e. the pattern of streamflow indices for the periods 1992–1999 and 2000–2007 was considerably different. For example, an abrupt increase in December flows after dam construction lasted until approximately year 1999, after which December flows were generally lower than in the pre-dam period. As with Włocławek Reservoir, there is no significant change in the studied parameters assessed using modelled flows, which shows that the natural climatic variability did not affect them. Visual comparison of pre-dam observed and simulated data shows on the other hand a moderate agreement between SWAT-based and gauge-based parameters. The effect of Siemianówka dam assessed by comparing post-impact observed and modelled data is different in different sub-periods. For example, high pulse number was

Table 1. Main features of investigated reservoirs.

Reservoir	River	Gauge	Catchment area [10^3 km^2]	Dam height [m]	Construction end year	Capacity [10^6 m^3]	Power plant capacity [MW]	Dominating functions
Włocławek	Wisła	Włocławek	171	20	1970	408	160	energy, recreation
Siemianówka	Narew	Bondary	1.10	9	1992	79.5	0.16	nature protection, irrigation, recreation
Czorsztyn	Dunajec	Sromowce Wyżne	1.3	56	1997	232	92	flood protection, energy

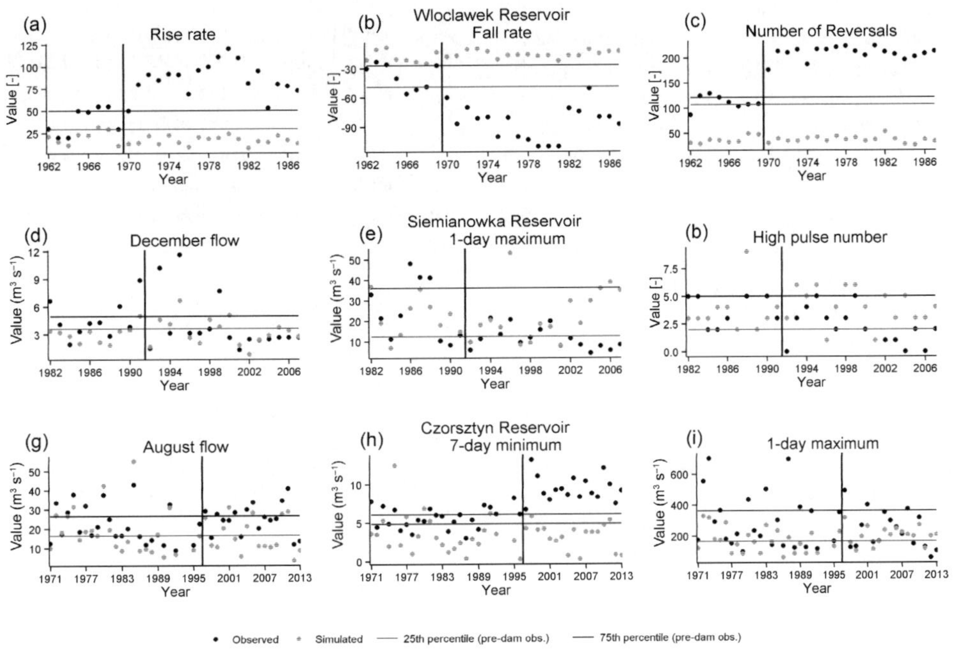

Figure 3. The effect of three dam reservoirs: Włocławek, Siemianówka and Czorsztyn on streamflow alteration assessed using observed and simulated flows. Black vertical line denotes the construction year of each reservoir. The 25th and 75th percentiles of the flow statistics were calculated based on pre-dam observed flows and delineate the likely interval of natural variability of a given statistics.

significantly lower for the observed flows than for the simulated flows in the sub-period 1992–1999, while in the later period they were similar.

3.3 Czorsztyn reservoir

Constructed in 1997, Czorsztyn Reservoir with a storage capacity of 232 millions m^3 is the third largest reservoir in Poland (with respect to capacity). In contrast to Włocławek and Siemianówka it is situated in the mountains, on the River Dunajec characterised by a more dynamic flow regime. Three parameters showing relatively large impacts related to dam construction were selected for illustrative purposes: mean August flow, 1-day maximum flow and 7-day minimum flow (Fig. 3g–i). The analysis based on gauged flows demonstrates that August mean flows and 7-day minimum flows increased and 1-day maximum flows decreased (including their variability) after dam construction. In contrast to two previous examples, the natural climatic variability seems to have affected streamflow as well. For example, simulated mean August flows were significantly lower, while simulated 1-day maximum flows had a considerably lower variability in the post-dam period as compared to the pre-dam period. Visual comparison of pre-dam observed and simulated data shows on the other hand a high agreement between SWAT-based and gauge-based August flows, and an underestimation of minimum and maximum flows by SWAT. The effect of Czorsztyn dam assessed in the CI design suggests a higher dam impact than when assessed in temporal comparison, in particular for August flow and 7-day minimum flow. While in the first case this statement seems to be true (because of good performance of SWAT in simulating August flows), in the latter it is probably wrong (because of underestimation of minimum flows by SWAT in the pre-dam period).

4 Discussion and outlook

The spatial calibration approach developed here in order to simulate natural discharge over two large river basins is applicable in other geographical settings, particularly for areas in which the degree of flow alteration is moderate (as in Poland), i.e. it is possible to identify both a subset of near-pristine benchmark catchments and heavily-regulated catchments. An approximate indication of such regions over the world can be made based on the global distribution of deficit watersheds of Milliman et al. (2008) or on the global distribution of aging of continental runoff in response to large reservoir impoundment (Vörösmarty and Sahagian, 2000). For river basins that are entirely near-pristine this approach would not bring much benefit, as the gauged hydrology is already natural. In contrast, for extremely regulated river basins, this approach would not work because of insufficient number of unaltered benchmark catchments.

This study showed that three different reservoirs caused alteration of different flow regime characteristics. The added value of using the large-scale high-resolution hydrological model for assessing flow alteration is manifold. The assessment using gauged flows is probably more accurate, but in many cases gauges can be situated far downstream from a dam (or another source of alteration, such as point source or abstraction), and thus the direct effect is dampened. A common issue with gauged flows are missing data, which can be supplemented by simulated streamflow. Finally, comparison of pre-dam and post-dam gauged flow data (BA design) carries the danger of ignoring the effect of natural climatic variability. In this respect, the modelled flows can serve as the proxy of a control gauge (CI design) within the full Before-After-Control-Impact (BACI) design, the preferred design of assessing flow alteration (Peñas et al., 2016). As shown in this study and elsewhere (Shrestha et al., 2014), care needs to be taken as far as the reliability of modelled flow statistics is concerned. If the model is able to predict the variability of a given parameter in the pre-impact period, it can be reliably used in the Control-Impact design. As with the August flow for Czorsztyn Reservoir, there was a good agreement for this parameter during the pre-impact period, and thus one could conclude that the dam effect assessed using the BA design was underestimated. The reason was that this design neglected the fact that post-impact streamflow was naturally lower than the pre-impact streamflow in August.

In Poland, environmental flows have been for a long time identified with minimum low flow thresholds. Due to the EU Water Framework Directive and its environmental flow policy (Acreman et al., 2009) this situation is gradually changing, and more importance is given to flow requirements of valuable ecosystems such as riparian wetlands (Piniewski et al., 2014). It is anticipated that designing environmental flows from dams will sooner or later become an important topic in Poland. Thus, a model simulating baseline (natural) hydrology in 90 % of the country may be a useful tool in guiding this process, since a similar analysis as performed here for only three reservoirs can be easily extended over all major sources of flow alteration.

Data availability

The underlying research data: (1) CHASE-PL Natural Hydrology dataset (CPL-NH), and (2) CHASE-PL Forcing Data: Gridded Daily Precipitation & Temperature Dataset 5 km (CPLFD-GDPT5), can be accessed for free at 3TU.Datacentrum repository (https://data.3tu.nl/repository/uuid:b8ab4f5f-f692-4c93-a910-2947aea28f42 and doi:10.4121/uuid:e939aec0-bdd1-440f-bd1e-c49ff10d0a07, respectively).

Acknowledgements. Support of the project CHASE-PL (Climate change impact assessment for selected sectors in Poland) of the Polish–Norwegian Research Programme is gratefully acknowledged. Help of Mateusz Szcześniak and other colleagues from the CHASE-PL project is highly appreciated. The author is grateful for support to the Alexander von Humboldt Foundation and to the Ministry of Science and Higher Education of the Republic of Poland.

References

Abbaspour, K., Rouholahnejad, E., Vaghefi, S., Srinivasan, R., Yang, H., and Klove, B.: A continental-scale hydrology and water quality model for Europe: Calibration and uncertainty of a high-resolution large-scale SWAT model, J. Hydrol., 524, 733–752, 2015.

Abbaspour, K. C., Johnson, C. A., and van Genuchten, M. T.: Estimating uncertain flow and transport parameters using a sequential uncertainty fitting procedure, Vadose Zone J., 3, 1340–1352, 2004.

Acreman, M., Aldrick, J., Binnie, C., Black, A., Cowx, I., Dawson, H., Dunbar, M., Extence, C., Hannaford, J., Harby, A., Holmes, N., Jarritt, N., Old, G., Peirson, G., Webb, J., and Wood, P.: Environmental flows from dams: the Water Framework Directive, Proceedings of the Institution of Civil Engineers – Engineering Sustainability, 162, 13–22, 2009.

Arnold, J. G., Srinivasan, R., Muttiah, R. S., and Williams, J. R.: Large-area hydrologic modeling and assessment: Part I. Model development, J. Am. Water Resour. As., 34, 73–89, 1998.

Berezowski, T., Szcześniak, M., Kardel, I., Michałowski, R., and Piniewski, M.: CHASE-PL Forcing Data – Gridded Daily Precipitation Temperature Dataset (CPLFD-GDPT5), Dataset on

3TU.Datacentrum, doi:10.4121/uuid:e939aec0-bdd1-440f-bd1e-c49ff10d0a07, 2015.

Berezowski, T., Szcześniak, M., Kardel, I., Michalowski, R., Okruszko, T., Mezghani, A., and Piniewski, M.: CPLFD-GDPT5: High-resolution gridded daily precipitation and temperature data set for two largest Polish river basins, Earth Syst. Sci. Data, 8, 127–139, doi:10.5194/essd-8-127-2016, 2016.

Donnelly, C., Yang, W., and Dahné, J.: River discharge to the Baltic Sea in a future climate, Climatic Change, 122, 157–170, 2014.

Gupta, H. V., Kling, H., Yilmaz, K. K., and Martinez, G. F.: Decomposition of the mean squared error and NSE performance criteria: Implications for improving hydrological modelling, J. Hydrol., 377, 80–91, 2009.

He, Y., Bárdossy, A., and Zehe, E.: A review of regionalisation for continuous streamflow simulation, Hydrol. Earth Syst. Sci., 15, 3539–3553, doi:10.5194/hess-15-3539-2011, 2011.

Huang, S., Krysanova, V., and Hattermann, F.: Projections of climate change impacts on floods and droughts in Germany using an ensemble of climate change scenarios, Reg. Environ. Change, 15, 461–473, 2015.

Mackay, S. J., Arthington, A. H., and James, C. S.: Classification and comparison of natural and altered flow regimes to support an Australian trial of the Ecological Limits of Hydrologic Alteration framework, Ecohydrology, 7, 1485–1507, 2014.

Mathews, R. and Richter, B. D.: Application of the Indicators of Hydrologic Alteration Software in Environmental Flow Setting, JAWRA Journal of the American Water Resources Association, 43, 1400–1413, 2007.

Milliman, J., Farnsworth, K., Jones, P., Xu, K., and Smith, L.: Climatic and anthropogenic factors affecting river discharge to the global ocean, 1951–2000, Global Planet. Change, 62, 187–194, 2008.

Pagliero, L., Bouraoui, F., Willems, P., and Diels, J.: Large-Scale Hydrological Simulations Using the Soil Water Assessment Tool, Protocol Development, and Application in the Danube Basin, J. Environ. Qual., 43, 145–154, 2014.

Peñas, F., Barquín, J., and Álvarez, C.: Assessing hydrologic alteration: Evaluation of different alternatives according to data availability, Ecol. Indic., 60, 470–482, 2016.

Piniewski, M., Okruszko, T., and Acreman, M. C.: Environmental water quantity projections under market-driven and sustainability-driven future scenarios in the Narew basin, Poland, Hydrol. Sci. J., 59, 916–934, 2014.

Piniewski, M., Szcześniak, M., Kardel, I., and Berezowski, T.: CHASE-PL Natural Hydrology dataset (CPL-NH), Dataset on 3TU.Datacentrum, doi:10.4121/uuid:b8ab4f5f-f692-4c93-a910-2947aea28f42, 2015.

Piniewski, M., Szcześniak, M., Kardel, I., Berezowski, T., Okruszko, T., Srinivasan, R., Vikhamar Shuler, D., and Kundzewicz, Z.: Modelling water balance and streamflow at high resolution in the Vistula and Odra basins, Hydrol. Sci. J., submitted, 2016.

Poff, N. L., Richter, B. D., Arthington, A. H., Bunn, S. E., Naiman, R. J., Kendy, E., Acreman, M., Apse, C., Bledsoe, B. P., Freeman, M. C., Henriksen, J., Jacobson, R. B., Kennen, J. G., Merritt, D. M., O'Keeffe, J. H., Olden, J. D., Rogers, K., Tharme, R. E., and Warner, A.: The ecological limits of hydrologic alteration (ELOHA): a new framework for developing regional environmental flow standards, Freshwater Biol., 55, 147–170, 2010.

Richter, B., Baumgartner, J., Wigington, R., and Braun, D.: How much water does a river need?, Freshwater Biol., 37, 231–249, 1997.

Shrestha, R. R., Peters, D. L., and Schnorbus, M. A.: Evaluating the ability of a hydrologic model to replicate hydro-ecologically relevant indicators, Hydrol. Process., 28, 4294–4310, 2014.

Vörösmarty, C. J. and Sahagian, D.: Anthropogenic Disturbance of the Terrestrial Water Cycle, BioScience, 50, 753–765, 2000.

Proc. IAHS, 373, 109–114, 2016
proc-iahs.net/373/109/2016/
doi:10.5194/piahs-373-109-2016

Regionalization of post-processed ensemble runoff forecasts

Jon Olav Skøien[1], **Konrad Bogner**[2], **Peter Salamon**[1], **Paul Smith**[3], and **Florian Pappenberger**[3]

[1]European Commission, Joint Research Centre (JRC), Institute for Environment and Sustainability, Ispra, 21027 (VA), Italy
[2]Swiss Federal Institute WSL, Mountain Hydrology and Mass Movements, Birmensdorf, 8903, Switzerland
[3]European Centre for Medium-Range Weather Forecasts, Reading, RG2 9AX, UK

Correspondence to: J. O. Skøien (jon.skoien@jrc.ec.europa.eu)

Published: 12 May 2016

Abstract. For many years, meteorological models have been run with perturbated initial conditions or parameters to produce ensemble forecasts that are used as a proxy of the uncertainty of the forecasts. However, the ensembles are usually both biased (the mean is systematically too high or too low, compared with the observed weather), and has dispersion errors (the ensemble variance indicates a too low or too high confidence in the forecast, compared with the observed weather). The ensembles are therefore commonly post-processed to correct for these shortcomings. Here we look at one of these techniques, referred to as Ensemble Model Output Statistics (EMOS) (Gneiting et al., 2005). Originally, the post-processing parameters were identified as a fixed set of parameters for a region. The application of our work is the European Flood Awareness System (http://www.efas.eu), where a distributed model is run with meteorological ensembles as input. We are therefore dealing with a considerably larger data set than previous analyses. We also want to regionalize the parameters themselves for other locations than the calibration gauges. The post-processing parameters are therefore estimated for each calibration station, but with a spatial penalty for deviations from neighbouring stations, depending on the expected semivariance between the calibration catchment and these stations. The estimated post-processed parameters can then be used for regionalization of the postprocessing parameters also for uncalibrated locations using top-kriging in the rtop-package (Skøien et al., 2006, 2014). We will show results from cross-validation of the methodology and although our interest is mainly in identifying exceedance probabilities for certain return levels, we will also show how the rtop package can be used for creating a set of post-processed ensembles through simulations.

1 Introduction

Ensemble modelling has a long history in meteorology, and is also increasingly used in hydrology, mainly using the meteorological ensembles as forcing. By perturbing the initial conditions or parameters of the model, an ensemble of forecasts is produced, assuming that this is a proxy of the uncertainty of the forecast. However, even if the perturbations are sampled from a probability distribution of the conditions or parameters, it is frequent that the resulting ensembles are both biased (the mean is systematically too low or too high) and wrongly dispersed (the ensemble variance indicates a too low or too high confidence in the forecast, compared with the observations afterwards).

It is therefore common to post-process the forecasts. Two commonly methods are frequently used in meteorology: Bayesian Model Averaging (Raftery et al., 2005), which mainly focuses on calibration, and optimization based on the Ensemble Model Output Statistics (Gneiting et al., 2005), referred to as EMOS. Mostly the EMOS-method is calibrated with the use of Continuous Ranked Probability Score (CRPS), which is an indicator which punishes both biases and dispersion errors.

We will here mainly focus on the EMOS-method. In the original contributions in meteorology, it was common to fit a regional set of parameters for the post-processing. From the post-processed distributions for each location, samples were drawn to generate a post-processed ensemble. These were spatially independent, but Berrocal et al. (2007) extended the methodology to use the spatial structure of the errors to generate a spatially structured covariance matrix which can be used to generate spatially consistent samples, based on the Geostatistic output perturbation technique (Gel et al., 2004). Their method is still using the same set of weights for all locations.

The application of these types of post-processing techniques in hydrology started later. Hemri et al. (2013) developed a method for postprocessing runoff forecasts for individual stations, using the methods of Berrocal et al. (2007) for incorporation of the correlation between lead times. This correlation is likely to be higher for runoff than for meteorological variables. Engeland and Steinsland (2014) presented a method which would fit different weights to different locations and lead times, but still assuming the same number of forecasts for all lead times.

Our application of ensemble forecasting is the European Flood Awareness System (EFAS, http://www.efas.eu), an operational service for flood forecasting in Europe. The forecasts are created by running the model with different meteorological deterministic and ensemble forecasts. Contrary to the previous applications, we would therefore expect some of the ensembles to be better for some regions, and we do not want a single parameter set for the complete modelling region. Additionally, not all ensembles are available for all lead times, and we would prefer a method which will assure temporal continuity between lead times. The data set is considerably larger than in previous studies, so we do for example not see the method of Engeland and Steinsland (2014) as feasible.

The previous applications in hydrology did not consider forecasts outside the calibration points, similar to what Berrocal et al. (2007) did for meteorological applications. In this paper we will present a methods which will make it easier to make predictions outside calibration points, and also for making simulations of the possible discharge.

2 Data

The analyses in this paper are based on a combination of meteorological forecasts and ensemble forecasts from ECMWF, DWD, COSMO-LEPS and UK Met Office. We use forecasts from a period of almost two years (8 January 2012–31 December 2013). For each day, the forecasts have up to 10 days lead time.

- ECMWF: The European Centre for Medium Range Weather produces forecasts for the next 10 days. The forecast from ECMWF is an ensemble with 51 members, in addition to a deterministic forecast
- DWD: The German Weather Service produces a deterministic forecast for the next 7 days.
- COSMO-Leps: The Cosmo consortium produces an ensemble forecast with 16 members.
- UK-MET: The UK Met office produces an ensemble of 24 members.

Each individual forecast has been used as input to the hydrological model LISFLOOD (Van Der Knijff et al., 2010; De Roo et al., 2000), giving an ensemble of runoff values for each forecast day and each lead time. LISFLOOD is a gridded model, which numerically predicts the runoff for each pixel in a 5×5 km grid. The extent of the forecasts and the hydrological model covers most of Europe.

As observed values, we are using simulated runoff at 701 stations. The runoff has been simulated from interpolated observed values, using the same model setup of LISFLOOD as for the forecasts. There are some additional stations in the original data set, but these were discarded from the analyses as the runoff appeared to be unreasonably high compared to the estimated basin size, or that some of the forecasts were not available for all lead times and models.

We are using simulated values instead of real observations for comparison, as these will have the same model errors as the forecasts, such as boundary errors and routing errors.

The simulated and forecasted runoff data is divided by catchment area in 1000 km^2. The normalization on area is to work with more area independent values, whereas the use of 1000 km^2 for normalization is to avoid some numerical issues.

3 Method

3.1 Post-processing

The post-processing method we are applying in this paper is based on the Ensemble Model Output Statistics method (Gneiting et al., 2005). Shortly described, the idea is that the mean and variance of a range of forecasts might be biased and wrongly dispersed, so we want to find a weighted mean of the ensemble, whereas the variance can be assumed to fit a regression equation. As we have a combination of deterministic and ensemble forecasts, we will use the deterministic forecasts and the mean of each ensemble forecast for the bias correction, i.e., for a particular station i and lead time l:

$$Y_{il} = a_{il} + b_{il1} X_{il1} + b_{il2} X_{il2} + \ldots + b_{iln} X_{iln} + e_{il} \quad (1)$$

where a_{il} is a constant, $b_{il1}, \ldots, b_{iln}$ are weights, e_{il} is an error term averaging to zero, and $X_{il1}, \ldots, X_{iln}$ are the forecasted variable for this location, deterministic or mean of the ensemble. The deterministic forecasts are from ECMWF and DWD, whereas we use the mean of the ensembles from ECMWF, COSMO and UK Met office, giving $n = 5$ for lead times 1–5. The forecast Y_{il} should then be unbiased, but we would also like to know the variance of e_{il}. This is modelled as a linear function of the variance of all ensembles:

$$\sigma_{il}^2 = \mathrm{Var}(e_{il}) = c_{il} + d_{il} S_{il}^2 \quad (2)$$

where S_{il}^2 is the variance of all individual ensemble members, and c_{il} and d_{il} are non-negative coefficients. This gives a Gaussian predictive distribution:

$$N\left(a_{il}+b_{il1}X_{il1}+b_{il2}X_{il2}+\ldots+b_{iln}X_{iln},\ c_{il}+d_{il}S_{il}^2\right) \quad (3)$$

This can be optimized by minimizing the continuous ranked probability score (CRPS), as described by Gneiting et al. (2005). For each day d in the calibration period, the CRPS-error for a certain station i and lead time l is defined as:

$$\mathrm{CRPS}\left(F_{ild}, y_{i,l+d}\right)=\int_{-\infty}^{\infty}\left[F_{ild}(t)-H\left(t-y_{i,l+d}\right)\right]^2 \mathrm{d}t \quad (4)$$

where $H(t-y_{i,l+d})$ is the Heaviside function, which is 0 for $t<y_{i,l+d}$ and 1 for $t \geq y_{i,l+d}$. If F is the CDF of a normal distribution with mean Y_{ild} and variance σ^2, the integral can be replaced with:

$$\begin{aligned}&\mathrm{CRPS}\left[N\left(Y_{ild},\sigma^2\right), y_{i,l+d}\right]\\&=\sigma_{ild}\left\{\frac{y_{i,l+d}-Y_{ild}}{\sigma_{ild}}\left[2\Phi\left(\frac{y_{i,l+d}-Y_{ild}}{\sigma_{ild}}\right)-1\right]\right.\\&\left.+2\phi\left(\frac{y_{i,l+d}-Y_{ild}}{\sigma_{ild}}\right)-\frac{1}{\sqrt{\pi}}\right\} \quad (5)\end{aligned}$$

where $z=\frac{y_{i,l+d}-Y_{ild}}{\sigma_{ild}}$ is the normalized prediction error and $\Phi(z)$ and $\phi(z)$ represents the CDF and the PDF of a N(0,1) distribution. Equations (4) and (5) can be seen as objective functions when summed over all instances used in the calibration. It is likely that F might violate the normal distribution assumption for runoff variables, however, we will for simplicity use this assumption here, and deal with deviations in future work.

3.2 Interpolation of weights

Our region of interest is Europe. We do therefore not expect one set of weights to be sufficient for the whole modelling domain. However, we have the ensembles for all grid cells along a river, and would like to be able to make post-processed forecasts also for other locations than the calibration locations. The solution is to interpolate the weights along the river network, to have unbiased predictions for each pixel where a prediction is wanted. For this we will use top-kriging (Skøien et al., 2006, 2014). Top kriging is a geostatistical method for interpolation between areas of different spatial support, such as observations along a river network. The method is well explained in the citations above and will only be summarized here for river related applications as follows:

- A sample variogram is estimated from the observations for each gauge, as a spatial average if the variable is a spatial aggregate such as runoff and most runoff statistics. The centre of the upstream contributing area is used to compute the distances. Variograms are binned according to the size of each of the catchments, not only distance.
- A variogram model is found by jointly fitting regularized variogram values to the binned sample variogram values.
- A covariance matrix of expected semivariances between observation catchments and between observation catchments and prediction catchments is found from the variogram model, based on the size and location of the catchments.
- Interpolation and cross-validation is performed as in normal kriging, based on the covariance matrices.

The most interesting features of this interpolation method is that it takes into account both network topology (2 locations which are connected on a river network usually gets higher weights than unconnected locations) and spatial proximity. The last feature takes into account both the size and the location of the catchments, not just the distance between the gauges or centres of gravity.

However, the fitting method in Eqs. (4) and (5) can give poorly correlated weights for neighbouring locations if two or more of the forecasts are highly correlated. For example, if we only had two forecasts and they were equal for a certain location, any combination of weights giving the same sum would give the same error. To force a certain correlation between weights, and variance coefficients between locations (all referred to as parameters below), we use an iterative procedure where we introduce a spatial penalty as a function of the modelled semivariance between two locations and the difference of all the m parameters:

$$S_{\mathrm{pen}}=P_{\mathrm{c}}\sum_{i=1}^{n}\frac{1}{\gamma_{0i}}\sum_{j=1}^{m}\frac{2\left|p_{0j}-p_{ij}\right|}{\left|p_{0j}\right|+\left|p_{ij}\right|} \quad (6)$$

Here the parameters of the calibration location is p_{0j} whereas p_{ij} is the parameter value for the n locations with the highest correlation to the calibration location. The expected semivariance γ_{0i} between the calibration location and the neigbouring locations is found from a regularized semivariogram for mean runoff. P_{c} is a penalty coefficient to scale the spatial penalty to the CRPS-error.

The calibration is done station by station. In the first iteration, no spatial penalty is added, as many neighbouring stations have not been computed yet. In the second iteration, P_{c} is set equal to 1. At the end of the second iteration, this coefficient is recomputed as two times the ratio between the CRPS-error and the spatial penalty. This P_{c} is used for the third iteration. In the calibration, the most recent parameter values are always used, i.e., if a neighbor has already been

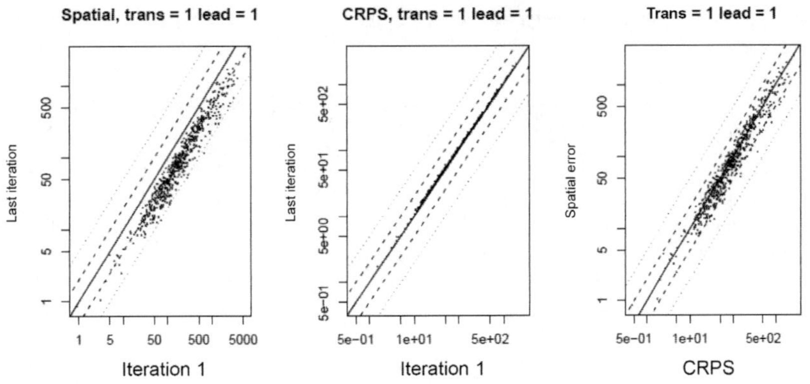

Figure 1. Left and central panel: Development of CRPS-errors and spatial errors from first to last iteration for lead time of one day. Right: Comparison of CRPS and spatial error after last iteration. Solid line represents 1 : 1, whereas dashed lines represent 2 : 1 and 1 : 2 and stapled lines 5 : 1 and 1 : 5.

updated, this value is used instead of the one from the previous iteration. The locations are visited in a random order for each iteration.

3.3 Simulations of runoff

A common usage of post-processing in meteorology is to create simulations of the variable of interest. This can also be done with the post-processing we are presenting here, based on the calibrated parameters and the semivariogram above. The simulation method is based on the Sequential Gaussian Simulation method (Deutsch and Journel, 1998), combined with Kriging with uncertain data (KUD) (de Marsily, 1986; Merz and Blöschl, 2005).

We start with the weighted mean and uncertainties for each calibration location.

In a random order, we visit all calibration locations and prediction locations, and do the following step for each of them:

1. For a new location, we predict the mean and the kriging variance, using the weighted mean for the calibration locations, and previously simulated locations as observations. For the KUD prediction, we use the weighted ensemble variance for the calibration locations.
2. Sample a value from the predictive distribution (traditionally assumed to be Gaussian) with the prediction as mean and the kriging variance as variance. Add this to the set of observed/simulated values. This simulated value will in the subsequent simulations have an uncertainty of zero in the KUD prediction.
3. Replace the weighted mean with a simulated value if the simulation concurs with a calibration location.

Simulation of runoff values implies some numerical challenges. First of all, when we have multiple points along a river segment, the contributing area of these points might be almost similar in many cases. This can create highly correlated neighbours, which can again create singular or close to singular covariance matrices. We have found some methods to automatically remove some of these neighbours, still there might be some numerical challenges in solving the kriging equations. There are therefore some cases where numerical issues can give a negative kriging variance, which is first of all physically impossible, second, makes it impossible to draw a value above. We are still in the process of finding robust solutions for these cases, in the meantime there will be some points where we are not able to make simulations.

A second issue is that runoff values are typically above zero. Using random sampling from a Gaussian distribution can give negative observations. We are therefore instead assuming a long-normal distribution in this case, log-transforming the predictive mean and variance with σ_l^2 as the logtransformed variance: $\sigma_l^2 = \log(\frac{\sigma^2}{\mu} + 1)$ before sampling. This ensures positive runoff values, but should be seen as an approximation to the correct solution and is likely to be slightly biased (Clark, 1998). We will further investigate better approaches for this case.

4 Results

4.1 Fitting of EMOS-parameters and effect of spatial penalty

The initial fitting of the parameters are done without the spatial penalty. We can therefore easily see how much the use of the penalty increases the CRPS-error and how we reduce the spatial errors as in Fig. 1. The CRPS-error increases marginally for all catchments, but the largest increase is less than 25 and 75 % of the catchments increase less than 5 %. The spatial penalty reduces considerably with the iterations, 55 % of the catchments see the spatial error reduced to less than 1/2. The last panel shows that the CRPS-error is dominating the total error for most catchments, and is consider-

Table 1. Cross-validation r^2-values for EMOS-parameters for different lead times. Bold numbers are below 0.9, italic numbers are below 0.8.

Lead	Var1 (c_{il})	Var2 (d_{il})	Intercept (a_{il})	DWD (b_{il1})	ECMWF (b_{il2})	ECMWF mean (b_{il3})	COSMO mean (b_{il4})	UK mean (b_{il5})
1	0.98	*0.74*	0.90	0.92	0.96	0.93	0.98	0.90
2	0.98	**0.88**	**0.85**	0.92	0.95	0.93	0.95	0.93
3	0.98	**0.89**	0.90	0.92	**0.87**	0.93	0.95	0.95
4	0.98	0.95	0.94	0.92	0.92	0.94	0.94	0.97
5	0.98	0.94	0.92	0.91	**0.88**	0.95	0.94	0.97
6	0.91	*0.57*	**0.88**	*0.76*	*0.65*	*0.79*	NA	**0.80**
7	0.96	0.97	0.96	0.94	0.91	0.94	NA	0.97
8	0.97	0.97	**0.85**	NA	0.94	0.94	NA	0.97
9	0.96	0.97	0.97	NA	0.96	0.95	NA	0.97
10	0.97	0.96	0.97	NA	0.94	0.94	NA	0.97

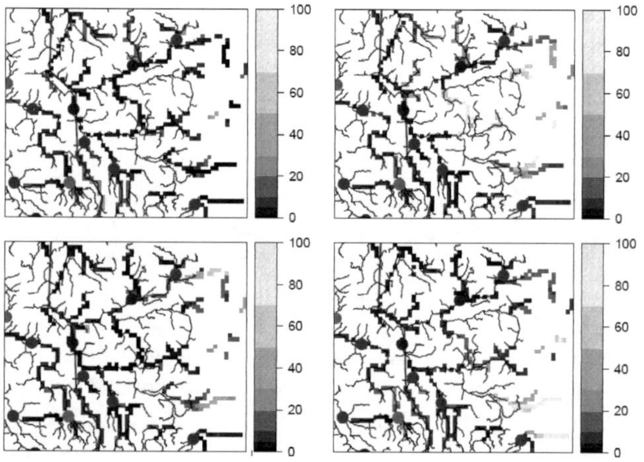

Figure 2. Predicted (dots) and simulated specific runoff (pixels) for one day lead time for a region on both sides of the German/Polish border (red line).

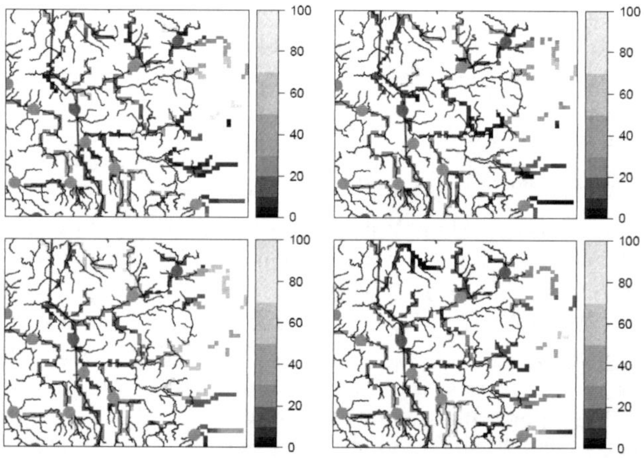

Figure 3. Predicted (dots) and simulated specific runoff (pixels) for 10 days lead time for a region on both sides of the German/Polish border (red line).

ably larger for a large group of them, whereas there are some (around 30 %) where the spatial error dominates.

We can also notice that there is quite a large range in the errors, both CRPS and spatial error in approximate range from 1–1000. We have not analyzed the reason for this, although it is likely not related to area. First, the runoff has been divided by catchment area, second, we have plotted (not shown) both errors and ratios between errors against catchment area, without finding any strong relationships.

4.2 Interpolation of EMOS-parameters

Table 1 gives an overview of r^2 of the cross-validated EMOS-parameters from Eqs. (1) and (2). The variable names in the equations are given in brackets for the column names. We can see that there is a good correspondence between the fitted parameters and the interpolated parameters. Some of the cells are given a color code, where the one red cell has $r^2<0.6$, there is one orange cell with $r^2<0.7$, three yellow cells with $r^2<0.8$ and 8 green cells with $r^2<0.9$. The remaining 59 parameters have $r^2>0.9$. This means that the top-kriging method can well be used for interpolating EMOS-parameters between different locations on the stream network, at least when the parameters have been fitted using a spatial penalty, as we have done in this manuscript. We have not yet examined in detail the reasons for the poorer results for a few locations and lead times.

4.3 Simulation of runoff fields

With the fitted parameters, we have an estimate of the predictive mean and uncertainty for each of the calibration locations. From these, we can simulate the specific runoff for each pixel along the river network. Figures (2) and (3) show the results from 4 simulations for the first and the 10th forecast day, respectively, based on forecasts from 17 February for a region on the German-Polish border. The forecast indicates a flood event to the end of this period (10th forecast day), so the predictions and the simulations are considerably higher for Fig. 3 than for Fig. 2. The dots show the predictive mean for the calibration locations based on the fitted EMOS-parameters, whereas the pixels show the simulated values based on the variogram and the predictive uncertainty from the calibration locations. We can see that the simulations are relatively close to the predicted mean for locations close to the calibration locations, whereas the deviations between simulations can be considerably larger in the smaller tributaries far from the calibration locations.

5 Conclusions

We have used the EMOS-method for post-processing of runoff predictions from an ensemble forecasting method. The results indicate that it is well possible to use top-kriging for interpolating the EMOS-parameters along the river network as long as the parameters have been fitted with a method which forces some degree of spatial continuity between the parameters.

We have also shown that it is possible to use top-kriging for simulation of runoff at uncalibrated locations, using the variogram and post-processed predictive distributions at the calibration locations. Using these simulations is a different approach than interpolating the EMOS-parameters to create uncorrelated predictive distributions for each locations along the river network. Such simulations have, as far as we know, not yet been used in hydrological forecasting, and the possible usages still need further analyses. One important aspect is that the uncertainty of the smaller tributaries will not only be based on meteorological uncertainty, as for ensemble modelling with a hydrological model with a single parameter setup, it will also include the modelling uncertainty.

We notice that there are still a few issues which have to be further improved in the analyses presented here. First of all, much of the theory is developed for variables with a normal distribution. However, runoff usually does not follow a normal distribution. We will in the near future analyse the possibilities for using transformations to be able to work with more normalized variables. We did use a lognormal transform for the simulations. However, the way it was done is not well founded in geostatistical theory, and will need further improvements. Some of the simulated values in the tributaries are extremely large, which can well describe the statistical uncertainty, but maybe not so much the meteorological uncertainty. Further comparisons of the simulations here and the results of each pixel from a distributed ensemble model will be necessary. Possible limitations when it comes to the use of models and scale have not yet been analysed, however, we do not see any reasons why the methodology could not be applied also for other ensemble models and hydrological models than the ones included in EFAS.

Acknowledgements. This work has been carried out within the framework of the European Flood Awareness System (EFAS) which is part of the Copernicus Emergency Management Services. We acknowledge the national hydrological services for providing observational data and ECMWF for the forecast data.

References

Berrocal, V. J., Raftery, A. E., and Gneiting, T.: Combining Spatial Statistical and Ensemble Information in Probabilistic Weather Forecasts, Mon. Weather Rev., 135, 1386–1402, 2007.

Clark, I.: Geostatistical estimation and the lognormal distribution, Geocongress, Pretoria, RSA, available at: http://uk.geocities.com/drisobelclark/resume/papers/Geocongress1998.zip (last access: 1 November 2006), 1998.

de Marsily, G.: Quantitative hydrogeology, Academic Press Inc., London, 1986.

De Roo, A. P. J., Wesseling, C. G., and Van Deursen, W. P. A.: Physically based river basin modelling within a GIS: The LISFLOOD model, in Hydrological Processes, John Wiley & Sons Ltd., 14, 1981–1992, 2000.

Engeland, K. and Steinsland, I.: Probabilistic postprocessing models for flow forecasts for a system of catchments and several lead times, Water Resour. Res., 50, 182–197, 2014.

Gel, Y., Raftery, A. E., and Gneiting, T.: Calibrated Probabilistic Mesoscale Weather Field Forecasting, J. Am. Stat. Assoc., 99, 575–583, 2004.

Gneiting, T., Raftery, A. E., Westveld, A. H., and Goldman, T.: Calibrated Probabilistic Forecasting Using Ensemble Model Output Statistics and Minimum CRPS Estimation, Mon. Weather Rev., 133, 1098–1118, 2005.

Hemri, S., Fundel, F., and Zappa, M.: Simultaneous calibration of ensemble river flow predictions over an entire range of lead times, Water Resour. Res., 49, 6744–6755, 2013.

Merz, R. and Blöschl, G.: Flood frequency regionalisation – Spatial proximity vs. catchment attributes, J. Hydrol., 302, 283–306, 2005.

Raftery, A. E., Gneiting, T., Balabdaoui, F., and Polakowski, M.: Using Bayesian Model Averaging to Calibrate Forecast Ensembles, Mon. Weather Rev., 133, 1155–1174, 2005.

Skøien, J. O., Merz, R., and Blöschl, G.: Top-kriging – geostatistics on stream networks, Hydrol. Earth Syst. Sci., 10, 277–287, doi:10.5194/hess-10-277-2006, 2006.

Skøien, J. O., Blöschl, G., Laaha, G., Pebesma, E., Parajka, J., and Viglione, A.: rtop: An R package for interpolation of data with a variable spatial support, with an example from river networks, Comput. Geosci., 67, 180–190, 2014.

Van Der Knijff, J. M., Younis, J., and De Roo, A. P. J.: LISFLOOD: a GIS-based distributed model for river basin scale water balance and flood simulation, Int. J. Geogr. Inf. Sci., 24, 189–212, 2010.

Proc. IAHS, 373, 115–118, 2016
proc-iahs.net/373/115/2016/
doi:10.5194/piahs-373-115-2016

A socio-hydrological comparative assessment explaining regional variances in suicide rate amongst farmers in Maharashtra, India

Nadja I. den Besten, Saket Pande, and Hubert H. G. Savenije

Department of Water Management, Delft University of Technolgy, Delft, the Netherlands

Correspondence to: Nadja I. den Besten (n.i.den.besten@student.tudelft.nl)

Published: 12 May 2016

Abstract. Maharashtra is one of the states in India that has witnessed one of the highest rates of farmer suicides as proportion of total number of suicides. Most of the farmer suicides in Maharashtra are from semi-arid divisions such as Marathwada where cotton has been historically grown. Other dominant crops produced include cereals, pulses, oilseeds and sugarcane. Cotton (fibers), oilseeds and sugarcane providing highest value addition per unit cultivated area and cereals and pulses the least. Hence it is not surprising that smallholders take risks growing high value crops without "visualising" the risks it entails such as those corresponding to price and weather shocks.

We deploy recently developed smallholder socio-hydrology modelling framework to understand the underlying dynamics of the crisis. It couples the dynamics of six main variables that are most relevant at the scale of a smallholder: water storage capacity (root zone storage and other ways of water storage), capital, livestock, soil fertility and fodder biomass. The hydroclimatic variability is accounted for at sub-annual scale and influences the socio-hydrology at annual scale. The model incorporates rule-based adaptation mechanisms (e.g., adjusting expenditures on food and fertilizers, selling livestocks) of smallholders when they face adverse conditions, such as high variability in rainfall or in agricultural prices.

The model is applied to two adjoining divisions of Maharashtra: Marathwada and Desh. The former is the division with relatively higher farmer suicide rates than the latter. Diverse spatial data sets of precipitation, potential evaporation, soil, agricultural census based farm inputs, cropping pattern and prices are used to understand the dynamics of small farmers in these divisions, and to attribute farmer distress rates to soil types, hydroclimatic variability and crops grown.

Comparative socio-hydrologic assessment across the two regions confirms existing narratives: low (soil) water storage capacities, no irrigation and poor access to alternative sources of incomes are to blame for the crisis, suggesting that smart indigenous solutions such as rain-water harvesting and better integration of smallholder systems to efficient agricultural supply chains are needed to tackle this development challenge.

1 Introduction

Smallholders contribute significantly to the total value of agricultural output in India, though their resilience to climate change and price volality make them susceptible to distress (Fan et al., 2013). The fragility of smallholders is very noticeable in farmer suicides rates in India. Between 1995 and 2012, approximately 284 673 farmers committed suicide in several regions in India, with a peak in 2006 and 2010 (Nagaraj, 2008; Mishra, 2014).

Understanding the challenges encountered by (small) farm holders requires a multi-disciplinary approach. The following research builds upon Pande and Savenije (2016) to explain smallholder dynamics between its natural resources and socio-economic situation within a water-centric approach. We zoom into the state of Maharashtra where one of the highest rates of suicide amongst farmers have been recorded

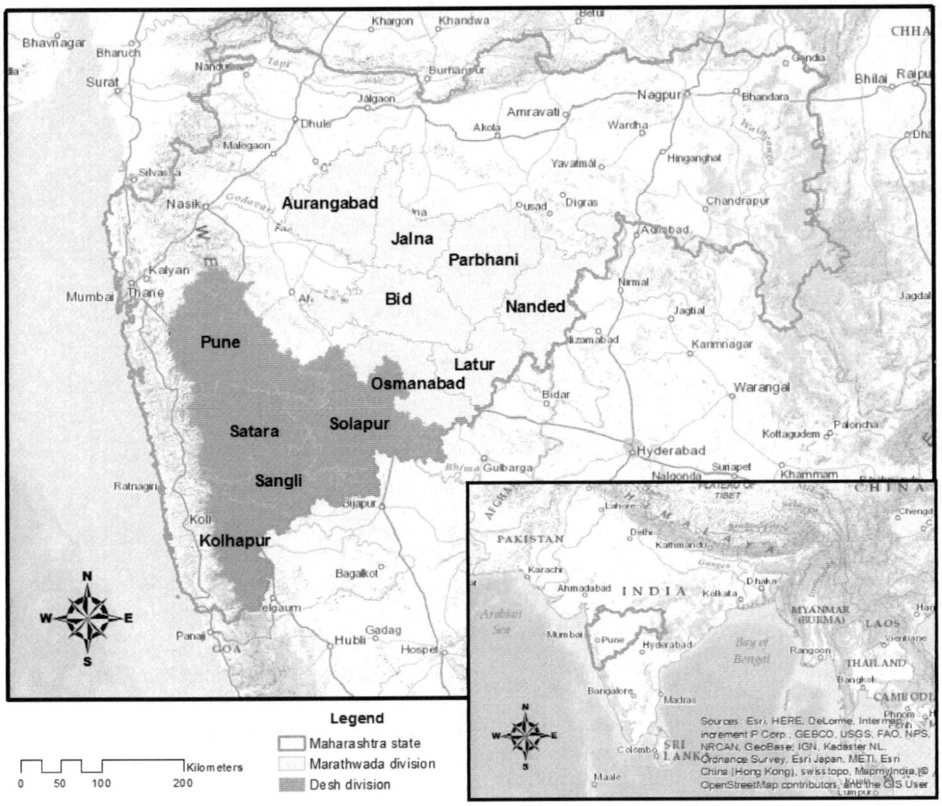

Figure 1. Location of simulated districts. Model simulations for the following eleven districts were conducted in Maharashtra, the districts in Marathwada: Aurangabad, Beed, Jalna, Latur, Osmanabad, Parbhani, and the districts in Desh: Pune, Sangli, Satara, Solapur and Kholapur. Spatial variable data sets on potential evaporation, soil properties and rainfall were weighted on district level.

(NCRB, 2014). The socio-hydrological modelling framework is applied to two regions of Maharashtra: Marathwada and Desh (see Fig. 1). We examine small farmers with 1–4 ha of ground. The agricultural income mainly depends on cotton and sugarcane production. Besides these crops, jowar (Sorghum), rice and soybean cultivation are also abundant. Near to sixty percent of the smallholders in the regions under study produce these five crops (Agriculture Census Division, 2011).

Marathwada division has experienced relatively higher farmer suicide rates than Desh (Ramanna, 2006; Khairnar et al., 2015). The aim of this paper is to explain this difference between the two sub-divisions through comparative assessment of smallholder socio-hydrological model simulations in terms of differences in hydro-climatic variability, soil heterogeneity and types of crops grown.

2 Methodology

The smallholder socio-hydrologic modelling framework proposed by Pande and Savenije (2016) is a dynamic model that can be made location specific. It consists of six socio-hydrological state variables: soil moisture, soil fertility, capital, livestock, fodder and labor availability. All these state variables are modelled by simple differential equations. Seven socio-hydrologic flux variables determine the interrelationships between these state variables, these are: crop production, livestock sales, expenditure, livestock production costs, crop production costs, labor factor and fertilizer factor. Climatic forcing and wage rate are external to the smallholder model system and influence the system as such.

To explain the modelling framework in a nutshell, consider a small farmer experiencing a year with disappointing crop yields due to poor rainfall, e.g. his/her capital encounters deficits as a result and therefore the farmer cuts down on his/her expenditure by selling livestocks, cutting down on investments, school fees and so on. The farmer stops cutting down on his/her expenditure when capital becomes positive again. These adjustments consequently affect the socio-hydrological state variables for the next year. The evolution of farmers capital thus depends on how much he or she is exposed to hydro-climatic variability (that affects crop yields) and price volatility (that affects crop income) and whether the farmer has access to hydrologic or financial instruments that can buffer these variabilities.

Table 1. Transpiration demand not met, gives the percentage of mean annual transpiration demand that is not met when rainfall is not enough to sustain crop water demand during growing season. The values in the brackets display the average yearly values of these deficits in [$mm\,yr^{-1}$] over the simulated time (1983–2009). % rainwater buffer: gives the percentage of years when the amount of annual rainfall was not enough to sustain transpiration demand, and consequently it gives the percentage of years when harvesting rainwater was not a sufficient strategy.

Crops	Marathwada		Desh	
	% transpiration demand deficit	% rainwater buffer	% transpiration demand deficit	% rainwater buffer
Sugarcane	84 [$-1080\,mm\,yr^{-1}$]	63	77 [$-945\,mm\,yr^{-1}$]	53
Jowar	31 [$-70\,mm\,yr^{-1}$]	29	22 [$-50\,mm\,yr^{-1}$]	15
Cotton	50 [$-220\,mm\,yr^{-1}$]	53	45 [$-200\,mm\,yr^{-1}$]	30
Rice	89 [$-285\,mm\,yr^{-1}$]	80	65 [$-180\,mm\,yr^{-1}$]	49
Soybean	44 [$-120\,mm\,yr^{-1}$]	35	37 [$-100\,mm\,yr^{-1}$]	19

2.1 Input data and parameters

The dominant soil types in the study regions were obtained through the Harmonized World Soil Database v1.2 (Fischer et al., 2008). All dominant soil textures throughout Marathwada and Desh are classified as loamy to clayey textures (USDA Texture classification) in the subsoil (Fischer et al., 2008). We assume that Available Water Storage Capacity (AWSC) is 175 $mm\,m^{-1}$ for loam and 200 $mm\,m^{-1}$ for clay. The soil water storages for every district are then calculated by weighted mean soil depth (India-WRIS WebGIS, 2016) of the districts multiplied by the previously mentioned AWSC.

Precipitation and potential evaporation data-series are created based on satellite products provided by Jones and Harris; Zomer et al. (2008). Freely available Global Potential Evapo-Transpiration (Global-PET) with a spatial resolution of 30 arcsec ($\sim$ 1 km at equator) is used to compute average potential evaporation on district level. The product used provides monthly averages over the years 1950–2000. These monthly averages are assumed to represent the potential evaporation over the simulated years. For precipitation a coarser product from the CRU-TS 3.0 Climate Database, with a 0.5° (30 arcmin) spatial resolution on land areas, is used.

Fertilizer and crop prices are obtained from The Worldbank (2010) with an exchange rate of Rs 45 = USD 1 to convert The Worldbank (2010) prices to Rs per kilogram. Crop specific yield coefficients are obtained as described by FAO (2015). Assumed application of fertilizers is as follows: 27 $kgN\,ha^{-1}$ for Jowar, 57 $kgN\,ha^{-1}$ for Paddy and 10 $kgN\,ha^{-1}$ applied for soybean production (FAO, 2005). The three crops are modelled as Kharif crops, growing from June until maximum October. All the crop factors have been obtained from FAO (FAO, 2015).

3 Results

Marathwada is climatologically different from Desh: 860 $mm\,yr^{-1}$ of rainfall falls on average in Marathwada, while Desh approximately receives 1000 $mm\,yr^{-1}$. The Solapur district within Desh that neighbours Marathwada is drier compared to other districts in Desh. Rain-fed agriculture is dominant in the districts under study (except under sugarcane production), therefore the water demands (crop transpiration) and supplies (rainfall) are analysed among all crops in the study area (Table 1).

The soil depths are highly variable within the study area. Shallow soils (0–25 mm) are primarily located in various districts of Marathwada, resulting in a disadvantage in average soil water storages for Marathwada compared to the region of Desh. The shallow soils limit the rooting depths of the crops grown in the area (Letey, 1985), hence decreasing estimated soil storage capacities.

The estimated soil water storages and district specific hydroclimatic forcing were used to simulate the effect on smallholder's capital and shown in Fig. 2.

4 Discussion

Marathwada is faced with higher suicide rate than the region of Desh. With the help of the water-centric approach within the socio-hydrological modelling framework, we tested a hypothesis that inclusion of spatial heterogeneity within the model can explain the differences in well-being of smallholders between the two regions. This is in addition to observed differences in hydro-climatic variability between the two regions. These outcomes suggest that high hydro-climatic variability, shallow soils and poor (risky) crop choices made by smallholders, can explain higher suicide rates in Marathwada. From Fig. 2 one can see that crop income falls in Marathwada when suicide rates start to rise. This indicates that Marathwada might contribute to the distress that farmers are facing in Maharashtra.

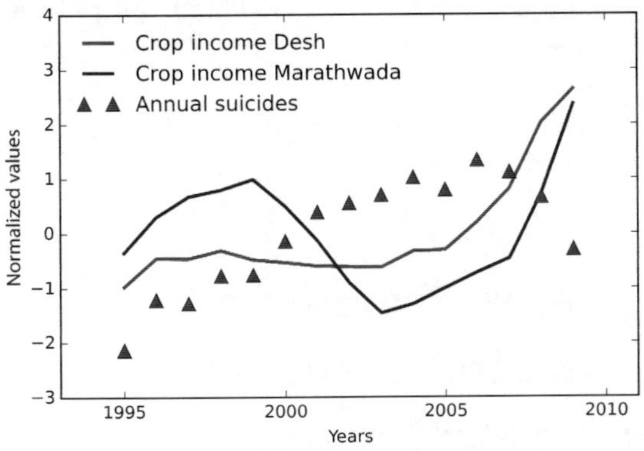

Figure 2. The capital (model) outcomes for all crops in all districts within a given sub-division (i.e. Desh and Marathwada) under study were aggregated based on agricultural census data on smallholders. Crop area weighted mean capital time series were estimated for the two divisions. Suicide rates at state level (Maharashtra) are plotted to indicate distress amongst farmers within the state. All 3 time series were normalized by substracting the mean and dividing by the standard deviation

Comparing hydro-climatological, soil characteristics and crops grown between the divisions one can see that the characteristics are unfavourable for Marathwada. Table 1 displays that the need to overcome crop transpiration demand deficits is higher in Marathwada than in the division of Desh. Further, even rainwater harvesting in Marathwada is not as successful as in Desh.

High water demanding crops such as cotton are generally favoured in Marathwada (24 % of the smallholders), while in Desh only 0.6 % of the smallholders choose to grow cotton primarily under rain-fed condition. Taking into account the observations in Table 1, creating local storages in Marathwada will not be enough to bridge monthly deficits during a growing season. A combination of solutions is indeed needed to outweigh unfavourable farming conditions in Marathwada. The policy and interventions in such regions could be steered in the direction of (1) preventing farmers from betting on a good monsoon and deciding to high risk crops in inadequate locations, and (2) to invest in local storages where biophysical characteristics (as soil depth e.g.) are adequate.

Data availability

The spatial distribution of soils was analyzed with a data set available at the repository of World Data Centre for Soils, re3data.org.

References

Agriculture Census Division: Agricultural Census Data Base, available at: http://agcensus.dacnet.nic.in/, last access: 20 December 2015, 2011.

FAO: Fertilizer use by crop in India, Tech. rep., Food and Agriculture Organization of the United Nations (FAO), Viale delle Terme di Caracalla, Rome, Italy, 2005.

FAO: Crop Water Information, available at: http://www.fao.org/nr/water/, last access: 11 January 2016, 2015.

Fan, S., Brzeska, J., Keyzer, M., and Halsema, A.: From subsistence to profit: Transforming smallholder farms, International Food Policy Research Institute (IFPRI), Tech. rep., 30 pp., 2013.

Fischer, G., Nachtergaele, F. O., Prieler, S., Teixeira, E., Tóth, G., van Velthuizen, H., Verelst, L., and Wiberg, D.: Global Agro-ecological Zones Assessment for Agriculture (GAEZ 2008), IIASA, Laxenburg, Austria and FAO, Rome, Italy, 2008.

India-WRIS WebGIS: Soil depth, available at: http://india-wris.nrsc.gov.in/soilapp.html, last access: 10 January 2016.

Jones, P. D. and Harris, I.: Climatic Research Unit (CRU) time-series datasets of variations in climate with variations in other phenomena, NCAS British Atmospheric Data Centre, Harwell Oxford, 2013

Khairnar, D. R., Bhosale, M. J., and Jadhav, M. A.: Lack of irrigation facilities, drought conditions and farmers suicides in Marathwada region, India, American Journal of Rural Development, 3, 74–78, 2015.

Letey, J.: Relationship between soil physical properties and crop production, in: Advances in soil science, Springer, 1, 277–294, doi:10.1007/978-1-4612-5046-3_8, 1985.

Mishra, S.: Farmers' suicides in India, 1995–2012: measurement and interpretation, Asia Research centre, London, 2014.

Nagaraj, K.: Farmers' suicides in India: magnitudes, trends and spatial patterns, 4th Edn., Bharathi puthakalayam, 421, Anna Salai, Teynampet, Chennai, 2008.

Pande, S. and Savenije, H. H.: A sociohydrological model for smallholder farmers in Maharashtra, India, Water Resour. Res., 52, doi:10.1002/2015WR017841, online first, 2016.

Ramanna, A.: Farmers rights in India: a case study, Tech. Rep. 4, The Fridtjof Nansen Institute, available at: http://www.mtnforum.org/sites/default/files/publication/files/4046.pdf, last access: 15 January 2016, 2006.

re3data.org: Registry of Research Data Repositories, re3data.org – doi:10.17616/R3X01J, last access: 26 April 2016.

Zomer, R. J., Trabucco, A., Bossio, D. A., and Verchot, L. V.: Climate Change Mitigation: A Spatial Analysis of Global Land Suitability for Clean Development Mechanism Afforestation and Reforestation, Agr. Ecosyst. Environ., 126, 67–80, 2008.

Proc. IAHS, 373, 119–123, 2016
proc-iahs.net/373/119/2016/
doi:10.5194/piahs-373-119-2016

Open Access
PIAHS

Modeling human-water-systems: towards a comprehensive and spatially distributed assessment of co-evolutions for river basins in Central Europe

Peter Krahe, Enno Nilson, Malte Knoche, and Anna-Dorothea Ebner von Eschenbach

Department Water balance, Forecasting and Predictions, Federal Institute of Hydrology, Koblenz, 56002, Germany

Correspondence to: Peter Krahe (krahe@bafg.de)

Published: 12 May 2016

Abstract. In the context of river basin and flood risk management there is a growing need to improve the understanding of and the feedbacks between the driving forces "climate and socio-economy" and water systems. We make use of a variety of data resources to illustrate interrelationships between different constituents of the human-water-systems. Taking water storage for energy production as an example we present a first analysis on the co-evolution of socio-economic and hydrological indicators. The findings will serve as for the development of conceptual, but fully coupled socio-hydrological models for selected sectors and regions. These models will be used to generate integrated scenarios of the climate and socio-economic change.

1 Introduction

The adequate availability of water is an important determinant for the socio-economic development – with water representing either an essential raw material or a threat e.g. in the case of flood events. At the same time the socio-economic development affects the water resources and the environment. Extensive reviews of challenges as well as guiding principles of developing integrated socio-hydrological models for (i) system understanding (ii) forecasting and prediction as well as (iii) policy and decision-making have recently been given by Sivapalan and Blöschl (2015) and Blair and Buytaert (2016). With this study we intend to assess the need and the feasibility of socio-hydrological approaches for solving practical management questions (cf. iii).

In the context of medium to long term river basin and flood risk management there is a growing need to improve the understanding of and the feedbacks between the water and the human component. Among others, scenarios of global and regional climate change need to be combined with those of developments of the society to come to "adaptive scenarios" instead of "climate driven impact scenarios". As a first step of an ongoing study we explore relevant human-water loops in a major part of Central Europe (Fig. 1).

In this paper (i) regions where human interference with the water cycle is particularly strong will be highlighted by a hydrological modelling approach, and (ii) time scales and potential drivers of a human-water system will be explored using the Swiss water power sector as a case study.

2 Study area

Our study area covers a major part of Central Europe (approx. 800 000 km^2), namely the German river catchments including their international upstream parts (Fig. 1). In a global perspective this area is usually not mentioned among the regions facing water resource problems. However, there are strong regional disparities of the natural water resources, which roughly follow a southwest-northeast gradient with the Alps in the south being an important "water tower". There is also a distinct seasonal contrast in natural water availability. Under the influence of the Alps (extending to the upper Rhine and southern tributaries of the Danube) runoff is usually higher in the summer months, while in the remainder of the area winter runoff is higher. Climate change may lead to stronger seasonal contrasts of water availability in the latter case (e.g. Schneider et al., 2013; Nilson et al., 2014).

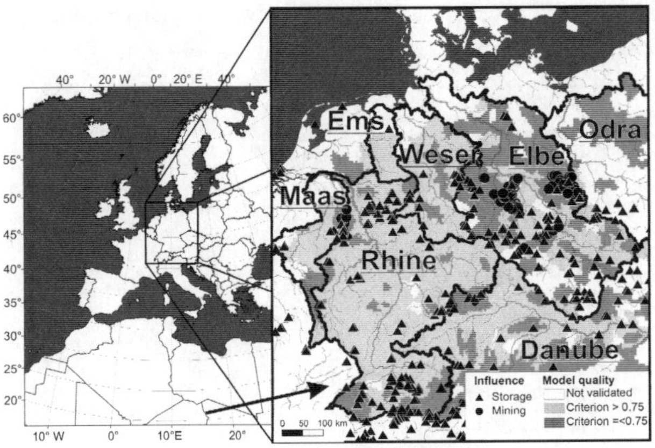

Figure 1. Model domain (left map) and validation results (right map) for river basins of the conceptual hydrological model LARSIM-ME. Grey scales in the right panel show classified values of the coefficient of determination (R^2) of daily flow values (November 1976 to October 2005). Locations of dams and mining were compiled from various sources (among others; Hydrological Atlas of Germany (BMU, 2003); Global Reservoir and Dam Database, GRanD; Lehner et al., 2011). Arrow marks the region selected as a case study.

Also, the water demand is unequally distributed regionally, being concentrated in a number of large agglomeration areas. The socio-economic systems in Central Europe are optimized to and dependent on relatively high and stable water resources. Consequently, the region is vulnerable to changes of water availability. Water resources and water use are thus regularly monitored on a national and continental level. Also, the "German Adaptation Strategy to Climate Change" introduces several indicators describing water resources (UBA, 2015).

3 Data and methods

Main challenges in socio-hydrological studies refer to the availability of long time series of data and indicators representing the socio-economic part of the human-water-system as well as the level of conceptualization and complexity at appropriate spatial and temporal scales that allow integrating them into hydrological models. We start assessing these challenges by identifying regions in our study area where human interference with the water cycle is particularly strong.

For that purpose, a hydrological model is applied for Central Europe ("LARSIM-ME") using a new homogeneous hydro-meteorological data product (DWD-BFG HYRAS) as input. The model has been parameterised by regionalization of model parameters from catchments that show no/weak human influence. We assume the simulated runoff to reflect natural conditions and compare them with observed runoff data and with an inventory of dams and mining areas in the model domain.

Starting from this version of the water balance model, human influence is currently successively being implemented. This will be done using operation plans of dams and water transfers, pumping volumes from open pit mines and water abstraction rules. Acquisition of this kind of information over a large region as the overall model domain is an intricate task because of the diversity of institutions and sectors involved. In some cases detailed information is often not available at all because it is retained due to privacy reasons.

Therefore, we evaluate methods to synthesize operation data for our model domain based on information from selected case studies that show an extraordinary good data situation. On top of data availability, each case study has to fulfil two prerequisites: (i) the sector has to be dependent on at least one component of the water balance (ii) the regional water balance is modified by the sector with at least one management measure.

In the following section we choose the case study "Water Power in Switzerland" as an example to highlight available data and to display time scales and drivers of change. Data resources are given in Table 1. Analyses with regard to co-evolution processes will be performed by comparing time series of these variables with water resources data as well as meteorological and hydrological data.

4 Results

4.1 Water balance simulations

Selected results of the large scale water balance simulation run with regionalized parameters for the period November 1976 to October 2005 are depicted in Fig. 1. As the model parameters were determined in catchments with no/weak human interference, the simulation can be regarded as representing natural flow conditions. The dark grey regions show particular strong deviations between simulations and observations. These regions are most frequent in the River Elbe catchment, the Alpine subcatchments of the Rivers Danube and Rhine, and in some catchments of tributaries of the Lower River Rhine. The spatial distribution of clusters of mining activities and water storages match well these regions (Fig. 1) indicating important modifications of the natural hydrological system in Central Europe (transfer systems are not shown). Obviously, water balance simulations aiming to match the observed runoff require additional representation of human activities.

4.2 Case study "Water Power in Switzerland"

Dams are mainly constructed and operated in Switzerland for energy supply. In the context of water resources monitoring, developments of water storage capacity of dams as well as of management rules are of main interest. Figure 2 shows a long term perspective on water storage capacities in Switzerland, runoff at River Rhine gauging station Basel near the

Table 1. Data used in case study "Water Power in Switzerland" (Δt = temporal resolution, d = daily, m = monthly, a = annual; as plotted in Figs. 2 and 3).

Theme	Δt	Source
Air temperature of river basin upstream of gauge Basel-Rheinhalle	a, m	CRU TS3.10 (Harris et al., 2014) EOBS8 (Haylock et al., 2008)
Runoff at gauge Basel-Rheinhalle	a, d	FOEN (2016)
Storage water levels	m	SFOE (2016b)
Power production from reservoirs	m	SFOE (2016a)
Electricity consumer price index	m	FSO (2016b)
Electricity consumption	m	SFOE (2016c)
Consumer price index electricity (type III)	m, a	FSO (2016c)
Population	a	FSO (2016a)
Gross domestic product GDP	a	< 1995: Zuercher (2010) > 1995 (change): FSO (2016d)
Storage capacity	a	SwissCOD (2016)
Water power production	a	< 1985: Swiss Energy Council (1987) ≥ 1985: SFOE (2016c)
Electricity consumption	a	< 1990: Swiss Energy Council (1987) ≥ 1990: SFOE (2016c)

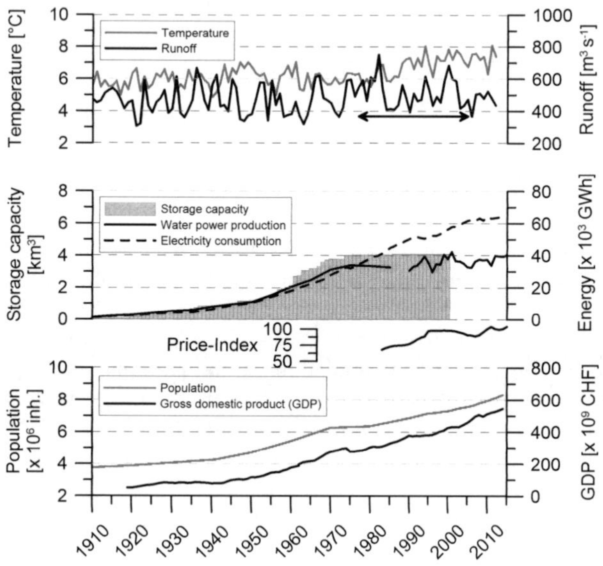

Figure 2. Annual data of water storage capacity of dams, water power generated (storage plus runoff power plants), and electricity consumed, hydrometeorologial conditions and socio-economic drivers in Switzerland since 1910. Runoff (lowest daily flow) and temperature refer to hydrological years (November–October). The arrow marks a period of a higher level of low flows after 1970. (Data sources: cf. Table 1)

Swiss/German border as well potential climatic and socio-economic drivers of change. Available data allows this assessment on an annual basis for the period 1910 to 2012.

At the beginning of the 20th century storage capacities and electricity consumption were very low. At this time water power was used more or less mechanically in the vicinity of the rivers (Swiss Energy Council, 1987). In the following decades storage capacities and water power production increased continuously until the 1970ies as a consequence of several developments: (i) a Swiss law targeting at the utilization of water power was released in 1918 (ii) technical innovations allowed the construction of larger dams and (iii) improved distribution of electricity over large distances. The latter also preconditioned the general electrification of the society and economy, which were both steadily growing. Since the mid of the 1970ies, there was no further increase of storage capacity and water power production. At the same time nuclear power was prioritized to meet the still growing electricity demand (Swiss Energy Council, 2015). It may be noteworthy, that after the mid 1970ies onwards the annual minimum runoff at gauge Basel hardly did ever drop below 400 $m^2 s^{-1}$ (cf. arrow in Fig. 2). This can to a large degree be attributed to the storage capacity implemented.

From 1995 onwards water power production data shows only minor long term increases (due to optimization of existing power plants), but a stronger year-to-year variability. This is at least in part due to a change of the data source and the way the data were calculated (cf. references given in Table 1). Nevertheless, these data may give an indication that – although socio-economical drivers dominated the centennial development – meteorological and hydrological conditions affect water power production variability on a year-to-year scale. This coincidence is, however not very clear on the annual scale.

In a monthly view (Fig. 3), the individual components of this human-water system show very systematic annual cycles. Storages are filled during the summer when precipitation and snow melt water runoff is high. The maximum level of water storage is usually reached in September or October.

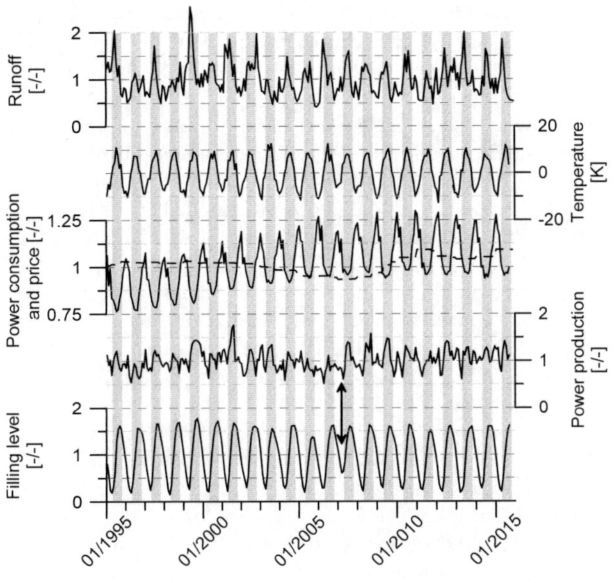

Figure 3. Time series of monthly water storage filling level, water power generation, electricity consumption/prices, and hydrometeorological conditions in Switzerland since 1995. All data were converted to anomalies with respect to their averages of the period 1995–2010. (dashed line: Electricity consumer price index; shading indicates hydrological seasons with white = winter; the arrow marks a singularity in winter 2006/07) (Data sources: cf. Table 1)

During the winter season, when runoff is typically low, water is released. The water power production shows a vague pattern with one weak peak during the high runoff period in summertime and one stronger peak during the release period in wintertime. The overall consumption of electricity shows – on top of a long term increase up to the year 2006 – a rather regular pattern with a single major peak during the heating season in winter. The amplitudes of most variables vary only moderately and the pattern persists remarkably stable over the years. One exception is the winter of 2006/2007 which was extraordinary warm. This coincides with higher than average minimum water levels of storages due to lower electricity consumption (less heating).

5 Discussion and conclusions

Clearly, potential water availability is the backbone of activity in the water power sector. However, technological developments were the main drivers for the onset and the cessation of the growth of water power production and the related modifications of the natural water system through reservoirs. The increase of water power production took place over three to four decades. Variability of natural drivers as reflected e.g. by air temperature and runoff manifests as second order variability in water power time series. This is visible on a year-to-year basis, but much more on a monthly timescale.

The steady seasonal pattern of reservoir operation represents a feature that could be described by comparatively simple models. This would allow us to generate storage data for the past as well as for climate impact studies. However, the long term socio-economy-driven developments tell us, that the pattern may be stationary only for limited time. Global climate change, a change of energy mix in Europe, and the ongoing deregulation of the energy market in Switzerland are possible candidates to cause changes in the operation of Swiss dams. Estimating the consequences of these so far unobserved phenomena remains a challenge.

Similar case studies are underway in other regions of our model domain and examining other sectors such as urban water supply, agriculture, navigation etc. We attempt to build simple conceptual models representing those sectors which interfere most with the water cycle in Central Europe. By combining these models with the water balance model we try to provide explicit information on spatial and temporal scales that are suitable for supra-regional water resources management.

Data availability

Most of the data used in this study are taken from freely available (open access) data sources listed under references. We acknowledge the E-OBS dataset from the EU-FP6 project ENSEMBLES (http://ensembles-eu.metoffice.com) and the data providers in the ECA&D project (http://www.ecad.eu). Locations of dams and mining taken from Hydrological Atlas of Germany are protected by copyright but published as a printed map (BMU, 2003). Digital data can be supplied for research activities by the author.

References

Blair, P. and Buytaert, W.: Socio-hydrological modelling: a review asking "why, what and how?", Hydrol. Earth Syst. Sci., 20, 443–478, doi:10.5194/hess-20-443-2016, 2016.

BMU (Bundesministerium für HAD Umwelt, Naturschutz und Reaktorsicherheit): Hydrologischer Atlas von Deutschland (Hydrological Atlas of Germany), Map 7.2 Water Supply, Bonn, 2003.

FOEN (Federal Office for the Environment, Switzerland): Hydrological Data Service for water courses and lakes, available at: http://www.bafu.admin.ch/wasser/13462/13494/15076/index.html?lang=en, 2016.

FSO (Swiss Federal Statistical Office): Bevölkerungsstand und -struktur – Detaillierte Daten (state and structure of population – detailed data), available at: http://www.bfs.admin.ch/bfs/portal/de/index/themen/01/02/blank/data/01.html, 2016a.

FSO (Swiss Federal Statistical Office): Schweizerische Preisindizes 1914–2014 (Price indices of Switzerland), available at: http://www.bfs.admin.ch/bfs/portal/de/index/themen/05/01/100.html, 2016b.

FSO (Swiss Federal Statistical Office): Landesindex der Konsumentenpreise (Consumer price index of state), available at: http://www.bfs.admin.ch/bfs/portal/de/index/themen/05/02/blank/data.html, 2016c.

FSO (Swiss Federal Statistical Office): Bruttoinlandprodukt – Daten, Indikatoren (Gross domestic product – data idicators), available at: http://www.bfs.admin.ch/bfs/portal/de/index/themen/04/02/01/key/bip_gemaess_produktionsansatz.html, 2016d.

Harris, I, Jones, P. D., Osborn, T. J., and Lister, D. H.: Updated high-resolution grids of monthly climatic observations – the CRU TS3.10 Dataset, doi:10.1002/joc.3711, 2014.

Haylock, M. R., Hofstra, N., Klein Tank, A. M. G., Klok, E. J., Jones, P. D., and New, M.: A European daily high-resolution gridded dataset of surface temperature and precipitation, J. Geophys. Res.-Atmos., 113, D20119, doi:10.1029/2008JD010201, 2008.

Lehner, B., Reidy Liermann, C., Revenga, C., Vorosmarty, C., Fekete, B., Crouzet, P., Döll, P., Endejan, M., Frenken, K., Magome, J., Nilsson, C., Robertson, J. C., Rodel, R., Sindorf, N., and Wisser, D.: Global Reservoir and Dam Database, Version 1 (GRanDv1): Dams, Revision 01. Palisades, NY: NASA Socioeconomic Data and Applications Center (SEDAC), doi:10.7927/H4N877QK, 2011.

Nilson, E., Krahe, P., Lingemann, I., Horsten, T., Klein, B., Carambia, M., and Larina, M.: Auswirkungen des Klimawandels auf das Abflussgeschehen und die Binnenschifffahrt in Deutschland, Schlussbericht KLIWAS-Projekt 4.01. KLIWAS-43/2014. BfG, Koblenz, doi:10.5675/Kliwas_43/2014_4.01, 2014.

Schneider, C., Laizé, C. L. R., Acreman, M. C., and Flörke, M.: How will climate change modify river flow regimes in Europe?, Hydrol. Earth Syst. Sci., 17, 325–339, doi:10.5194/hess-17-325-2013, 2013.

SFOE (Swiss Federal Office of Energy): Electricity Statistics, Wochenstatistik Elektrizitätsbilanz (Weekly statistic of electricity balance), available at: http://www.bfe.admin.ch/themen/00526/00541/00542/00630/index.html?lang=en&dossier_id=00767#, 2016a.

SFOE (Swiss Federal Office of Energy): Electricity Statistics, Füllungsgrad der Speicherseen (filling level of reservoirs, available at: http://www.bfe.admin.ch/themen/00526/00541/00542/00630/index.html?lang=de&dossier_id=00766, 2016b.

SFOE (Swiss Federal Office of Energy): Electricity Statistics, Gesamte Erzeugung und Abgabe elektrischer Energie in der Schweiz (total generation and release of electricity in Switzerland), available at: http://www.bfe.admin.ch/themen/00526/00541/00542/00630/index.html?dossier_id=00769, 2016c.

SwissCOD (Swiss Committee On Dams): Dams in Switzerland, available at: http://www.swissdams.ch/index.php/en/swiss-dams/dams-in-switzerland, 2016.

Sivapalan, M. and Blöschl, G.: Time scale interactions and the coevolution of humans and water, Water Resour. Res., 51, 6988–7022, doi:10.1002/2015WR017896, 2015.

Swiss Energy Council (Ed.): Energiestatistik der Schweiz 1910–1985, Zürich 1987, available at: http://www.worldenergy.ch/Seiten/Energiestatistik/uebersicht/Einleitung/?oid=1482&lang=de, 1987, 2015.

Umweltbundesamt (UBA) (Ed.): Monitoringbericht 2015 zur Deutschen Anpassungsstrategie an den Klimawandel, Bericht der Interministeriellen Arbeitsgruppe Anpassungsstrategie der Bundesregierung, Dessau-Roßlau, available at: http://www.umweltbundesamt.de/publikationen/monitoringbericht-2015, 2015.

Zuercher, B.: Das Wachstum der Schweizer Volkswirtschaft seit 1920, Die Volkswirtschaft 1/2-2010, 9–13, 2010.

Proc. IAHS, 373, 125–129, 2016
proc-iahs.net/373/125/2016/
doi:10.5194/piahs-373-125-2016

Open Access
PIAHS

Institutions in transitioning peri-urban communities: spatial differences in groundwater access

Sharlene L. Gomes and Leon M. Hermans

Department of Technology, Policy, and Management, Delft University of Technology, Delft, 2628BX, the Netherlands

Correspondence to: Sharlene L. Gomes (s.l.gomes@tudelft.nl)

Published: 12 May 2016

Abstract. Urbanization creates challenges for water management in an evolving socio-economic context. This is particularly relevant in transitioning peri-urban areas like Khulna, Bangladesh where competing demands have put pressure on local groundwater resources. Users are unable to sufficiently meet their needs through existing institutions. These institutions provide the rules for service provision and act as guidelines for actors to resolve their water related issues. However, the evolving peri-urban context can produce fragmented institutional arrangements. For example in Khulna, water supply is based on urban and rural boundaries that has created water access issues for peri-urban communities. This has motivated local actors to manage their groundwater needs in various ways. General institutional theories are well developed in literature, yet little is known about institutions in transitioning peri-urban areas. Institutions that fail to adapt to changing dynamics run the risk of becoming obsolete or counter-productive, hence the need for investigating institutional change mechanisms in this context. This paper examines peri-urban case studies from Khulna using the Institutional Analysis and Development framework to demonstrate how institutions have contributed to spatial differences in groundwater access with local actors investing in formal and informal institutional change as a means of accessing groundwater.

1 Introduction

The global south is expected to witness significant levels of urbanization in the coming decades which will have implications for the management of water resources (UN DESA, 2014). We see evidence of this in the Ganges Delta, where rapid and largely uncontrolled urban expansion has brought growing concerns of water scarcity, competition, and inequity among users (Kumar et al., 2011; Thissen et al., 2013). The situation is particularly severe in peri-urban areas surrounding Khulna city, Bangladesh, where groundwater is the main source for domestic and livelihood purposes.

Peri-urban areas are dynamic, representing the transition zones in processes of urban expansion with a heterogeneous composition (Allen, 2003; Narain, 2010). In such a context, the management of groundwater resources requires a coordinated approach among actors, based on a shared set of rules. Institutions are defined as formal and informal "rules" that structure interaction and behavior in society (North, 1990). They offer a means for resolving social dilemmas by offering prescriptive guidelines. However, peri-urban villages in Khulna have witnessed the failure of existing institutions in managing water resources in a sustainable and equitable way (Thissen et al., 2013). Achieving effective institutional arrangements in this context, however, is complex. Institutions are expected to evolve with the changing context, yet this is not always possible due to various actor constraints that prevents this. This can result in the creation of a fragmented mix of rural and urban institutions. This research explores how peri-urban institutions and institutional change has contributed to spatial differences in groundwater access in peri-urban Khulna.

2 Conceptual framework for peri-urban institutional analysis

The adapted Institutional Analysis and Development (IAD) framework in Fig. 1 focuses on key aspects relevant to the study of institutions and institutional change. The analysis

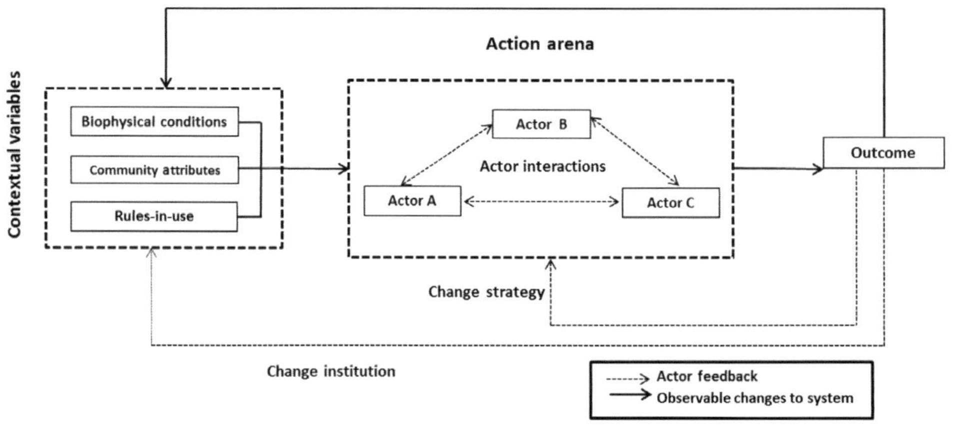

Figure 1. IAD Framework (adapted from Ostrom, 2005).

centres around the action arena, consisting of interactions between multiple actors with an ability to influence their system of interest. In this case, the action arena refers to peri-urban groundwater management. Strategic behaviour within this arena and its resulting outcome is based on contextual variables such as the institutions or "rules in use", community attributes such as socio-economic conditions, and biophysical conditions (Ostrom, 2005)

The institutional context includes consciously designed and codified formal laws, constitutions, or property rights as well as informal, socially constructed norms and values that together comprise a society's culture (North, 1990; Williamson, 1998). They define the participants, their position, resources, possible actions, and the costs and benefits of their associated outcomes within the action arena (Ostrom, 2005). In society, these institutions are organized in a nested structure from local operational rules to increasingly embedded higher order collective choice, constitutional, and meta-constitutional rules (North, 1990; Ostrom, 2005; Williamson, 2000).

Feedback loops from the outcomes are an important part of this framework offering a means to study institutional change. North (1996) describes institutional change to arise through a process of actor learning from the outcomes of their strategic decisions. Here, negative outcomes are a signal to change strategies or alter the institutions in order to achieve their objective. Although institutions are expected to evolve with actor's needs in this manner, this is not always the case. Theories highlight that the embeddedness of rules brings a higher transaction costs for actors investing in institutional change (North, 1990). As a result, institutions may gradually become eroded over time, replaced, or ignored altogether without purposive action (Scharpf, 1997). Moreover, institutions are not socially efficient, so institutional change also depends on the bargaining power of the actors involved (North, 1990, 1996). In Fig. 1, system feedback resulting in observed changes to the biophysical and socio-economic context is differentiated from actor feedback that produces institutional change, given this is a key research objective.

3 Methodology

Institutional aspects of groundwater access in peri-urban Khulna is analyzed using the framework in Fig. 1. The villages of Hoglandanga and Matumdanga are selected as case studies given the peri-urban characteristics they display and evidence of groundwater issues. They are currently witnessing urban expansion and are formally located outside Khulna city jurisdiction. Multi-actor interactions in groundwater access, an issue highlighted by both communities, is examined within the contextual variables. Field research reveals three key stages in the evolution of this issue representing the past, present, and future scenario of ground water access. In each stage, the study examines the form of institutional change invested in to manage groundwater access.

Primary data from government websites and published reports is verified and supplemented through field discussions with 18 respondents from government agencies and water users in both villages (Gomes, 2015). Interviewees were selected through structured and snowball sampling methods (Harrell and Bradley, 2009). Interview methods include focus group discussions and key informant interviews. Field data gathered in May–June (2015) was subsequently verified through a de-briefing meeting with local actors or email communication. Formal and informal institutions are analyzed based on their level of operation in the nested structure and the action arena variables they define (Ostrom, 2005).

4 Results

4.1 Initial situation: rural service provision

Both peri-urban villages fall under the jurisdiction of rural service providers such as the Department of Public Health Engineering (DPHE) and the Bangladesh Agricultural Development Corporation (BADC) for domestic and irrigation purposes respectively. Tube wells are provided via water and sanitation (WATSAN) or Irrigation committees at the sub-district level, headed by the sub-district chairman with representatives from local unions, BADC, and DPHE (BADC, 1985; Local Government Division, 1998). Villages may apply for a tube-well licence through written application at their local union after which the committee decides the allocation of licences among the villages. Allocation is based on the following selection criteria: aquifer conditions, distance from neighbouring tube wells, beneficial area, effect on existing tube wells, and suitability of sites as per formal rules (BADC, 1985; Gomes, 2015). However, this quota system does not appear to be fully operational in practice with insufficient tube wells provided in the past 4–5 years (Gomes, 2015). Public tube wells from DPHE are provided at a subsidized cost, making this an attractive option for those unable to invest in a private tube well. The resulting outcome is that of water scarcity due to the failure of formal institutions. Community discussions give the impression that their remote location away from the administrative centre limits their voice in the sub-district committee (Gomes, 2015)

4.2 Current situation: informal service provision

Given this outcome, the villages adopted a new strategy by creating informal institutions to access groundwater. Presently, local private tube well owners sell groundwater to marginalized irrigators. Tariffs are agreed upon by both parties, although it is unclear if there are negotiations or if the prices are elastic given the seasonal scarcity highlighted in both communities. In Matumdanga, water from shallow aquifers was sold in 2015 at a rate of 33–45 EUR 0.5 ha^{-1} per season for unlimited use while in Hoglandanga the shrimp farmers pay a rate of approximately 0.5 EUR h^{-1} to cover fuel and electricity costs (Gomes, 2015). Meanwhile, domestic users have informal rules for sharing existing tube wells such as a first come first serve and queue system to collect water. Women in Matumdanga, typically responsible for household water collection follow rules in the Muslim culture by avoiding water collection from their local Mosque during hours of prayer (Gomes, 2015).

4.3 Future situation: urban service provision

A potential future scenario of institutional change results as a by-product of urbanization. Khulna Water Supply and Sewerage Authority (KWASA) caters to the water and sanitation needs of Khulna city. Currently, 95 % of the water supply is from groundwater resources via private and shared tube-wells, however, KWASA plans to extend coverage through surface water projects funded by donor agencies (Gomes, 2015). Urban administration is managed by the Khulna City Corporation (KCC) who in 2007 and 2014 submitted proposals to extend city boundaries from 45 to 114 km^2 based on the Khulna master plan (Gomes, 2015). Once approved, the plan would bring Matumdanga and Hogladanga villages under KCC jurisdiction, making water supply the responsibility of KWASA. However, KWASA currently faces a supply gap, with unserved areas presently relying on private tube wells (Gomes, 2015). Thus, it is unclear if groundwater access could improve in these villages under urban administration in the future.

5 Discussion

These results show how peri-urban water access in Khulna is influenced by institutions. Formal institutions in the initial stage produced spatial differences in groundwater access between areas with and without tube well allocations. Although policies specify rules for allocation, lack of adequate enforcing and monitoring mechanisms can be a deterrent to its functioning at the local level. Monitoring and enforcement of rules can help prevent opportunistic behavior by placing a higher risk for those considering defecting (Ostrom, 2005). The need for enforcement, however, depends on whether there is a known willingness for local committees to deviate from rules. There appear to be potential payoffs for some elected members within the committee to support deviation. One could be internal reciprocal norms to appease the constituencies that elect them, which influences their decision. Or it may be the result of conflicting rules such as the Bangladesh Water Act (2013) promoting the right to water and others stating that communities should play a greater role in managing the costs for water services (Local Government Division, 1998; Ministry of Law, Justice and Parliamentary Affairs, 2013; Ministry of Water Resources, 1999). Greif (2006)'s theory on legal pluralism highlights that rules may be complementing, reinforcing, or conflicting to one another. Furthermore, limited understanding of the context in the design of rules or combination of rules can produce unintended or disastrous outcomes (Ostrom, 2005). Thus, research needs to look beyond local allocation rules to higher order rules or rules from other sectors and how they work in relation to one another.

The undesirable allocation outcome prompted local actors to invest in institutional change in order to meet their water needs. In the current situation, we observe that institutional change depends on one's resources. The location of both villages away from rural decision centers, give users limited bargaining power to seek a more equitable distribution of licenses. Thus, actors chose to invest in informal institutional change given the high transaction costs. Here, we see different sets of informal rules for irrigators and domestic users. While irrigators opted for a market system to access groundwater, domestic users share existing tube wells. The latter bears less financial implications but has costs in terms of time, distance travelled, and convenience. Thus, even with informal institutions we see differences in the options between peri-urban users.

In the future, constitutional level rules could again influence groundwater access through urbanization. Formal change by replacing rural with urban service provision creates some uncertainties for users in these communities. For domestic users there is the risk that existing gaps in urban water supply will continue, once again creating spatial difference in access within urban areas. For irrigators, future groundwater dependency is uncertain given the existing changes in land use (Gomes, 2015). When peri-urban villages come under KCC jurisdiction, the fate of peri-urban land will depend upon the master plans. So it is likely that land use will be a bigger uncertainty than water access.

6 Conclusion

Problems in peri-urban Khulna show that institutions fail to effectively manage groundwater access during urban transitions. It also highlights the challenges with institutional design in this context. The changing needs in peri-urban areas mean existing institutions across a rural-urban divide become ineffective over time causing spatial patterns in water access. Addressing inefficient rules needs to consider how they are applied in practice. Understanding actor motivations and objectives can help in this regard. In Khulna, we observe a shift from formal to informal institutions with the future bringing yet another formal change. Theories suggest that institutions work if they are stable in the long term but this is not possible given the dynamics of peri-urban areas. Thus peri-urban areas are challenged with balancing functionality with stability, the former addressed in this case through informal means. This research signals that improving water management during urban transitions requires focusing on the institutional context, for which the IAD framework offers a means to operationalize key concepts in the peri-urban context.

Acknowledgements. This research was funded by the Urbanising Deltas of the World programme of the Netherlands Organisation for Scientific Research (grant no. W 07.69.104), "Shifting Grounds: Institutional transformation, enhancing knowledge and capacity to manage groundwater security in peri-urban Ganges Delta systems". Thanks are due to Wil Thissen and Vishal Narain for their critical reflections on the ideas presented. Also to A. T. M. Zakir Hossain, Sk. Nazmul Huda, and Md. Riad Hossain from Jagrata Juba Shangha and Md. Rezaul Hasan from Bangladesh University of Engineering and Technology for valued assistance during the fieldwork in Bangladesh.

References

Allen, A.: Environmental planning and management of the peri-urban interface: perspectives on an emerging field, Environ. Urban., 15, 135–148, doi:10.1177/095624780301500103, 2003.

BADC: The Groundwater Management Ordinance, available at: http://bdlaws.minlaw.gov.bd/pdf_part.php?id=686 (last access: 31 January 2016), 1985.

Gomes, S. L.: Pre-scoping Visit-Field Report, Khulna, Delft University of Technology, Delft, the Netherlands, 2015.

Greif, A.: Institutional Trajectories: How Past Institutions Affect Current Ones, in: Institutions and the Path to the Modern Economy: Lessons from Medieval Trade, Cambridge University Press, New York, USA, 2006.

Harrell, M. C. and Bradley, M. A.: Data Collection Methods: Semi-Structured Interviews and Focus Groups, RAND Corporation, Santa Monica, California, USA, 2009.

Kumar, U., Khan, M. S. A., Rahman, R., Mondal, M. S., and Huq, H.: Water Security in Peri-Urban Khulna: Adapting to Climate Change and Urbanization, Discussion Paper No. 2, SACI WaterS, Secunderabad, Andhra Pradesh, India, 2011.

Local Government Division: National Policy for Safe Water Supply and Sanitation, available at: http://www.dphe.gov.bd/pdf/National-Policy-for-Safe-Water-Supply-&-Sanitation-1998.pdf (last access: 31 January 2016), 1998.

Ministry of Law, Justice and Parlimentary Affairs: Bangladesh Water Act, available at: http://www.ielrc.org/content/e1313.pdf (last access: 31 January 2016), 2013.

Ministry of Water Resources: National Water Policy, available at: http://www.buet.ac.bd/itn/publications/sector-documents/documents/National_Water_Policy.pdf (last access: 31 January 2016), 1999.

Narain, V.: Periurban water security in a context of urbanization and climate change: A review of concepts and relationships, SaciWATERS, available at: http://saciwaters.org/periurban/idrcperiurbanreport.pdf (last access: 22 October 2014), 2010.

North, D. C.: Institutions, Institutional Change and Economic Performance, Cambridge University Press, Cambridge, UK, 1990.

North, D. C.: Epilogue: Economic Performance Through Time, in: Emperican Studies in Institutional Change, Cambridge University Press, Cambridge, USA, 342–355, 1996.

Ostrom, E.: Understanding Institutional Diversity, Princeton University Press, Princeton, NJ, USA, 2005.

Scharpf, F. W.: Games Real Actors Play: Actor- Centered Institutionalism in Policy Research, Westview Press, Boulder Colorado, USA, 1997.

Thissen, W. A. H., Hermans, L. M., Prakash, A., Banerjee, P., Khan, M. S. A., Salehin, M., Narain, V., Kempers, R., Hossain, A. T. M. Z. P., and Banerjee, P. S.: Shifting Grounds: Institutional transformation, enhancing knowledge and capacity to manage groundwater security in peri-urban Ganges Delta systems, Project Proposal for the Urbanizing Deltas of the World programme of the Netherlands Organization for Scientific Research, Delft University of Technology, Delft, the Netherlands, 2013.

UN DESA: World Urbanization Prospects: The 2014 Revision, Highlights, No. ST/ESA/SER.A/352, United Nations, New York, USA, 2014.

Williamson, O. E.: Transaction cost economics: how it works; where it is headed, De Economist, 146, 23–58, 1998.

Williamson, O. E.: The new institutional economics: taking stock, looking ahead, J. Econ. Lit., 38, 595–613, 2000.

Proc. IAHS, 373, 131–136, 2016
proc-iahs.net/373/131/2016/
doi:10.5194/piahs-373-131-2016

Open Access
PIAHS

Human signatures derived from nighttime lights along the Eastern Alpine river network in Austria and Italy

Serena Ceola[1], Alberto Montanari[1], Juraj Parajka[2], Alberto Viglione[2], Günter Blöschl[2,3], and Francesco Laio[4]

[1]Dipartimento di Ingegneria Civile, Chimica, Ambientale e dei Materiali, Università di Bologna, Bologna, 40136, Italy
[2]Institute of Hydraulic Engineering and Water Resources Management, Vienna University of Technology, Vienna, 1040, Austria
[3]Centre for Water Resource Systems, Vienna University of Technology, Vienna, 1040, Austria
[4]Dipartimento di Ingegneria dell'Ambiente, del Territorio e delle Infrastrutture, Politecnico di Torino, Torino, 10129, Italy

Correspondence to: Serena Ceola (serena.ceola@unibo.it)

Published: 12 May 2016

Abstract. Understanding how human settlements and economic activities are distributed with reference to the geographical location of streams and rivers is of fundamental relevance for several issues, such as flood risk management, drought management related to increased water demands by human population, fluvial ecosystem services, water pollution and water exploitation. Besides the spatial distribution, the evolution in time of the human presence constitutes an additional key question. This work aims at understanding and analysing the spatial and temporal evolution of human settlements and associated economic activity, derived from nighttime lights, in the Eastern Alpine region. Nightlights, available at a fine spatial resolution and for a 22-year period, constitute an excellent data base, which allows one to explore in details human signatures. In this experiment, nightlights are associated to five distinct distance-from-river classes. Our results clearly point out an overall enhancement of human presence across the considered distance classes during the last 22 years, though presenting some differences among the study regions. In particular, the river network delineation, by considering different groups of river pixels based on the Strahler order, is found to play a central role in the identification of nightlight spatio-temporal trends.

1 Introduction

Anthropogenic closeness to rivers is of fundamental relevance for socio-hydrological purposes, including flood and drought management, water pollution and exploitation, but also the human pressure on river ecosystems. In order to analyse human signatures in the proximity of streams and rivers, census data and satellite data such as Landscan could be easily employed (Small, 2004; Kummu et al., 2011), even though they cannot provide spatially and temporally detailed information about human presence. While census data are usually provided on a yearly basis at subnational level (i.e. coarse spatial resolution), spatially detailed Landscan data at 1 km resolution do not allow for a temporal trend analysis. To overcome this weakness, recent studies (Elvidge et al., 1997; Ceola et al., 2014, 2015) used fine scale remotely sensed data, as nighttime lights, to support global and local analyses of human presence and water related issues.

We explore here new opportunities offered by nightlight data to better decipher the interactions between human and water systems. More specifically, we provide insights about the spatio-temporal evolution of human presence along the river network in the European Eastern Alpine region and in its immediate proximity, by adopting satellite images of nighttime lights, available as yearly snapshots from 1992 to 2013 at a high spatial resolution (i.e., nearly 1 km at the equator). In particular, nighttime lights are associated to the river network position and to additional four distance classes.

2 Materials and methods

2.1 Satellite Nighttime Light Data Set

Nighttime light time series, collected by the US Air Force Weather Agency under the Defense Meteorological Satellite Program (DMSP) – Operational Linescan System (OLS), are provided as freely available digital products by the National Geophysical Data Center from the National Oceanic and Atmospheric Administration (NOAA) at http://ngdc.noaa.gov/eog/dmsp/downloadV4composites.html.

Nightlight data, produced on a yearly basis from 1992 to 2013, represent cloud-free nocturnal luminosity from sites with protracted lighting (i.e., cities, towns, gas flares). Sunlit and moonlit data and observations from ephemeral phenomena like fires are excluded from the data set. Nightlight values, expressed as an adimensional digital number DN_{obs}, range from 0 to 63, corresponding to conditions characterized by absence of lights and pronounced luminosity, respectively. Nightlights thus represent a valuable proxy for the presence of human settlements and economic activity and they have been widely employed for demographic, economic, and environmental purposes (Elvidge et al., 1997, 2009; Small, 2004; Chand et al., 2009; Chen and Nordhaus, 2011; Bennie et al., 2014). Recently, Ceola et al. (2014, 2015) identified from nightlights data a higher human concentration in the vicinity of the river network, as well as a clear correlation between nightlights and economic losses due to flood events, showing that the more illuminated areas presented higher flood damages.

Nightlights cover almost the entire world (180° W–180° E longitude, 75° N–65° S latitude) and they are available as raster products at a very detailed spatial resolution, i.e., 30 arcsec, corresponding to nearly 1 km at the equator. Six different satellites collected the nighttime light data set, as reported in Table 1.

2.2 River Network Data Set

Regarding the location of the river network, the JRC's Catchment Characterisation and Modelling (CCM, version 2.1) open database is used (http://ccm.jrc.ec.europa.eu/php/index.php?action=view&id=23). The CCM2 database covers the entire European continent, and it is based on a pan-European DEM mosaic with a 3 arcsec grid-cell resolution, updated land cover and climate data, and improved algorithms for data analysis.

2.3 Study Region Administrative Boundaries Data Set

In order to analyse the spatio-temporal evolution of human presence in the Eastern Alpine region, freely available vector files for Austria (http://www.naturalearthdata.com/downloads/10m-cultural-vectors/10m-admin-0-countries/) and three North-Eastern Italian Regions (i.e., Trentino Alto Adige, Veneto and Friuli Venezia Giulia, http://www.istat.it/it/archivio/24613) are employed.

2.4 Preparation of Repurposed Satellite Nightlight Data

In order to get a unique and representative nightlight value for each pixel and for each year, for those years presenting two different satellites operating simultaneously (i.e., 1994, 1997, 1998, 1999, 2000, 2001, 2002, 2003, 2004, 2005, 2006, 2007, see Table 1), a new nightlight product is first obtained as the average nocturnal luminosity from the two overlapped data sets.

Second, because raw nightlight data are not on-board calibrated and cannot be compared among the 22 year period, a preliminary intercalibration procedure is required. Therefore, a well-established empirically based intercalibration technique is employed (see e.g., Elvidge et al., 2009; Chen and Nordhaus, 2011). This technique uses F121999 as the reference satellite composite, because it is characterized by the highest DN_{obs} values that correspond exactly to the brightest areas of urban centres. The reference region for the intercalibration procedure is Sicily, since it presents a wide spread of DN_{obs} values with a more sharply defined diagonal clusters of points, but also it does not show significant changes in nightlights over time. Thus, by applying the intercalibration procedure, satellite nightlight composites from Sicily are compared among years to find suitable regression coefficients by using:

$$DN = C_0 + C_1 \cdot DN_{obs} + C_2 \cdot DN_{obs}^2, \quad (1)$$

where DN refers to intercalibrated nightlight values, DN_{obs} represents raw nightlight data and C_0, C_1 and C_2 are the empirical coefficients.

Third, all data associated to gas flares are excluded from the nightlight database, because they are deemed to be irrelevant for the analysis, and finally, a 22-year time series of nightlights for each of the considered study regions is extracted.

2.5 Preparation of Repurposed River Network Data and Identification of Distance Classes

To identify how the human presence is settled across streams and rivers, five distinct distance-from-rivers classes are defined. The original vectorial CCM river network is converted to a raster file and then pixels are classified by adopting the euclidean distance approach. Distance-0 pixels represent river network pixels, distance-1 pixels identify all pixels adjacent to streams and rivers, while distance-2, distance-3, and distance-4 pixels are defined from farther concentric zones (i.e., distance classes 1, 2, 3 and 4 correspond to 1, 2, 3 and 4 km far from the river network, respectively).

Table 1. Nighttime lights data set, satellite number and observation year.

	Satellite number					
	F10	F12	F14	F15	F16	F18
1992	F101992					
1993	F101993					
1994	F101994	F121994				
1995		F121995				
1996		F121996				
1997		F121997	F141997			
1998		F121998	F141998			
1999		F121999	F141999			
2000			F142000	F152000		
2001			F142001	F152001		
2002			F142002	F152002		
2003			F142003	F152003		
2004				F152004	F162004	
2005				F152005	F162005	
2006				F152006	F162006	
2007				F152007	F162007	
2008					F162008	
2009					F162009	
2010						F182010
2011						F182011
2012						F182012
2013						F182013

The performed analysis focuses on the following different groups of river pixels and associated distance-from-river pixels (see Fig. 1): CCM river pixels with Strahler order (i) ≥ 1 (i.e., all original river pixels are considered, where Strahler order $= 1$ represents small streams); (ii) ≥ 2; (iii) ≥ 3; (iv) ≥ 4 (Strahler order $= 4$ identifies major rivers). Given that the CCM river network spatial resolution is more detailed than the nightlight one (i.e., 3 arcsec versus 30 arcsec), we decided to consider all these different categories of river pixels in order to examine how the human signatures, as derived from the spatio-temporal nightlight trends, are influenced by the river network definition. As a consequence, when the original CCM river network is considered, river pixels correspond to nearly 40 % of the total number of pixels across Austria, Veneto and Friuli Venezia Giulia, while they are about 60 % for Trentino Alto Adige (see Fig. 1). In the case of CCM river pixels with Strahler order ≥ 2, ≥ 3 and ≥ 4, the associated river pixel percentage reduces to nearly 24 % (for Austria, Veneto and Friuli Venezia Giulia) and 33% (for Trentino Alto Adige), 13 % (for Austria, Veneto and Friuli Venezia Giulia) and 17 % (for Trentino Alto Adige) and, 7 % (for Austria, Trentino Alto Adige, Veneto and Friuli Venezia Giulia), respectively (see Fig. 1).

2.6 Temporal and Spatial Distribution Analysis of Repurposed Nightlights for the Study Regions

To account for the human presence as derived from nightlights, for each study region x (Austria, Trentino Alto Adige, Veneto and Friuli Venezia Giulia) and for each year t (from 1992 to 2013), we first analyse nocturnal light values gathered in correspondence of each of the five distance-from-river classes j (where j ranges from 0 to 4): for each pixel i laying on the selected distance class, a DN value for all considered years is obtained, labelled as $\mathrm{DN}_{i,j}(x,t)$.

In order to visualize tendencies in time, for each region x, for each year t and for each distance-from-river class j, we evaluate the empirical frequency distribution of $\mathrm{DN}_{i,j}(x,t)$ values. In addition, we estimate a characteristic DN value, $\mathrm{DN}_j(x,t)$, by averaging values measured in pixels located along each distance class j as:

$$\mathrm{DN}_j(x,t) = \frac{\sum_{i=1}^{n_j} DN_{i,j}(x,t)}{n_j(x)}, \tag{2}$$

where n_j represents the pixel abundance for distance class j. Thus, regional average values are analysed from both a spatial and a temporal perspective to detect the possible presence of trends in space and time.

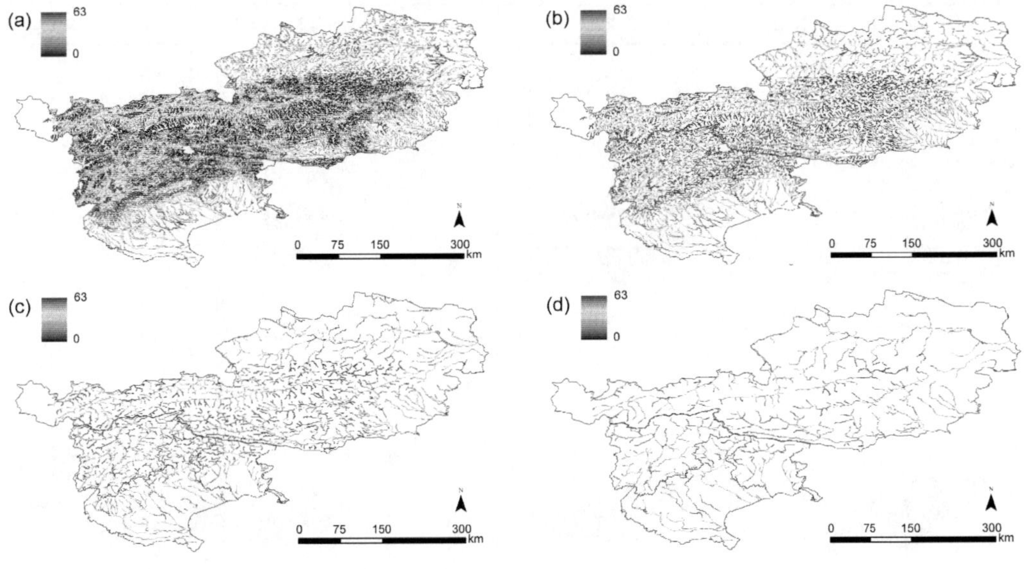

Figure 1. Repurposed nighttime lights (expressed as DN values) for year 2013 masked along the river network in the Eastern Alpine region (Austria and North Eastern Italian regions – Trentino Alto Adige, Veneto and Friuli Venezia Giulia): CCM river pixels with Strahler order **(a)** ≥ 1; **(b)** ≥ 2; **(c)** ≥ 3 and **(d)** ≥ 4 are considered.

3 Results and discussion

In order to identify human signatures across the European Eastern Alpine region, for each group of river pixels and associated distance-from-river pixels, we analysed repurposed nightlight data. Figure 1 shows, as an example, 2013 nightlights in correspondence of the river network pixels as defined from the aforementioned groups. As outlined in the legend, red pixels refer to areas with maximum DN values, which correspond to extensively urbanized areas (significant anthropogenic presence and associated economic activity), while blue pixels represent areas without nocturnal luminosity, likely characterized by the absence of human settlements and economic activities. Spatio-temporal trends of averaged nightlights $\mathrm{DN}_j(x, t)$ within each study region x for all the considered distance-from-river classes j (i.e., j ranges from 0 to 4) and years t (from 1992 to 2013) are then reported in Fig. 2.

From the temporal perspective, regardless of the group of river pixels identified by the Strahler order, a clear enhancement of nightlights in time is revealed for each study region and distance-from-river class, thus clearly proving that the human presence close to streams and rivers consistently increased in the Eastern Alpine region from 1992 to 2013.

From the spatial perspective, the nightlight trends across distance classes between the considered groups of river pixels reveal completely different behaviours. More specifically, river pixels with Strahler order ≥ 1 and ≥ 2 identify an extremely dense river network, which includes very small river reaches. Indeed, as shown in Fig. 1a, b, it seems that almost all pixels are streams or rivers. These two classes are dominated by upstream minor river reaches, which are likely not associated to human presence. Indeed, these pixels mainly lie in the natural environment of the Eastern Alps, not particularly influenced by anthropogenic activities. When moving far from the river network, average nightlights within these two classes tend to increase: this is valid for Austria, Veneto and Friuli Venezia Giulia, characterized by variable altitudes (including mountainous and plain areas), while it does not apply to Trentino Alto Adige, mainly constituted by mountains (Fig. 2, column 1 and 2).

Conversely, river pixels with Strahler order ≥ 3 and ≥ 4, which represent by definition only major streams and rivers, may provide a more reliable representation of human signatures. For these pixels, higher average nightlights are found for distance-0 pixels, while a decreasing luminosity trend is found in correspondence of higher distance-from-river classes (valid for all study regions, except for Strahler order ≥ 3 in Friuli Venezia Giulia, see Fig. 2, column 3 and 4). These results confirm the main outcomes of Ceola et al. (2014, 2015), where the HydroSHEDS river network was adopted.

Concerning the ranges of variability of average $\mathrm{DN}_j(x, t)$ values, when comparing the different river network groups the following interesting features have emerged. For all the considered study regions, distance-0 pixels present higher nocturnal luminosity values with increasing Strahler order river pixels. For Austria and Trentino Alto Adige, as the Strahler order increases, nightlights raise as well in all distance-from-river classes. In particular, while the lower DN values do not differ among the river network groups, there is an increment in the upper range DN value. Conversely, for

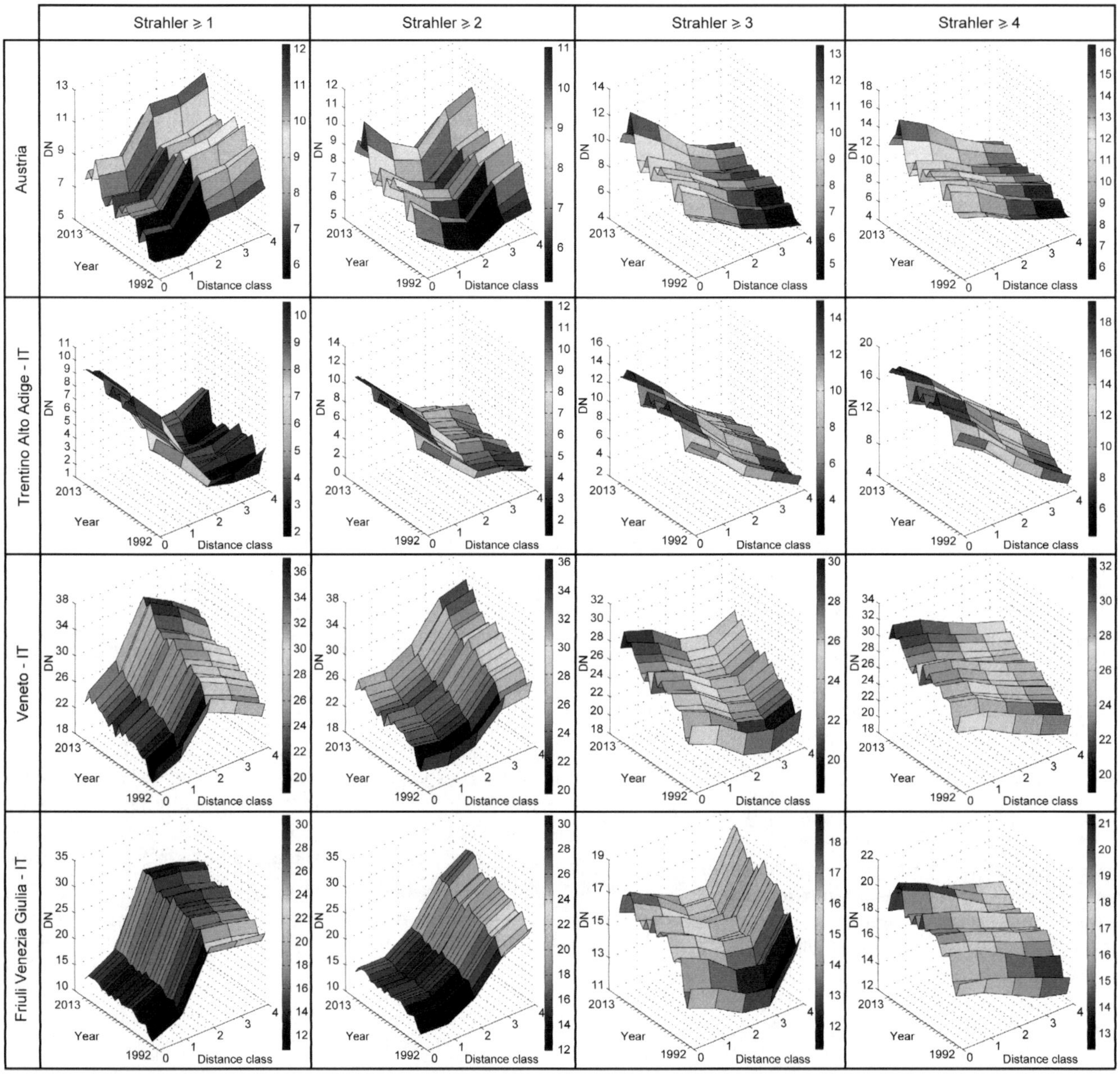

Figure 2. Spatial and temporal trend of yearly nightlight values in Austria (first row), Trentino Alto Adige (second row), Veneto (third row) and Friuli Venezia Giulia (forth row). Columns 1, 2, 3 and 4 refer to CCM river pixels with Strahler order ≥ 1, ≥ 2, ≥ 3 and ≥ 4, respectively. Distance classes 0, 1, 2, 3 and 4 correspond to: river pixels, 1, 2, 3 and 4 km far from the river network, respectively.

Veneto and Friuli Venezia Giulia, the maximum average luminosity evaluated for each distance class decreases with increasing Strahler order river pixels.

4 Conclusions

The spatially and temporally extensive analysis of human signatures from nightlights performed so far (i.e., by using detailed information on the river network location and nearby pixels) provide valuable insights into the evolution of human presence. Our results clearly point out an overall enhancement of human presence across the considered distance classes during the last 22 years, though presenting some differences among the study regions. From 1992 to 2013 average nightlights increased with respect to their initial value of about 30 % for Austria, 15 % for Trentino Alto Adige and 12 % for Veneto and Friuli Venezia Giulia. In addition, the river network delineation is found to play a central role in the identification of nightlight spatio-temporal trends, as highlighted by our results. Perspective work envisages to compare these outcomes with demographic data, as well as the European Settlement Map (i.e., a spatial raster data set that is mapping human settlements in Europe based on SPOT5 and SPOT6 satellite imagery, produced with Global Human Settlement Layer technology by the European Commission, Joint Research Centre, Institute for the Protection and Security of the Citizen, Global Security and Crisis Management Unit). Furthermore, linkages between nightlights and drought issues due to enhanced human water demands are also planned to be uncovered.

Acknowledgements. The research leading to these results has received funding from the European Union Seventh Framework Programme (FP7/2007-2013) under grant agreement no. 603587, project "SWITCH-ON" (Sharing Water-related Information to Tackle Changes in the Hydrosphere – for Operational Needs).

References

Bennie, J., Davies, T. W., Duffy, J. P., Inger, R., and Gaston, K. J.: Contrasting trends in light pollution across Europe based on satellite observed night time lights, Sci. Rep., 4, 3789, doi:10.1038/srep03789, 2014.

Ceola, S., Laio, F., and Montanari, A.: Satellite nighttime lights revealing increased human exposure to floods worldwide, Geophys. Res. Lett., 41, 7184–7190, doi:10.1002/2014GL061859, 2014.

Ceola, S., Laio, F., and Montanari, A.: Human-impacted waters: New perspectives from global high resolution monitoring, Water Resour. Res., 51, 7064–7079, doi:10.1002/2015WR017482, 2015.

Chand, T., Badarinath, K., Elvidge, C., and Tuttle, B.: Spatial characterization of electrical power consumption patterns over India using temporal DMSP-OLS night-time satellite data, Int. J. Remote Sens., 30, 647–661, doi:10.1080/01431160802345685, 2009.

Chen, X. and Nordhaus W.: Using luminosity data as a proxy for economic statistics, P. Natl. Acad. Sci. USA, 108, 8589–8594, doi:10.1073/pnas.1017031108, 2011.

Elvidge, C., Baugh, K., Kihn, E., Kroehl, H., and Davis, E.: Mapping city lights with nighttime data from the DMSP operational linescan system, Photogramm. Eng. Rem. S., 63, 727–734, 1997.

Elvidge, C., Sutton, P., Ghosh, T., Tuttle, B., Baugh, K., Bhaduri, B., and Bright, E.: A global poverty map derived from satellite data, Comput. Geosci., 35, 1652–1660, doi:10.1016/j.cageo.2009.01.009, 2009.

Kummu, M., de Moel, H., Ward, P., and Varis, O.: How close do we live to water? A global analysis of population distance to freshwater bodies, PLoS ONE, 6, e20578, doi:10.1371/journal.pone.0020578, 2011.

Small, C.: Global population distribution and urban land use in geophysical parameter space, Earth Interact., 8, 1–18, 2004.

Proc. IAHS, 373, 137–141, 2016
proc-iahs.net/373/137/2016/
doi:10.5194/piahs-373-137-2016

Hydrological and hydraulic models for determination of flood-prone and flood inundation areas

Hafzullah Aksoy, Veysel Sadan Ozgur Kirca, Halil Ibrahim Burgan, and Dorukhan Kellecioglu

Department of Civil Engineering, Hydraulics Division, Istanbul Technical University, Istanbul, 34469, Turkey

Correspondence to: Hafzullah Aksoy (haksoy@itu.edu.tr)

Published: 12 May 2016

Abstract. Geographic Information Systems (GIS) are widely used in most studies on water resources. Especially, when the topography and geomorphology of study area are considered, GIS can ease the work load. Detailed data should be used in this kind of studies. Because of, either the complication of the models or the requirement of highly detailed data, model outputs can be obtained fast only with a good optimization. The aim in this study, firstly, is to determine flood-prone areas in a watershed by using a hydrological model considering two wetness indexes; the topographical wetness index, and the SAGA (System for Automated Geoscientific Analyses) wetness index. The wetness indexes were obtained in the Quantum GIS (QGIS) software by using the Digital Elevation Model of the study area. Flood-prone areas are determined by considering the wetness index maps of the watershed. As the second stage of this study, a hydraulic model, HEC-RAS, was executed to determine flood inundation areas under different return period-flood events. River network cross-sections required for this study were derived from highly detailed digital elevation models by QGIS. Also river hydraulic parameters were used in the hydraulic model. Modelling technology used in this study is made of freely available open source softwares. Based on case studies performed on watersheds in Turkey, it is concluded that results of such studies can be used for taking precaution measures against life and monetary losses due to floods in urban areas particularly.

1 Introduction

Floods are among the most common natural hazards that cause several damages to the properties, injuries and losses of lives. It occurs in different regions all over the World extending to more than 1/3 of the area and 82 % of population (Dilley, 2005). Urbanization and continuous industrial activities trigger the flood and increase the flood damage in flood plains of hydrological watersheds. It is therefore very important to delineate flood-prone areas well in advance to be able to take preventive measures to minimize any damage the flood may cause.

Floods are studied by hydrological, hydraulic and topographical inputs to be analyzed at both spatial and temporal scales. For this aim, numerical models have been developed to calculate flood discharge due to precipitation of a given return period. With the help of the recent modelling technology, flood forecasting becomes more accurate and requires less time than before. Geographical Information Systems (GIS) contributes as an important and very useful tool in this type of hydrological and hydraulic models.

In this study, a regional-scale hydrological model was developed first for the delineation of the flood-prone areas in hydrological watersheds. Two topographical indexes were used in the hydrological model. Once flood prone areas are determined, a hydraulic model was implemented at local-scale, for calculating the flow depth and velocity in risky areas along the river for which the freely available HEC-RAS hydraulic model was used. Case study from a watershed in Northern part of Turkey shows the applicability of the models in delineating the flood-prone areas and the flow depth in risky parts of the river in terms of flood risk maps for given return periods of flood discharges.

2 Method

The proposed approach to assess flood hazard has a two-step process, the first step involves the detection and mapping of flood-prone areas by using GIS support and considering the topography of the watershed after which areas at risk are studied with more details by implementing the HEC-RAS hydraulic model to assess flooding parameters (Aksoy et al., 2014). This step is basically a flood susceptibility assessment process and is used to prioritize the flood-prone areas in terms of the importance of the assets at risk. The second step is in fact the flood hazard assessment stage, leading to results which can support decision makers regarding the selection of preventive measures. In the literature, no study exists dealing with the problem as in the way used here other than a case study performed very recently by Samela et al. (2016) for delineation of flood-prone areas in ungauged basins.

2.1 Regional scale hydrological modelling by wetness indices

In the first step of regional-scale analysis that accommodates the hydrological model, flood-prone areas in the watershed are identified. For this aim, two softwares, Quantum GIS (QGIS) and System for Automated Geoscientific Analyses GIS (SAGA GIS), fully supported by GIS were used. Both GIS softwares are freely available and user-friendly. The methods to map flood-prone areas at regional scale are based on the Topographic Wetness Index (TWI) approach and its variant, the SAGA WI, respectively.

TWI used in QGIS was suggested by Beven and Kirkby (1979) as

$$\mathrm{TWI} = \ln\left(\frac{A_s}{\tan\beta}\right) \tag{1}$$

where A_s is the (upslope) flow accumulation area (or drainage area) per unit contour length (Wilson and Gallant, 2000) and β is the angle of the slope. TWI has found a wide range of application in hydrology (Moore et al., 1991; Quinn et al., 1995; Sørenson et al., 2006). Although TWI assumes that the soil in the watershed is homogeneous and isotropic as a constraint, it was found that topographical changes become much more important, therefore, the difficulty in this assumption becomes negligible compared to the change in the topography of the watershed. This allows one to use TWI in the hydrological analysis in this study.

The SAGA WI used in SAGA GIS is based on a modified catchment area calculation which does not consider the flow as a very thin film. As a result, it predicts, for cells situated in valley floors with a small vertical distance to a channel, a more realistic and with higher potential soil moisture as compared to TWI. This fact is translated to a wider area calculated as potentially covered with water during a flood event.

The TWI can be effectively used to reveal the flooding susceptibility by mapping the flood prone areas. The procedure to calculate the TWI and the SAGA WI in order to define the susceptibility to flooding (the flood-prone areas) is very straight forward since the selected open source software Quantum GIS (QGIS) incorporates the respective routines built into the SAGA GIS.

2.2 Local scale hydraulic modelling in HEC-RAS

The hydraulic analysis is performed to determine hydraulic characteristics of water flow, such as flow depth, flow velocity, forces due to water flowing on a surface or at a hydraulic structure, etc. The aim of this part in this study is to determine the characteristics of flow and potential flooding in streams and areas that have suffered flooding in the past taking into account the precipitation and surface runoff draining from the basin.

Hydraulic modelling and analysis have shown extensive research effort to simulate flood propagation in rivers. This is expressed by numerous 1- and 2-dimensional simulation models including HEC-RAS, Mike11, FLO-2D etc. The outcomes of these models, in riverine flood hazard assessment, include the level of inundation (flood water level), the intersection of the flood level with the terrain (ground surface) to create the flood plain extent, the difference between the flood level and the terrain, which is used to calculate the depth and the flow velocity.

Hydraulic analysis for assessing hydraulic behavior for the needs of this study was implemented with the HEC-RAS software (USACE, 2010) that can be downloaded from the official Hydrologic Engineering Center, River Analysis System (HEC-RAS) web site freely.

3 Study area and watershed

As mentioned previously, the pilot implementation of flood hazard assessment includes two phases: a regional implementation and a local implementation. The regional implementation was done in Ikizce watershed located in Samsun and Ordu provinces, the Middle Black Sea region in northern part of Turkey (Fig. 1). The local implementation for the flood hazard assessment was done for the Akçay Creek in the Ikizce watershed. Akçay Creek is located within the Terme district of Samsun, which is 56 km to the East of the Samsun city center. All rivers including the Akçay Creek are known to be critical in terms of flash floods, causing losses, even casualties in the watershed.

Samsun province exhibits the Northwestern climate of Turkey, which covers the coastal Black Sea area including the Northern faces of the mountains and Northeastern part of Marmara region. The mean annual precipitation in Samsun is 683.2 mm. Maximum precipitation is seen in October, November and December with the values of 88.8, 82.5 and 72.9 mm, respectively. It may be of particular interest

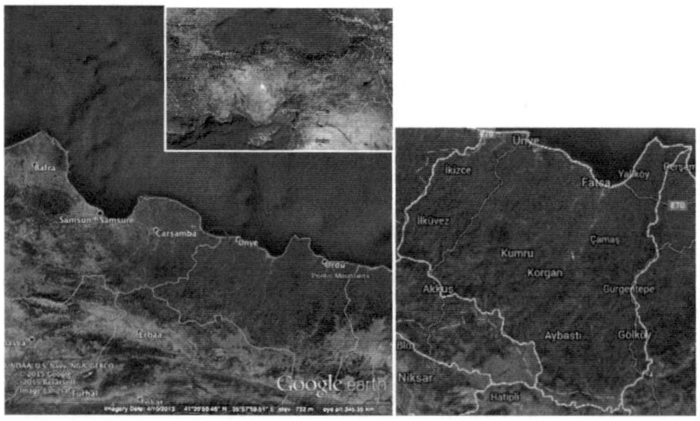

Figure 1. Geographic location and terrain view of Samsun (courtesy of GoogleEarth®) and Ikizce watershed.

to mention that the maximum daily precipitation recorded in Samsun is 113.2 mm during 32 years of measurements. The temperature difference between summer and winter is not sparse. Summers are relatively cool whereas winters are relatively warmer in coastal lowlands and colder in higher ground with precipitation of snow. The mean annual temperature is 14.2 °C, with the mean monthly summer temperatures of 23.3, 23.2 and 20.0 °C in August, July and June, respectively. In the winter, the mean monthly values drop down to 6.6, 7.0 and 7.8 °C in February, January and March, respectively. Natural plant cover is humid and wide-leaved forest, turning to coniferous trees in higher lands. In the coastal areas along the river banks, there is an intense population of reeds. In the coastal region of Samsun (including Terme) the major type of encountered soil is clayey sand. The mean infiltration capacity of this soil type is measured to be 9 mm h^{-1}, a quite low value.

Akçay Creek discharges directly to Black Sea (Fig. 1). The downstream region of the creek is generally covered by agricultural area, as well as industrial facilities. There is an LNG power plant, an ancient bridge and a primary school building in the close vicinity of this zone. The major intercity highway which connects Samsun to Ordu in the East passes across the Akçay Creek with a new bridge (Fig. 2).

Figure 2. A detailed view of Akçay Creek in Ikizce watershed.

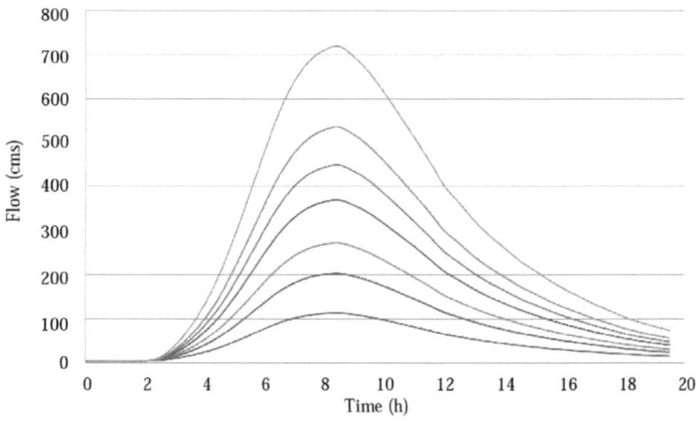

Figure 3. Flood hydrograph in Akcay Creek for 2, 5, 10, 25, 50, 100 and 500-year return periods.

4 Data

The data for the regional-scale implementation were compiled from the maps. The Ikizce watershed was delineated from the website of Ministry of Forestry and Water Affairs of Turkey http://geodata.ormansu.gov.tr/. For the hydrological model, 1/25 000 scale maps were used as the topographical data obtained from Turkish Army General Command of Mapping.

For the local-scale model, cross-sections of the river were produced from 1/500 scale digital maps. For using in the local-scale implementation, the 2, 5, 10, 25, 50, 100 and 500-year return period flood hydrographs of the Akçay Creek were obtained from State Hydraulic Works (DSI) of Turkey as given in Fig. 3. Peak discharges of the flood hydrographs are tabulated in Table 1. A steady-state flow with the flood discharge is assumed in the hydraulic analysis. Cross sections at every 20 m longitudinal distance were used to define the flow geometry in HEC-RAS. These cross-sections were extracted from the DEM with a grid resolution of 2 m.

Data availability

The local-scale maps were made available for implementation of this particular study but not to share with a third party or publicly.

5 Implementation

5.1 Regional-scale implementation

For the regional scale, the previously explained methodology was used which was implemented in the QGIS and SAGA GIS. For this purpose, the following procedure is applied:

Table 1. The flood hydrograph of Akçay Creek.

Q_2	Q_5	Q_{10}	Q_{25}	Q_{50}	Q_{100}	Q_{500}
			($m^3\,s^{-1}$)			
109.57	199.02	267.87	365.30	445.09	531.01	713.10

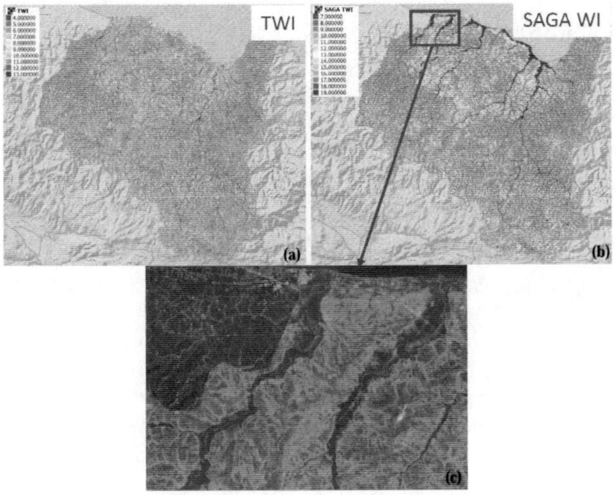

Figure 4. Regional-scale analysis of Ikizce watershed, **(a)** TWI, **(b)** SAGA WI; **(c)** a close-up view of downstream area of Akçay Creek.

- Using the sub-basin templates in http://geodata.ormansu.gov.tr, the DEM files are trimmed into sub-basins, such that each file covers a single sub-basin.
- Once the sub-basin DEMs are generated, they are processed with the integrated computational steps using QGIS interface.
- The aspects, unit catchment area, TWI and SAGA WI are calculated.
- These indexes are converted into maps which show the flood-prone areas as the result of regional-scale flood hazard assessment.

Although there is not a strict universal criterion to differentiate the flood-prone areas solely on the isolated values of the aforementioned indexes, the distribution of the indexes across the river basins give very crucial qualitative information about the flood susceptibility of any location of interest within the basin. Figure 4a and b shows the TWI and SAGA WI for the Akçay Creek in Ikizce basin. In Fig. 4c, a close-up view of the SAGA WI is shown.

5.2 Local-scale implementation

The area chosen for the local scale flood hazard assessment is the Akçay Creek in the Akçay village, which is located in the border of Samsun and Ordu provinces. In the output of the regional scale flood hazard assessment model, the downstream area of Akçay Creek came out to be a critical flood-prone zone based on wetness indexes in Fig. 4a and b. The close-up view of this location in Fig. 4c shows that the downstream section of the river is the most susceptible zone in the watershed against flooding. Therefore, it was decided to model this region at local scale by means of the 1-D hydraulic modelling using the HEC-RAS software.

As stated above, for the hydraulic analysis, cross-sections of the river were produced from 1/500 scale-digital maps at every 20 m along the river. The HEC-RAS model was run for the 2-, 5-, 10-, 25-, 50-, and 100-year return period flood hydrographs under steady-state subcritical flow conditions.

Results are given in Fig. 5 for floods with the above mentioned return periods. As can be seen, when the peak discharge exceeds the 10 year-return period extreme value, the creek tends to overflow from its bed. This draws a parallel picture with what happens in reality. Especially at the last 300 m of the creek (i.e. downstream of the highway bridge, see Fig. 2) the inundation area is considerably high. In the modelling, it was seen that the highway bridge is flooded with return periods higher than 100 years. A flood recorded in July 2012 was found to be roughly the 85 year-return period flood in the watershed (Onoz et al., 2012). If the storm surge that rises the sea level at the downstream (say, at the order of 0.5–1 m) is taken into account, the observation in July 2012 comes to an agreement with the model results.

6 Conclusions

Following conclusions can be drawn from the regional scale and local scale flood hazard assessment modelling studies implemented:

- The proposed regional scale methodology is simple, easy and inexpensive (free software and minimum amount of data requirement); yet it is very effective in terms of pinpointing the flood-prone locations in hydrological watersheds.
- 1/25 000 scale maps are very convenient for the application of the regional model to yield high resolution results. However, experience shows that even the ASTER DEM maps (freely downloadable from http://www.jspacesystems.or.jp/ersdac/GDEM/E/index.html) can yield satisfactory results on revealing flood-prone areas.
- An additional point, based on several case studies performed by Ermiş (2015) (not covered in this study), is that the regional scale flood hazard assessment model can be operated both with small basins (in size of a few hundred km^2) and relatively large basins (in size of a few thousand km^2). In any case, operating with smaller

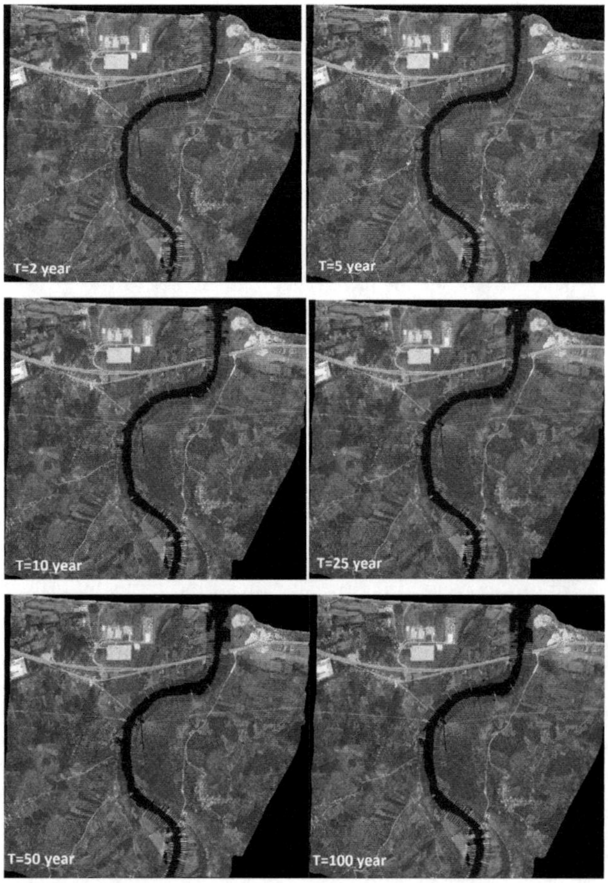

Figure 5. Results of local scale hydraulic model for different return period floods.

basins may be more practical since the precision of the results would be presumably higher.

- The local scale flood hazard assessment model (the 1-D hydraulic model) implemented by use of HEC-RAS yielded precise results for the probable flood inundation areas for different return periods of flood occurrences.

- As a specific result of the case study here, it was seen that the Akçay Creek overflows its bed for return periods larger than 10 years and the events tend to become more catastrophic for extreme events larger than 100 year-return period, since the highway bridge is inundated leading to further flooding upstream.

It should additionally be noted that any possible sea level rise in Black Sea (i.e. due to storm surge, wind or wave setup) was not accounted for in the flood modelling. Such rises in the sea level are likely to increase the flooded areas due to the downstream-controlled flow regime in the river.

Acknowledgements. This study is based on the research project "A Scientific Network for Earthquake, Landslide and Flood Hazard Prevention – SciNetNatHazPrev" (no. TR11C1.01-02/309) which is implemented under the Black Sea Basin Joint Operational Program, co-funded by the European Union and national funds of the participating countries.

References

Aksoy, H., Kirca, V. S. O., and Papatheodorou, K.: Flood hazard assessment and modelling practices in Turkey, Proceedings of the Mediterranean Meeting on "Monitoring, modelling and early warning of extreme events triggered by heavy rainfalls", University of Calabria, Cosenza, Italy, 26–28 June 2014, 205–214, 2014.

Beven, K. J. and Kirkby, M. J.: A physically-based variable contributing area model of basin hydrology, Hydrol. Sci. Bull., 24, 43–69, 1979.

Dilley, M.: Natural disaster hotspots: A global risk analysis, Vol. 5, World Bank Publications, Washington, DC, 2005.

Ermiş, I. S.: Determination of flood-prone areas with topographic wetness indices in the river basins, Istanbul Technical University, Institute of Science and Technology, Istanbul, June 2015, 65 pp., 2015 (in Turkish).

Moore, I. D., Grayson, R. B., and Ladson, A. R.: Digital terrain modeling: A review of hydrological, geomorphological and biological applications, Hydrol. Process., 5, 3–30, 1991.

Onoz, B., Kabdasli, S, Varol, O.E., Yuce, A., and Bora, O.: Report on assessment of flood characteristics and risks due to Akçay River for OMV Gas & Power, Technical Report by ITU and CEC, NO. C-1209047, 2012.

Quinn, P., Beven, K., and Lamb, R.: The ln(αs/tanβ): how to calculate it and how to use it within the TOPMODEL framework, Hydrol. Process., 9, 161–182, 1995.

Samela, C., Manfreda, S., Paola, F. D., Giugni, M., Sole, A., and Fiorentino, M.: DEM-based approaches for the delineation of flood-prone areas in an ungauged basin in Africa, J. Hydrol. Eng., 21, 06015010, doi:10.1061/(ASCE)HE.1943-5584.0001272, 2016.

Sørensen, R., Zinko, U., and Seibert, J.: On the calculation of the topographic wetness index: evaluation of different methods based on field observations, Hydrol. Earth Syst. Sci., 10, 101–112, doi:10.5194/hess-10-101-2006, 2006.

USACE: HEC-RAS Users Manual for version 4.1, CPD-68, Davis, CA (downloadable from http://www.hec.usace.army.mil/software/hec-ras/documentation/HEC-RAS_4.1_Users_Manual.pdf), 2010.

Wilson, J. P. and Gallant, J. C.: Digital terrain analysis. Terrain analysis: Principles and applications, 1–27, 2000.

Proc. IAHS, 373, 143–146, 2016
proc-iahs.net/373/143/2016/
doi:10.5194/piahs-373-143-2016

Open Access
PIAHS

Vehicles instability criteria for flood risk assessment of a street network

Chiara Arrighi[1], Nicolas Huybrechts[2,3], Abdellatif Ouahsine[2], Patrick Chassé[3], Hocine Oumeraci[4], and Fabio Castelli[1]

[1]Department of Civil and Environmental Engineering, University of Florence, Florence, Italy
[2]Laboratoire Roberval, Sorbonne Universités, Université de Technologie de Compiègne, CNRS, Centre de Recherches de Royallieu, Compiègne Cedex, France
[3]CEREMA-DTecEMF, Margny Lès Compiègne, France
[4]TU Brauschweig, Leicthweiss Institute for Hydraulic Engineering and Water Resources, Braunschweig, Germany

Correspondence to: Chiara Arrighi (chiara.arrighi@dicea.unifi.it)

Published: 12 May 2016

Abstract. The mutual interaction between floods and human activity is a process, which has been evolving over history and has shaped flood risk pathways. In developed countries, many events have illustrated that the majority of the fatalities during a flood occurs in a vehicle, which is considered as a safe shelter but it may turn into a trap for several combinations of water depth and velocity. Thus, driving a car in floodwaters is recognized as the most crucial aggravating factor for people safety. On the other hand, the entrainment of vehicles may locally cause obstructions to the flow and induce the collapse of infrastructures. Flood risk to vehicles can be defined as the combination of the probability of a vehicle of being swept away (i.e. the hazard) and the actual traffic/parking density, i.e. the vulnerability. Hazard for vehicles can be assessed through the spatial identification and mapping of the critical conditions for vehicles incipient motion. This analysis requires a flood map with information on water depth and velocity and consistent instability criteria accounting for flood and vehicles characteristics. Vulnerability is evaluated thanks to the road network and traffic data. Therefore, vehicles flood risk mapping can support people's education and management practices in order to reduce the casualties. In this work, a flood hazard classification for vehicles is introduced and an application to a real case study is presented and discussed.

1 Introduction

According to global statistics, in the last decade floods affected the largest number of people with respect to other natural hazards (EM-DAT, 2012). In developed countries the majority of fatalities occurs in vehicles (Jonkman and Kelman, 2005; Maples and Tiefenbacher, 2009; Fitzgerald et al., 2010; Kellar and Schmidlin, 2012). On one hand, vehicles can become unstable by losing traction even for very low water depths, or they become buoyant, so that they are swept away by floodwaters and possibly produce a debris. On the other hand, several fatalities occur for inappropriate high-risk behaviours, like driving in flooded streets. These are the reasons why vehicles are usually recognized as one of the most aggravating factors in urban floodings (Rodriguez et al., 2006; Franklin et al., 2014). Although the problem of vehicle mobilization is so crucial, only a few experimental investigation on the incipient motion conditions of small-scale vehicle models have been carried out so far (Shu et al., 2011, Xia et al., 2011, 2014). The experimental data on vehicles instability, expressed as pairs of water depth and velocity, are affected by a large scatter, which depends on the different characteristics (i.e. weight, shape, height of the planform) of the considered vehicle model.

Moreover, the different contribution of the hydrodynamic forces and the influence of flow regime on the mechanism of the onset of motion has been numerically studied by Arrighi et al. (2015).

The aim of the work is to introduce hazard criteria for vehicles at rest in flooded streets easily applicable to inundation maps, which account for both flood and vehicles characteristics, in order to overcome the scatter of existing experimental data.

2 Dimensionless instability diagram for flooded vehicles

Incipient motion for sliding occurs when the drag force exerted by the water just exceeds the friction force, which is the product of the friction coefficient μ and immersed weight of the car (Eq. 1). The onset of motion by sliding of a vehicle is modeled using the geometric scheme shown in Fig. 1a. The flow direction has an orientation β with respect to the longitudinal axis of the car and the bed slope is assumed null, the wheels are considered as locked.

$$\frac{1}{2}\rho C_{\mathrm{D}} U^2 A_{\mathrm{D}} > \mu\Big[\rho_{\mathrm{c}} g (H_{\mathrm{V}} - h_{\mathrm{c}}) l L - \rho g (H - h_{\mathrm{c}}) l L - \frac{1}{2}\rho C_{\mathrm{L}} U^2 A_{\mathrm{L}}\Big] \quad (1)$$

where ρ is the water density, ρ_{c} is the car density, U is the mean flow velocity, H is the water depth, H_{V} is the height of the vehicle, h_{c} is the height of the planform, g is the acceleration of gravity, l and L are the car width and length and C_{D} and C_{L} are the drag and lift coefficients (official IS of units). The reference areas for drag (A_{D}) and lift (A_{L}) forces are the projection of the full area of the car normal to the flow (Eq. 2) and the planform (Eq. 3) which are respectively

$$A_{\mathrm{D}} = (l \cdot \cos\beta + L \cdot \sin\beta) \cdot (H_{\mathrm{V}} - h_{\mathrm{c}}) \quad (2)$$

$$A_{\mathrm{L}} = L \cdot l \quad (3)$$

The manipulation of Eq. (1) here modified to account for any given flow orientation after Arrighi et al. (2015) yields a relation between the square of Froude number $\frac{U^2}{gH}$ and a dimensionless group of variables which is called mobility parameter θ_{V}

$$\theta_{\mathrm{V}} = \frac{2L}{(H_{\mathrm{V}} - h_{\mathrm{c}})} \cdot \alpha \cdot \left(\frac{\rho_{\mathrm{c}} \cdot (H_{\mathrm{V}} - h_{\mathrm{c}})}{\rho \cdot (H - h_{\mathrm{c}})} - 1\right) \quad (4)$$

$$\alpha = \frac{l}{l \cdot \cos\beta + L \cdot \sin\beta} \quad (5)$$

where the first term is a shape factor, α accounts for the angle of flow incidence and the latter accounts for the immersed weight of the car.

The mobility parameter has been calculated for three experimental datasets (Shu et al., 2011, Xia et al., 2011, 2014) including seven different car models and densities and three model scales (1 : 14, 1 : 18, 1 : 43). The values of the calculated mobility parameter θ_{V} plotted against Froude number identify a unique threshold of incipient motion (Fig. 1c),

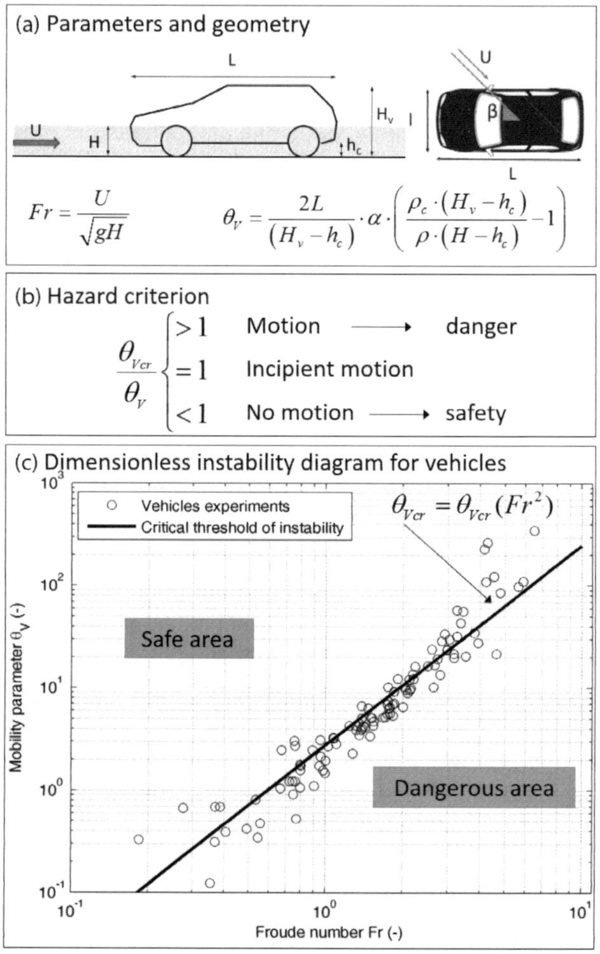

Figure 1. Sketch of the car geometry and definition of the mobility parameter **(a)**, definition of hazard criterion **(b)** and dimensionless instability diagram for flooded vehicles with identification of the critical threshold **(c)** (experimental data by Shu et al., 2011; Xia et al., 2011, 2014).

which separates safe conditions (above the curve) from dangerous conditions (below the curve) where the vehicle mobilization is likely to occur. Thus, a hazard criterion can be defined as the ratio between the critical threshold $\theta_{\mathrm{V_{cr}}}$ and the mobility parameter θ_{V} for a predefined reference vehicle and flow regime (i.e. water depth and velocity of an inundation map) as shown in Fig. 1b.

3 The case study of Ajaccio (France)

The study area considered to show the application of the vehicles hazard criterion derived from the mobility parameter, is the Les Cannes district in Ajaccio, situated in south-western Corse (France). The flood simulation has been carried out with TELEMAC-2D (Galland et al., 1991) solving the shallow water equations over an unstructured grid of triangular elements. The results have been provided by CEREMA (Centre d'études et d'Expertise sur les Risques, l'Environment, la Mobilité et l'Aménagement) and are described in detail in a project report for the Direction Départementale de l'Equipment Corse du Sud (Sogreah, 2006).

The Cannes district is located downstream of three small catchments (Arbitrone, Moulins Blancs and Arbajola), whose streams are culverted in the urbanized area up to the gulf (Fig. 2, top). The cadastral map 1 : 5000 has been used to extract the built footprint and a detailed topographic survey has been carried out to measure the terrain elevation (max 30 m a.s.l.). The mesh is composed of 10 511 nodes and 18 486 elements, with edge size ranging from 0.5 to 25 m in the sea.

Although the area is not densely urbanized, it hosts two educational facilities and some commercial activities and it is connected to the port area through an important traffic artery along the coastline, the Cours Jean Nicoli.

Benchmark inundation events date back to 23 January 2003 and 25–26 November 1990 and allowed the validation of the numerical model (Sogreah, 2006).

The maximum water depth reached in the Cannes district for the simulated 100 years scenario is about 2 m (Fig. 2, top) and the maximum velocity is above $3.0\,\mathrm{m\,s^{-1}}$. The dark blue area at the bottom of the top panel in Fig. 2 represents the sea level used as a southern boundary condition. The hazard level is calculated for a Ford Focus and $\beta = 90°$ (i.e. a medium size passenger vehicle oriented normally to the flow) combining in each computational cell, for each time step, the variables obtained during the simulation (i.e. water depth and velocity). Figure 2 (bottom) shows the extraction of the maximum value of the vehicle hazard over the simulation time. The colour scale goes from 0.25 (green colour, very low hazard) up to 1.5 (dark violet, very high hazard). Red areas are those likely to be prone to incipient motion of vehicles (high hazard) and violet areas can be considered as extremely dangerous for parked vehicles since the hazard level is higher than one. Moreover, since many open areas are used as car parking for the commercial areas, it is likely that vehicles can be swept away by floodwaters for a similar scenario.

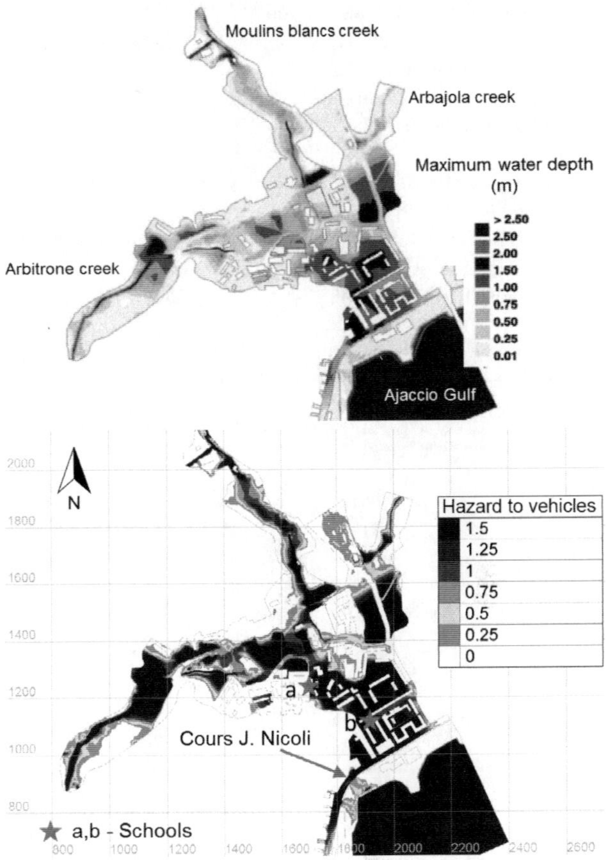

Figure 2. Flood depth map for the simulated 100 years scenario (top) and temporal maximum of the hazard for vehicles with identification of vulnerable buildings and main traffic artery (bottom).

during several time slots and represents a vulnerability factor. In order to prevent people's fatalities, hazard maps for vehicles can be adopted to implement non-structural flood risk mitigation measures (i.e. people education, traffic management) or emergency actions (roads closure).

4 Conclusions

The mobility parameter allows defining hazard criteria capable of accounting for both flood and vehicles characteristics. The hazard map here presented can be drawn for different probability scenarios and can be used to assess the risk for vehicles if the parking/traffic density is accounted for. From this preliminary test, the main traffic artery shows very high hazard levels for the simulated 100 years flood and can be significantly affected in case of event.

The available visual documentation of the 2003 flood event demonstrate that vehicles mobilization occurred in the area. The presence of the schools is a further traffic-attractor

References

Arrighi, C., Alcèrreca-Huerta, J. C., Oumeraci, H., and Castelli, F.: Drag and lift contribution to the incipient motion of partly submerged flooded vehicles, J. Fluids Structures, 57, 170–184, 2015.

EM-DAT: Disasters in numbers (October 2000), available at: http://www.emdat.be/publications, 2012.

Fitzgerald, G., Du, W., Jamal, A., Clark, M., and Hou, X. Y.: Flood fatalities in contemporary Australia (1997–2008), Em. Med. Australasia, 22, 180–186, 2010.

Franklin, R. C. King, J. C., Aitken, P. J., and Leggat, P. A.: "Washed away"-assessing community perceptions of flooding and prevention strategies: A North Queensland example, Nat. Haz., 73, 1977–1998, 2014.

Galland, J. C., Goutal, N., and Hervouet, J. M.: TELEMAC: A new numerical model for solving shallow water equations, Adv. Water Res., 14, 138–148, 1991.

Jonkman, S. N. and Kelman, I.: An analysis of the causes and circumstances of flood disaster deaths, Disasters, 29, 75–97, 2005.

Kellar, D. M. M. and Schmidlin, T. W.: Vehicle-related flood deaths in the United States, 1995–2005, J. Flood Risk Manage., 5, 153–163, 2012.

Maples, L. and Tiefenbacher, J.: Landscape, development, technology and drivers: The geography of drownings associated with automobiles in Texas floods, 1950–2004, Appl. Geogr., 29, 224–234, 2009.

Rodriguez, H., Quarantelli, E. L., and Dynes, R. R.: Handbook of Disaster Research, Springer Science, 2006.

Sogreah: Direction Départementale de l'Equipment Corse du Sud. Evaluation du risqué pluvial sur la commune d'Ajaccio-Etude hydraulique dans les bassins versant d'Arbitronne, San Remedio at la Madonnuccia, No. 2 74 0190, 2006.

Shu, C., Xia, J., Falconer, R., and Lin, B.: Estimation of incipient velocity for partially submerged vehicles in floodwaters,J. Hydrol. Res., 49, 709–717, 2011.

Xia, J., Teo, F. Y., Lin, B., and Falconer, R.: Formula of incipient velocity for flooded vehicles, Nat. Haz., 58, 1–14, 2011.

Xia, J., Falconer, R. A., Xiao, X., and Wang, Y.: Criterion of vehicle stability in floodwaters based on theoretical and experimental studies, Nat. Haz, 70, 1619–1630, 2014.

Proc. IAHS, 373, 147–151, 2016
proc-iahs.net/373/147/2016/
doi:10.5194/piahs-373-147-2016

Improving predictions of the effects of extreme events, land use, and climate change on the hydrology of watersheds in the Philippines

Rubianca Benavidez[1], **Bethanna Jackson**[1], **Deborah Maxwell**[1], **and Enrico Paringit**[2]

[1]School of Geography, Environment and Earth Sciences, Victoria University of Wellington, Wellington, 6012, New Zealand
[2]Disaster Risk and Exposure Assessment for Mitigation Program, Quezon City, 1101, Philippines

Correspondence to: Rubianca Benavidez (rubianca.benavidez@vuw.ac.nz)

Published: 12 May 2016

Abstract. Due to its location within the typhoon belt, the Philippines is vulnerable to tropical cyclones that can cause destructive floods. Climate change is likely to exacerbate these risks through increases in tropical cyclone frequency and intensity. To protect populations and infrastructure, disaster risk management in the Philippines focuses on real-time flood forecasting and structural measures such as dikes and retaining walls. Real-time flood forecasting in the Philippines mostly utilises two models from the Hydrologic Engineering Center (HEC): the Hydrologic Modeling System (HMS) for watershed modelling, and the River Analysis System (RAS) for inundation modelling. This research focuses on using non-structural measures for flood mitigation, such as changing land use management or watershed rehabilitation. This is being done by parameterising and applying the Land Utilisation and Capability Indicator (LUCI) model to the Cagayan de Oro watershed (1400 km^2) in southern Philippines. The LUCI model is capable of identifying areas providing ecosystem services such as flood mitigation and agricultural productivity, and analysing trade-offs between services. It can also assess whether management interventions could enhance or degrade ecosystem services at fine spatial scales. The LUCI model was used to identify areas within the watershed that are providing flood mitigating services and areas that would benefit from management interventions. For the preliminary comparison, LUCI and HEC-HMS were run under the same scenario: baseline land use and the extreme rainfall event of Typhoon Bopha. The hydrographs from both models were then input to HEC-RAS to produce inundation maps. The novelty of this research is two-fold: (1) this type of ecosystem service modelling has not been carried out in the Cagayan de Oro watershed; and (2) this is the first application of the LUCI model in the Philippines. Since this research is still ongoing, the results presented in this paper are preliminary. As the land use and soil parameterisation for this watershed are refined and more scenarios are run through the model, more robust comparisons can be made between the hydrographs produced by LUCI and HEC-HMS and how those differences affect the inundation map produced by HEC-RAS.

1 Introduction

The Philippines is regularly struck by tropical cyclones. An annual average of twenty enter the region, of which about nine make landfall and often cause destructive floods (Lasco et al., 2009). In 2011, Typhoon Washi caused heavy flooding and destruction in the Cagayan de Oro region, leaving 1268 casualties and PHP 2 billion in damages ($\sim$ USD 46 million in 2011) (National Disaster Risk Reduction Management Council, 2012a). Since the Philippines is located within the typhoon belt, the country is vulnerable to the effects of climate change due to possible increases in typhoon frequency and intensity (Intergovernmental Panel on Climate Change, 2012). The vulnerability of the country to tropical cyclones and climate change underscores the need for more proactive disaster risk management, which to date is mainly carried out through flood forecasting and structural flood protection projects.

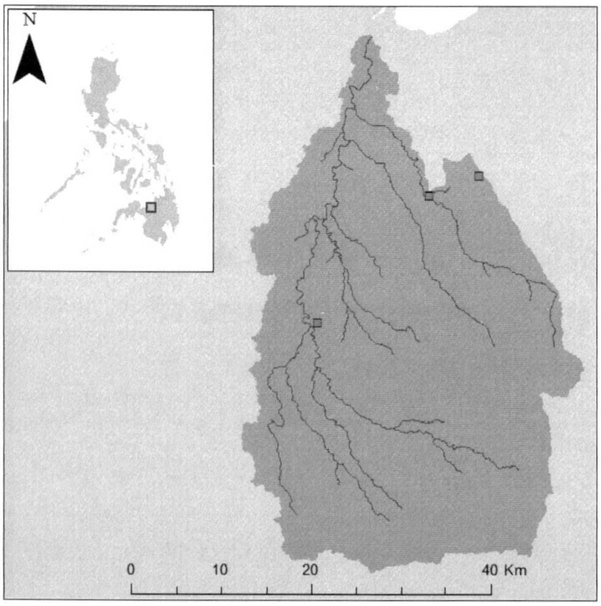

Figure 1. The Cagayan de Oro watershed and rain gauges relative to the Philippines.

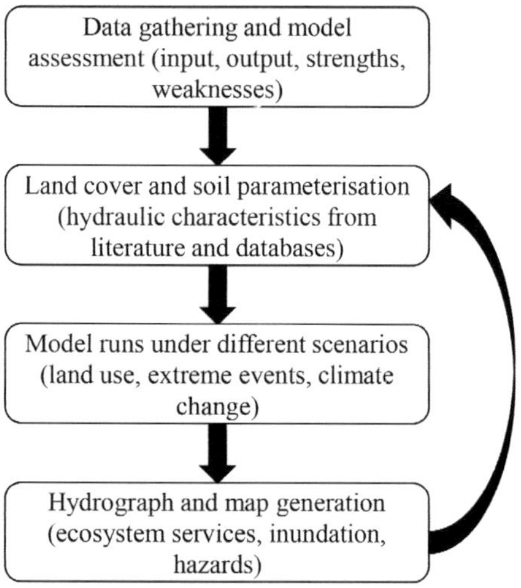

Figure 2. Methodology of the study.

The official programme of disaster mitigation in the Philippines is Project NOAH (Nationwide Operational Assessment of Hazards). This programme mainly uses two models from the United States Hydrologic Engineering Center (HEC): the Hydrologic Modeling System (HMS) for watershed modelling, and the River Analysis System (RAS) for inundation modelling (Brunner, 2010; Scharffenberg, 2013). Project NOAH uses these models for real-time flood forecasting and produces flood hazard maps available on a public website (Santillan et al., 2013). These models have also been used to analyse the watershed response to extreme events and changing land use, estimating peak flow and occurrence time, and modelling the floodplain inundation (Mabao and Cabahug, 2014; Santillan et al., 2011).

Although these models can assist in decision-making through running different land use scenarios, they do not identify areas to target for potential management interventions. Reforestation activities in the Cagayan de Oro (CDO) watershed (Fig. 1) tend to occur in easily-accessible areas, which may not be the areas that would optimally benefit from management interventions (Center for Environmental Studies and Management, 2014). This research applies a model developed to aid in scientifically-sound decision-making regarding land management: the Land Utilisation and Capability Indicator (LUCI). This is being used to identify areas providing existing ecosystem services, areas where management interventions can enhance or degrade ecosystem services, trade-offs between services, and impacts of changing land cover (Jackson et al., 2013). This type of ecosystem services modelling has not been carried out in the Philippines to date, and the LUCI model can aid in more sustainable land management in the CDO watershed. This research is also the first application of LUCI in a tropical catchment, which will allow testing of how well the model is able to represent hydrological processes in such regions.

This research is still on-going, so the results presented in this paper are preliminary. As the soil and land use parameterisation is refined, the models will be run under more scenarios of changing land use, extreme events, and extreme events affected by climate change. The main aims of this research are: (1) to apply the LUCI model to the Cagayan de Oro catchment to help identify priority areas for land management; (2) assess how the LUCI model fits into the existing disaster risk management framework in the Philippines, and (3) assess how the hydrological response of the watershed changes under different scenarios of land use, extreme events, and climate change.

2 Methodology

An overview of the methodology is shown in Fig. 2. First, thematic datasets of land cover, soil, and topography were gathered from various sources and at different scales (local, national, global). Next, land cover and soil parameterisation for the LUCI model was carried out through correlating local soil names with the United States Department of Agriculture (USDA) soil taxonomy, and using the Soil Water Characteristics (SWC) model to estimate the soils' hydraulic characteristics (Saxton and Rawls, 2006). The watershed and floodplain model in HMS and RAS were made and parameterised by the Disaster Risk and Exposure Assessment for Mitigation Program (DREAM) (2015).

In LUCI, the land cover and soil dataset was used to produce ecosystem services maps in order to identify which ar-

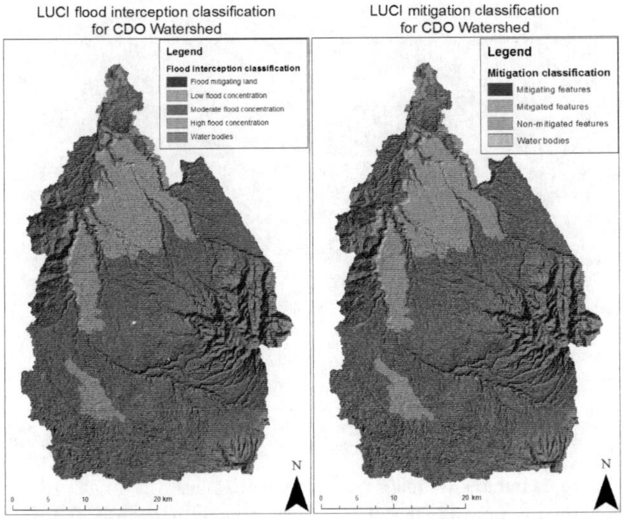

Figure 3. Flood interception and mitigation classification for the CDO watershed.

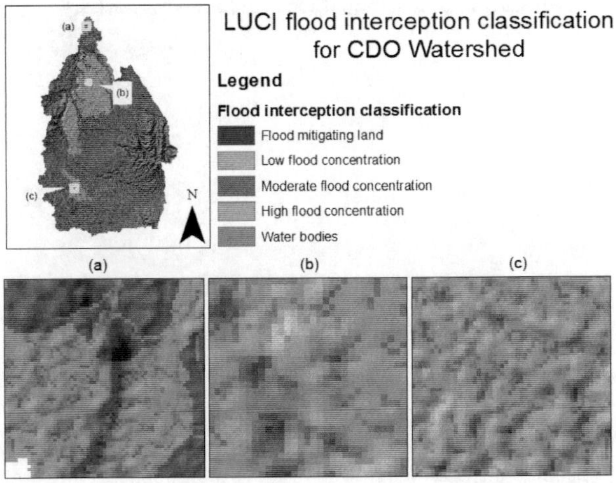

Figure 4. A zoom of some areas of the watershed to show flood interception at a smaller scale.

eas could be targeted for management interventions to improve flood mitigation.

Although this study aims to run the models under different scenarios, this paper presents one extreme event scenario: Typhoon Bopha (local name: Pablo). In December 2012, Typhoon Bopha hit the region and caused a total of 1067 casualties and estimated PHP 36 billion in damages (~ USD 870 million in 2012) (NDRRMC, 2012b). In this scenario, the land cover and soil dataset was the same dataset used to construct the current flood forecasting model for the CDO watershed. The rainfall data was taken from three different rain gauges: Bubunawan, Libona, and Talakag, and was provided by the DREAM Program (2015).Using the Typhoon Bopha rainfall data, rainfall-runoff modelling was performed in both LUCI and HEC-HMS to produce hydrographs for comparison. Both hydrographs were then used as input into HEC-RAS for inundation modelling, and the resulting inundation maps were prepared in ArcMap 10.3.1.

3 Results and discussion

This section presents the preliminary results of this research: the initial LUCI results for the CDO watershed, the hydrographs from LUCI and HEC-HMS, and the resulting inundation maps from HEC-RAS.

3.1 Initial LUCI results

The LUCI model is used to assess the existing and potential impacts of changes in land use management on the following ecosystem services: habitat connectivity, flood mitigation, erosion/sediment delivery, carbon sequestration, and agricultural productivity (Jackson et al., 2013). The flood mitigation service is the main focus for this paper, but this LUCI application aims to model the rest of the ecosystem services and trade-offs as the land cover and soil parameterisation is further refined. The model makes use of a "traffic light" system for its output maps: red means to stop or that changes in land use may degrade the ecosystem service, yellow/orange means that careful management interventions can be carried out, and green areas are priority areas since management interventions have the potential to improve the service (Jackson et al., 2013).

Figure 3 shows the initial LUCI results for the study site for the ecosystem service of flood mitigation. For the flood interception map, red areas are those providing good flood mitigation and any land management changes are likely to adversely affect its capability for flood mitigation. Conversely, green areas are areas of high and moderate flood concentration and can be targeted for management interventions such as rehabilitation or reforestation. The yellow/orange areas of flood interception are areas of low flood concentration, but can be targeted for management interventions since the LUCI algorithm identifies them as non-mitigated features.

At the watershed scale, the green areas or areas that are possible targets for management interventions are not immediately obvious. Figure 4 shows that the model has identified areas for management interventions at the smaller scale, identifying areas where flow accumulates before reaches streams and rivers. Based on the initial LUCI results, small-scale and targeted management efforts have the potential to be more useful to improving flood mitigation services than choosing large portions of land for reforestation. This is important because limited funding for watershed rehabilitation was one of the problems identified in the development of the watershed management plan, so a smaller-scale but more targeted approach may be more beneficial (CESM, 2014). As

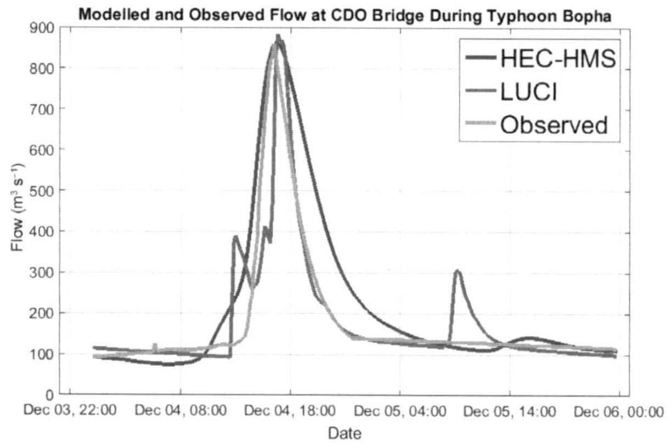

Figure 5. The modelled hydrographs from HEC-HMS and LUCI compared to the observed flow during Typhoon Bopha.

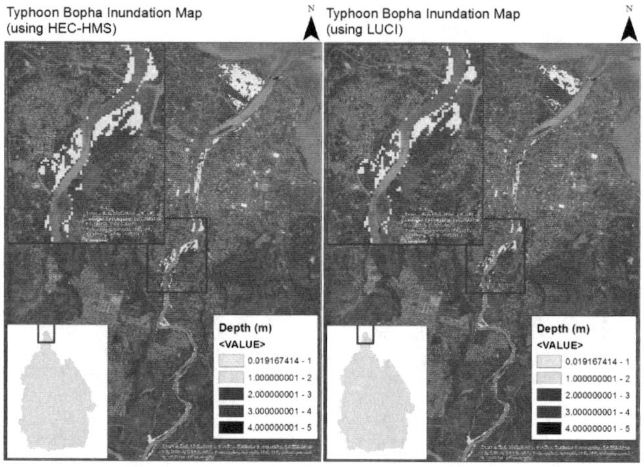

Figure 6. Inundation maps produced in HEC-RAS using the HEC-HMS hydrograph and the LUCI hydrograph, with the inset showing the flood water encroaching on the river banks and buildings.

the soil and land cover parameterisation in LUCI is refined and iterative model runs are done, this application can produce a much clearer picture of priority rehabilitation areas in the watershed and be used to assess the potential result of the local government's plans for rehabilitation or utilisation.

3.2 Typhoon Bopha results: LUCI, HEC-HMS, and HEC-RAS

Under the scenario of Typhoon Bopha, the hydrographs of HEC-HMS and LUCI compare reasonably well with the observed flow at the Cagayan de Oro Bridge (Fig. 5). The peak flows were: 864 679 $m^3 s^{-1}$ (HEC-HMS), 883 963 $m^3 s^{-1}$ (LUCI), and 858.3 $m^3 s^{-1}$ (observed). Both hydrographs from LUCI and HEC-HMS show a similar peak flow volume and occurrence time, as well as a smaller peak after the main flood event. The LUCI hydrograph shows a flashier watershed response and a later peak flow occurrence time compared to the HEC-HMS hydrograph. However, the LUCI hydrograph shows a more similar falling limb when compared to the observed flow. Another similarity between the LUCI hydrograph and the observed flow is the flood volume, as LUCI is able to capture the more abrupt rise and fall of the flood peak, while the HEC-HMS hydrograph shows a more gradual rise and fall, implying an over-prediction of the flood volume.

When both hydrographs are run through HEC-RAS, the produced inundation maps are relatively similar (Fig. 6). The maximum inundation depths were: 4.82 (HEC-HMS) and 4.63 m (LUCI). Although the inundation extent using the HEC-HMS hydrograph is larger compared to the inundation extent produced by the LUCI hydrograph, the inundation occurs in the same areas: at the mouth of the river and areas where the river meanders. Based on this initial comparison, LUCI is able to predict the watershed's response with similar performance to the HEC-HMS model, and the inundation map based on the LUCI hydrograph is also similar to the inundation map based on the HEC-HMS hydrograph.

LUCI, HEC-HMS, and HEC-RAS can be used in combination to complement each other's strengths and weaknesses. For example, LUCI has the capability to go down to significantly finer spatial scales than HEC-HMS. Given its fine resolution, LUCI can run reasonably efficiently at the watershed scale, which has advantages for flood forecasting and uncertainty estimation. However, HEC-HMS takes less processing time to run over the entire watershed. The current configuration of HEC-HMS used by the disaster risk programme uses the curve number method to determine the sub-basin runoff. This method uses one composite number to represent the different soil and land use combinations present within a sub-basin (Scharffenberg, 2013). While this allows the programme to take less processing time, it cannot capture the explicit spatial configuration of elements within the watershed. This makes it less useful for assisting in decision-making regarding land use management. By contrast, the LUCI model uses the soil and land use data to account for the storage and permeability of the different elements within the watershed, which allows the model to perform watershed modelling and represent possible management impacts at fine spatial scales (Jackson et al., 2013). This method allows LUCI to take spatial configuration into account and identify areas that would either benefit from management interventions, or where changing land use would degrade an ecosystem service. This would make the LUCI model more useful to stakeholders and local government making management decisions, or choosing areas for rehabilitation or development.

The models address different areas of disaster risk management: the HEC models for short-term flood forecasting and LUCI for long-term land use planning. The expedience of the HEC models allows the DREAM Program to automate the process of watershed modelling and inundation modelling at 10-min intervals to produce maps that are accessible through a public website (Santillan et al., 2013). The HEC models have also been well-applied in the Philippines as this automated system is being used for flood forecasting in eighteen major river basins in the country (Lagmay, 2012). Through running scenarios of local government plans for rehabilitation and development, the LUCI model can complement the existing system by assessing the potential impacts of land use changes, and identifying priority areas for management interventions. By coupling LUCI with HEC-RAS, the impacts of management decisions can be elucidated through inundation modelling and comparing the resulting maps with maps from the baseline scenarios.

4 Conclusions and further research

Given the vulnerability of the Philippines to typhoons and their resulting floods, it is important to invest in proactive measures for disaster risk mitigation. Three of the ways this can be done is through robust hydrological modelling for flood forecasting, structural measures for flood protection, and through making environmentally-sound land management decisions within the watershed. The initial LUCI application in the Cagayan de Oro watershed seeks to address the component of land management decisions through modelling the ecosystem service of flood mitigation in the CDO watershed. The model was able to identify areas to target for management interventions to improve the ecosystem service of flood mitigation. The priority areas are areas before the flow reaches streams and rivers, suggesting that the small-scale targeted rehabilitation approach may be more useful to improving flood mitigation compared to reforesting large areas within the watershed. Under the Typhoon Bopha scenario, LUCI and HEC-HMS produced similar hydrographs in terms of peak flow and occurrence time, and similar inundation maps. The next steps for this research are to refine the soil and land cover parameterisation to produce LUCI maps of the different ecosystem services, identify trade-offs, and to run the model under different land cover scenarios. After that, the rainfall-runoff modelling capability of the LUCI model can be compared with the HEC models to assess how the LUCI model is able to accurately represent watershed hydrological response.

Data availability

This research used secondary data provided by the DREAM Program (https://dream.upd.edu.ph/).

Acknowledgements. R. Benavidez acknowledges the support and guidance of her supervisors: B. Jackson, D. Maxwell, and E. Paringit. Thanks to the fellow hydrology researchers at the Victoria University of Wellington and the University of the Philippines. Thanks to the DREAM Program for providing thematic datasets, rainfall datasets, and the basin and floodplain model. Thanks to the Cagayan de Oro City Hall, the Department of Public Works and Highways, the Regional Development Council, the Department of Agriculture, and the Philippine Atmospheric, Geophysical and Astronomical Services Administration for providing datasets and watershed reports.

References

Brunner, G.: HEC-RAS User's Manual, 2010.

Center for Environmental Studies and Management: Formulation of an Integrated River Basin Management and Development Master Plan for Cagayan de Oro river Basin, San Juan City, Metro Manila, Philippines, 2014.

Disaster Risk and Exposure Assessment for Mitigation Program: HEC-RAS and HEC-HMS Model Parameterised for Cagayan de Oro, Philippines, University of the Philippines Diliman, Quezon, City, 2015.

Intergovernmental Panel on Climate Change: Managing the Risks of Extreme Events and Disasters to Advance Climate Change Adaptation, Cambridge, UK, and New York, USA, 2012.

Jackson, B., Pagella, T., Sinclair, F., Orellana, B., Henshaw, A., Reynolds, B., and Eycott, A.: Polyscape: A GIS mapping framework providing efficient and spatially explicit landscape-scale valuation of multiple ecosystem services, Landscape Urban Plan., 112, 74–88, doi:10.1016/j.landurbplan.2012.12.014, 2013.

Lagmay, A. M. F.: Disseminating near real-time hazards information and flood maps in the Philippines through Web-GIS, DOST-Project NOAH Open-File Reports, 1, 28–36, ISSN 2362-7409, 2012.

Lasco, R. D., Pulhin, F. B., Jaranilla-Sanchez, P. A., Delfino, R. J. P., Gerpacio, R., and Garcia, K.: Mainstreaming adaptation in developing countries: The case of the Philippines, Climate and Development, 1, 130, doi:10.3763/cdev.2009.0009, 2009.

Mabao, K. and Cabahug, R. G.: Assessment and Analysis of the Floodplain of Cagayan De Oro River Basin, Mindanao Journal of Science and Technology, 12, 147–170, 2014.

National Disaster Risk Reduction and Management Council: Final Report on the Effects and Emergency Management re Tropical Storm "SENDONG" (Washi), 2012a.

National Disaster Risk Reduction and Management Council: Update re Effects of Typhoon "PABLO" (Bopha), 2012b.

Santillan, J. R., Makinano, M., and Paringit, E.: Integrated Landsat Image Analysis and Hydrologic Modeling to Detect Impacts of 25-Year Land-Cover Change on Surface Runoff in a Philippine Watershed, Remote Sensing, 3, 1067–1087, doi:10.3390/rs3061067, 2011.

Santillan, J. R., Ramos, R. V., Recamadas, S., David, G., Paringit, E. C., Espanola, N. C., and Alconis, J.: Use of geospatial technologies and numerical modeling to monitor and forecast flooding along Marikina River, Philippines, Proceedings of the 12th SEASC: Geospatial Cooperation Towards a Sustainable Future, 2013.

Saxton, K. E. and Rawls, W. J.: Soil Water Characteristic Estimates by Texture and Organic Matter for Hydrologic Solutions, Soil Sci. Soc. Am. J., 70, 1569, doi:10.2136/sssaj2005.0117, 2006.

Scharffenberg, W. A.: HEC-HMS: User's Manual, Washington, D.C., 2013.

Proc. IAHS, 373, 153–159, 2016
proc-iahs.net/373/153/2016/
doi:10.5194/piahs-373-153-2016

Open Access
PIAHS

Visualising DEM-related flood-map uncertainties using a disparity-distance equation algorithm

S. Anders Brandt[1,*] and Nancy J. Lim[1,*]

[1]Department of Industrial Development, IT and Land Management, University of Gävle, 80176 Gävle, Sweden
[*]These authors contributed equally to this work.

Correspondence to: S. Anders Brandt (sab@hig.se)

Published: 12 May 2016

Abstract. The apparent absoluteness of information presented by crisp-delineated flood boundaries can lead to misconceptions among planners about the inherent uncertainties associated in generated flood maps. Even maps based on hydraulic modelling using the highest-resolution digital elevation models (DEMs), and calibrated with the most optimal Manning's roughness (n) coefficients, are susceptible to errors when compared to actual flood boundaries, specifically in flat areas. Therefore, the inaccuracies in inundation extents, brought about by the characteristics of the slope perpendicular to the flow direction of the river, have to be accounted for. Instead of using the typical Monte Carlo simulation and probabilistic methods for uncertainty quantification, an empirical-based disparity-distance equation that considers the effects of both the DEM resolution and slope was used to create prediction-uncertainty zones around the resulting inundation extents of a one-dimensional (1-D) hydraulic model. The equation was originally derived for the Eskilstuna River where flood maps, based on DEM data of different resolutions, were evaluated for the slope-disparity relationship. To assess whether the equation is applicable to another river with different characteristics, modelled inundation extents from the Testebo River were utilised and tested with the equation. By using the cross-sectional locations, water surface elevations, and DEM, uncertainty zones around the original inundation boundary line can be produced for different confidences. The results show that (1) the proposed method is useful both for estimating and directly visualising model inaccuracies caused by the combined effects of slope and DEM resolution, and (2) the DEM-related uncertainties alone do not account for the total inaccuracy of the derived flood map. Decision-makers can apply it to already existing flood maps, thereby recapitulating and re-analysing the inundation boundaries and the areas that are uncertain. Hence, more comprehensive flood information can be provided when determining locations where extra precautions are needed. Yet, when applied, users must also be aware that there are other factors that can influence the extent of the delineated flood boundary.

1 Introduction

1.1 Background

In a time of climate change, many countries now require production of flood risk maps when planning and managing built-up areas. Hence, numerous maps have been produced and many of them constitute important documents to hinder or mitigate the negative consequences big floods may bring with them. However, as the maps usually show a potential flood inundation area corresponding to a water discharge so big it has never been experienced before, it is virtually impossible to get an exact, or even near, match between the model output and the future actual flood. This calls for sensitivity and uncertainty modelling (cf. for example Merwade et al., 2008; Pappenberger et al., 2008, for general treatises on these issues). Among the different methods Monte Carlo simulations and probabilistic methods have been commonly used for analysing uncertainties in flood models (e.g., Pappenberger et al., 2005; Werner et al., 2005a). The techniques are applied to test the sensitivity of the results, in terms of the produced water surface elevation, discharge and the flood's spatial extent, to hydraulic model inputs by randomly altering the parameters or parameter sets used. The robustness of the model is often attributed to the number of realisations (from a few hundreds to thousands) that the models are fed with, which in return are used for computing statistics that can quantify the uncertainty.

Provided that the correct water discharge has been selected, examples of issues causing uncertainties include: ground- and river bottom friction parametrisation (Schumann et al., 2007; Werner et al., 2005b), buildings (Koivumäki et al., 2010), cross-section spacing (for 1-D models) (Castellarin et al., 2009; Cook and Merwade, 2009), the modeller's and end users' skills and competences, etc. Nevertheless, the quality of the digital elevation model (DEM) has been shown to have a profound effect on the correctness of the inundation boundary delineation, something which will be looked into deeper in this paper.

1.2 DEM-related uncertainties

The influence of DEM resolution on flood modelling accuracy has been acknowledged by a number of authors (e.g., Brandt, 2005; Casas et al., 2006; Cook and Merwade, 2009). Unlike input sensitivities, the quantification of ambiguities produced by the DEM are most often represented by comparisons of flood extents or water surface elevation produced by using various DEM resolution. Hence, the implementation of Monte Carlo simulations for digital elevation models is more limited because they require alterations, for instance, of the resolution or accuracy of the DEM for each test, prior to the simulation process. Also, depending on the quality of the DEM to be tested, the hydraulic model used, as well as the size of the test site, the modelling time may increase dramatically.

Previous research has also rather consistently (cf. Brandt, 2016) concluded that DEMs of between 4 and 10 m resolution (or point spacing) are enough to model flood boundaries with sufficient accuracy (e.g., Werner, 2001; Casas et al., 2006; Raber et al., 2007). However, those recommendations are based on the total characteristics of the model results; i.e. the total areal extent may not differ much between the model and the true inundation areal extent. This means that local terrain conditions have not been accounted for, and hence, there may be quite large discrepancies at some locations. Although those may be few in number, they may still severely impact the suitability for planning, constructing, or managing urban areas. A first attempt to see whether the disparities between modelled and real flood boundaries could be linked via the quality of the DEM was reported in Brandt (2009), using two particular river reaches of the Eskilstuna River, Sweden. This was further elaborated in Brandt and Lim (2012), and later quantified by Brandt (2016) through an empirically derived equation together with an algorithm capable of producing uncertainty zones of varying width. As this empirical equation is based on only one single river, there is a need for comparative studies to verify the approach of modelling and visualising DEM and slope-dependent uncertainties.

1.3 Aims and objectives

The aim of this paper is to illustrate the relation between the DEM quality and the accuracy of flood modelling. Using the Testebo River, Sweden, as test area for the empirical-based disparity-distance equation and algorithm (from Brandt, 2016), that considers both DEM resolution and terrain slope to create prediction-uncertainty zones around resulting inundation extents of a one-dimensional (1-D) hydraulic model, the study's objectives are to:

- assess whether the equation is applicable on rivers with different characteristics than the Eskilstuna River;
- relate the DEM and slope contributions as part of the total flood boundary delineation's uncertainty;
- identify both advantages and limitations of the method in accounting for model uncertainty based on DEM resolution and the slope characteristics of the floodplain.

2 Methods

To evaluate the applicability of the disparity-distance equation and algorithm by Brandt (2016) in deriving uncertainty boundaries that account for both the DEM resolution and the slope characteristics of the area, Lim's (2011) earlier modelling results for a part of the Testebo River (i.e. Forsby and Varva), north of Gävle, Sweden, were utilised as test cases. This area is characterised by flat floodplains, composed mainly of pasture lands. The results were derived from two different TIN elevation models. One from Lantmäteriet's (the Swedish mapping, cadastral and land registration authority) 50 m data, and one from filtered laser-scanned data (from SWECO), with an equivalent resolution of 2.1 m. The TIN models, which served as the main topographic data, were created from each elevation data set, in combination with river bathymetry comprising of echo-sounded and interpolated channel points (Lim, 2009). The hydraulic modelling was performed with HEC-RAS steady-flow model, using a peak discharge of $160\,m^3\,s^{-1}$. This was equivalent to the 100-year flow in the area, and also to the flooding that happened in 1977, to which the validation data for the study were based on.

Because model results, especially from 1-D models such as HEC-RAS, can be affected by both the spacing and number of cross sections in defining the topographic characteristics of the area (Cook and Merwade, 2009), and how a modeller assigns parameters in the hydraulic model, these impacts on the delineation of the uncertainty zones were also looked at in this present study. Modelling results of Testebo River by both Brandt and Lim (cf. Brandt and Lim, 2012)

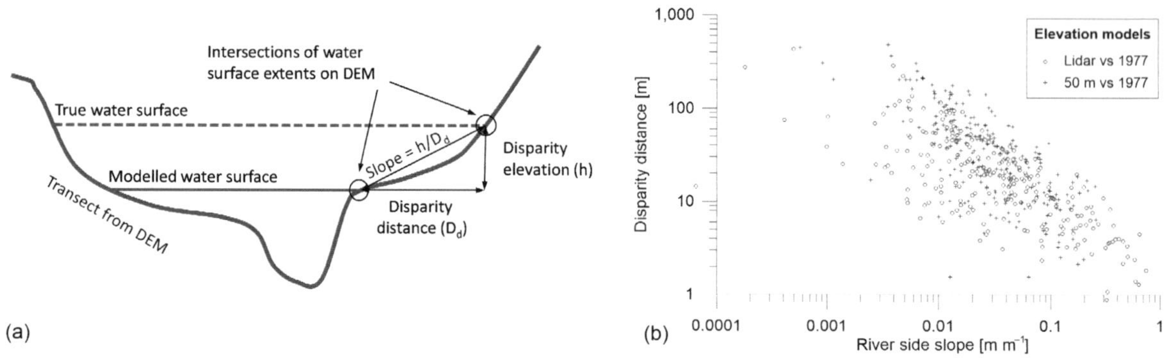

Figure 1. **(a)** Calculation of disparity distances. **(b)** Relation between disparity distance and slope in the Testebo River (values from Lim, 2011).

were used in determining uncertain areas and compared with each other. Although both were based on the same elevation model (i.e. LiDAR data combined with river bathymetry), cross sections, roughness coefficient and boundary conditions used varied between them.

2.1 Disparity calculations

The location of the modelled flood boundary will almost always not be located at the true flood boundary. In most cases they will be close to each other, but at some, they may be placed far away from each other. In order to quantify these differences, the distances and slopes (taken from the locations where the hydraulic model's cross sections intersect the flood boundary lines) were measured according to Fig. 1a. After the distance disparities were plotted against the slopes, a point cloud with a distinctive falling tendency of the highest disparities as the slope increases will appear (Fig. 1b). In Fig. 1b, it can also be seen that a DEM with poorer quality in general will produce higher disparities for the same slopes.

Based on the appearance of the point cloud, it follows that the disparity distance (D_d) can be expressed as a function of slope (S) [m m^{-1}] perpendicular to the flow direction, resolution or cell size of the DEM (δ) [m], and percentile of interest or confidence level (P) [%] for estimating the uncertainties. Taking S as the main dependent variable, it can be expressed as:

$$D_\mathrm{d} = f(S, \delta, P) = cS^z, \quad (1)$$

where c and z are coefficient and exponent, respectively, containing δ and P. Based on the results from the Eskilstuna River, the quantile regression for a number of hydraulic simulations with different DEM resolutions yielded the following empirical equation (Brandt, 2016):

$$D_\mathrm{d} = \left[\delta^{0.970} 0.000792 P^{1.303}\right] S^{[0.1124 \ln(\delta) + 0.0709 \ln(P) - 1.0064]}, \quad (2)$$

2.2 Algorithm

To translate the results from the D_d equation to uncertainty zones around the modelled flood boundary line, the algorithm presented in Brandt (2016) was used. In short, the following procedure is carried out (Fig. 2): (1) The modelled flood polygon, the hydraulic model's cross sections, and the DEM are used as input. (2) In an iterative process (i.e. going from the cross-section/modelled flood-boundary intersection both towards the centre of the river and away from the river), first, calculate the slope and distance between the node and intersection. Next, check whether the distance exceeds the D_d distance for the corresponding slope. (3) When D_d is exceeded, inner and outer uncertainty elevation values are recorded. (4) The cross sections are populated with the uncertainty elevation values and from these, two water surface elevation models are created: one with inner uncertainty elevations and another with outer uncertainty elevations. (5) Finally, the water surface elevation models are compared with the DEM to produce three distinctive areas: almost certain to be flooded, uncertain to be flooded, and almost certain not to be flooded.

2.3 GIS implementation

The algorithm was implemented in GIS using the following data: (1) the flood polygon from the HEC-RAS modelling, (2) the river cross sections with the water surface elevation, (3) stream centreline, which was used for dividing the cross section into left and right parts (while looking downstream the channel), and (4) a DEM over the area.

Two point layers were generated. The first layer consisted of points extracted at the intersection of the cross section and the flood boundary produced from model simulation. The x and y coordinates for these points were derived, together with the water surface elevation at the given location. The points were coded according to the reference number of the cross section where they were extracted, and if they were located at the right or left part of the channel (Fig. 3).

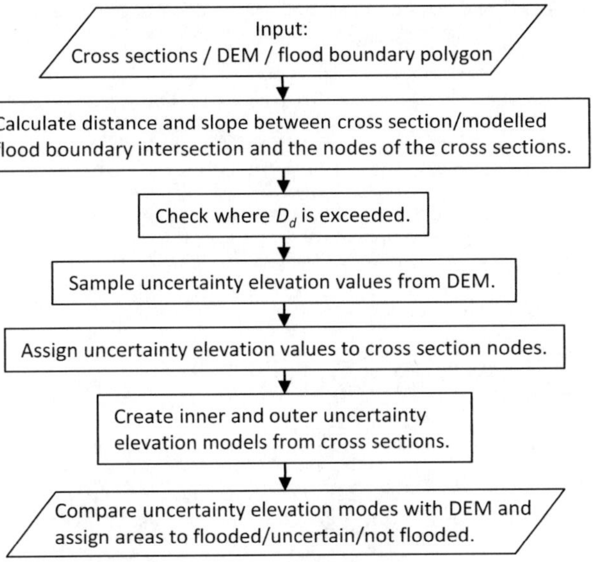

Figure 2. Simplified flowchart of work steps.

The second layer were nodes sampled along each cross section that intersected the edges of the TIN model (Fig. 3). Similar to the first set of points where the flood extent and cross section intersected, these nodes were coded according to the cross-section number and their location at the channel. Furthermore, it was also determined if they were located inside or outside the modelled flood extent. Their x and y coordinates, together with the elevation information were extracted for these points.

To determine the uncertainty boundaries for the inner and outer flooded extents, four new elevation values have to be derived per cross section: an outer and inner elevation for the left and right parts of the channel, respectively. Beginning with the left-hand channel's first cross section and the sampled nodes, which were inside the flooded extent, their disparity from the cross-section/flood-polygon intersection was determined using Eq. (2), with 95 % confidence interval. This was computed node-by-node, starting from the node closest to the flood boundary, going towards the river's centre. The computation was stopped when the disparity was exceeded by the D_d value from Eq. (2). The elevation information from this node was taken and assigned as the inner elevation for all the sampled nodes at the left part of the cross section.

In getting the outer flood boundary elevation for the left part of the channel, the disparity computation was repeated for the sampled points outside the flood boundary, beginning from the node nearest the intersection of cross section and modelled flood boundary. Again, once the disparity was exceeded by the computed disparity from Eq. (2), the computation was stopped and the elevation value at this node was used as the outer elevation for that cross section's nodes at the left channel. These steps were repeated for the sampled

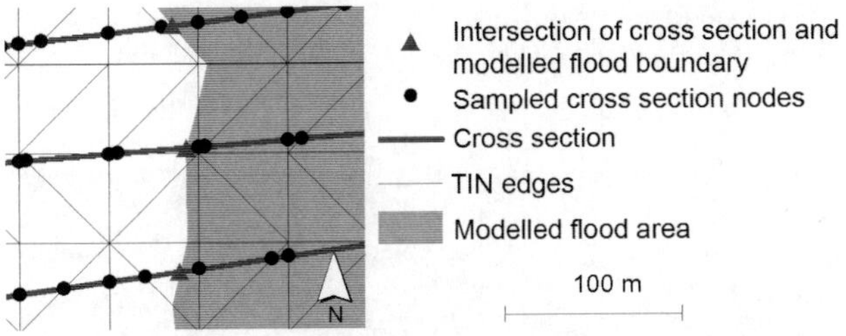

Figure 3. Points sampled at the cross sections intersecting the flood boundary, and nodes at the cross sections intersecting the edges of the TIN model.

nodes at the right part of the cross section to get the inner and outer elevation boundaries. Then the same procedures were followed for the sampled nodes in the succeeding cross sections.

After determining the inner and outer elevations for all sampled nodes, two new point datasets were generated: one containing all points assigned with the outer elevation, and another having the inner elevation information. These point datasets were mapped and used for producing new uncertainty TIN models with the new elevation values. Each of the TIN models was then subtracted from the DEM to delineate areas that are flooded, uncertain, and not flooded.

2.4 Agreement between model and actual flood

The digitised extent based on the actual flooding in 1977 was compared with the result of the generated inner and outer uncertainty boundaries, in addition to the original modelling results. This was measured according to the feature agreement statistic for the overlap (FA) used in Raber et al. (2007):

$$\mathrm{FA} = \frac{F_{\mathrm{Obs}} \cap F}{(F_{\mathrm{Obs}} \cap F) + (F_{\mathrm{Obs}} F)} \times 100, \qquad (3)$$

where the areal size of the overlap between the actual observation (F_{Obs}) and the modelled results (i.e. F) is divided by the total areal size. The latter is computed from the sum of the size of the overlap, and the sizes of the areas that are overestimated (i.e. areas that are supposed to be dry but flooded in the modelled results) and underestimated (i.e. areas that are supposed to be flooded but are dry) in the modelled results. Based on this equation, the percentage of areas that were overestimated and underestimated were also computed to see how big part of the total they constitute.

3 Results

3.1 Predicted uncertain and flooded areas based on the disparity-distance equation

The uncertainty zones produced using the modelled results from the 50 m data and the LiDAR data are represented by the red regions in Fig. 4, while those that will most likely be

Figure 4. Estimated uncertain and flooded areas using the disparity-distance equation algorithm. Results were based on DEM resolutions of 50 and 2.1 m (LiDAR).

flooded are located within the blue areas. The black and dark blue lines are the original boundaries derived from the modelling results and the extents of the actual flooding in 1977, respectively. The remaining areas are those that are foreseen to be dry by the equation.

Both the flooded and the uncertain areas were larger using the 50 m DEM, compared with LiDAR (Table 1). When the results from two different modellers are compared, the estimated size of the uncertain zone was the same (i.e. 0.22 km^2), but not for the flooded areas; Brandt's flood area was 0.13 km^2 bigger than Lim's. If the flood area from the

Table 1. Total sizes of areas predicted to be uncertain and flooded using the disparity-distance equation.

Predicted status	Total area size (km^2) 50 m	LiDAR (Brandt)	LiDAR (Lim)
Flooded	1.02	0.84	0.71
Uncertain	0.93	0.22	0.22

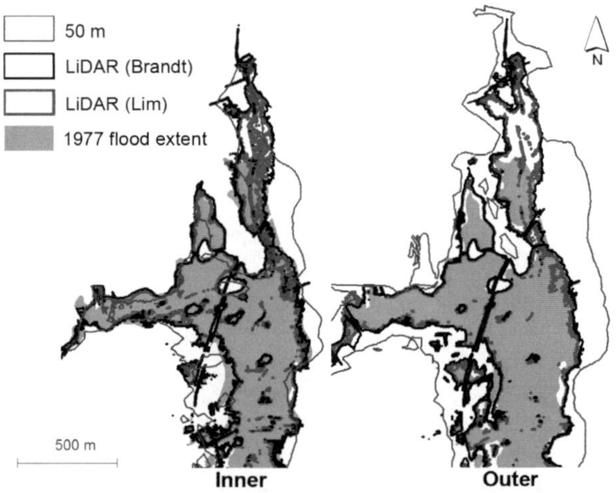

Figure 5. Differences in the flooded areas within the generated inner and outer boundaries from the equation, in comparison with the real flooding data.

50 m data is compared with both Brandt and Lim's results, the difference is 0.18 and 0.31 km^2, respectively.

The width of the uncertainty zones also depended on the characteristics of the terrain. In flat terrains, they were broader, while in areas that are confined with steep side slopes, they were narrower and closer to the original modelling result's extent. For laser-scanned data in relatively flat locations, the outer uncertainty boundary often extends until it reaches a location where the flow of water will be restricted by an elevated ground.

3.2 Comparison of predicted flooding extents with the validation data

Deviation from the 1977-flood data was most evident on the results based on the 50 m data. Almost certain to be flooded areas (i.e. inner uncertainty boundary) derived from the equation did not fit the observed data (Fig. 5). However, if the outer extent of the uncertainty zone will be considered, almost the entire 1977-flood area is inside the extent.

For the LiDAR data, the largest discrepancy between the modelled and the actual flooding extent existed mainly at the northern part of the study area (Fig. 5). In both modelling results, these areas were predicted to be flooded rather than

Table 2. Results of feature agreement statistics for the different modelling results derived from DEMs having different data resolution, and as produced by different modellers, compared with the flood of 1977.

		Percent (%) of total		
		50 m	LiDAR (Brandt)	LiDAR (Lim)
Original model	Overlapping	56.68	74.99	74.33
	Overestimated	42.13	15.56	13.18
	Underestimated	1.19	9.45	12.48
Inner	Overlapping	59.49	74.15	70.39
	Overestimated	29.30	14.63	9.34
	Underestimated	11.21	11.21	20.27
Outer	Overlapping	41.52	70.89	74.37
	Overestimated	58.42	25.90	19.04
	Underestimated	0.06	3.21	6.59

dry. The disparity from the observed data for the two results at this location was greater than 500 m.

In the south western portion of the river, there is an area supposed to be flooded that was modelled dry by both Brandt and Lim (Fig. 4). Despite the disagreement, the uncertainty in this area was relatively low, as manifested by narrower uncertainty zone sizes, because it bordered steeper terrain. The actual flood line was located above, on an elevated ground.

To quantify how much of the predicted flooding within the newly estimated uncertainty extents match the flooding of 1977, feature agreement statistics were computed for all the results, i.e. original modelled results and flooded areas within the inner and outer extents, as shown in Table 2. Overlapping areas with the actual data from 1977 were higher with the LiDAR data (almost 75 %), as revealed by the original results of Brandt and Lim, compared with the 50 m DEM (56.68 %). Although the overlap results varied for the inner and outer boundaries produced by both modellers (wherein higher overlap was computed from Brandt's inner boundaries, while the outer uncertainty boundaries from Lim's modelling results produced higher percentage), they registered between 70 to 75 % of the total, which were still higher than those produced from the 50 m elevation.

For the 50 m data, the assessed agreement of the flooded areas within the inner boundary was higher (59.49 %) than the original modelling result (56.68 %), and the areas within the outer extent (41.52 %). This was attributed to a significantly larger resulting inundation area. Even the outer uncertainty area was overestimated by 58.42 % (cf. Fig. 5), and together with the lowest underestimation (11.21 %), it is clear that most of the original inundation zones were already within the limits.

4 Discussion and conclusion

By using the disparity-distance equation, the underlying resolution of the DEM affected how certain or uncertain areas are to be flooded. It was evident in the results that poorer resolution data, such as the 50 m DEM, produced bigger uncertainties than the LiDAR-based elevation models. Because the details in the terrain were lost with the 50 m data, already flat terrain areas became even flatter. With the laser-scanned data, the effect of flat terrain was also demonstrated in terms of the thickness of the uncertainty zones. The modelled result's uncertainty becomes higher as the slope becomes flatter, while minimal uncertainty was attained in areas with steep-sided slopes. This is because when there is a significant change in the elevation value, the equation and algorithm prevent the uncertainty to go further in the perpendicular direction of the main flow. It must also be noted that all terrain elevation data used in the study were supplemented with bathymetric data. Thus, the estimated uncertainty using the equation may be higher, particularly in flat areas, if the data had not been supplemented with river bottom elevations. Lack of river bathymetry can generate gentler slopes, making the modelled water flow over the floodplain, and extend the water surface extents (Lim, 2009), as well as making the inner uncertainty zone broader towards the river.

This study also shows that the effect of resolution is more significant than the effect brought about by the different modellers using different cross sections, roughness values, and boundary conditions. Based on the feature agreement assessment, the values derived for both modellers did not differ much from each other, as when compared with using the 50 m and the LiDAR data. Their overestimated area was in the flat part of the floodplain, with slope values around 0.002 m m^{-1}. Whether this area will really be flooded or not is difficult to determine, as there may be other factors that also affect the lateral flow, adding to the uncertainty of the flood delineation. On the other hand, the underestimation in both their models occurred in locations where water flow was limited by steep side slopes. In this case, the accuracy of the validation data became in question.

The relatively small differences among the modelled total flooded areas (1.02, 0.84, and 0.71 km^2 for 50 m, Brandt's LiDAR, and Lim's LiDAR, respectively) compared with the striking difference in uncertain areas (0.93, 0.22, 0.22 km^2) between DEM resolutions (but not between modellers), indicate that the effect on the total area is mostly related to parameters such as roughness, whereas the size of the uncertainty zone will be more related to the resolution of the DEM. It did not matter that the modelled boundary of Brandt was generally situated at a distance outside Lim's boundary.

The discrepancy between the 1977 flood and the uncertainty zones produced makes it clear that DEM and slope-dependent uncertainties alone do not account for the total uncertainty. Uncertainties in models have been recognised in different literature (e.g. Pappenberger et al., 2005, 2008;

Schumann et al., 2007; Werner et al., 2005a) to be affected by other factors such as the roughness values, boundary conditions and even the presence of structures (both in the river and the floodplain), which could have also affected this discrepancy derived in the produced results. The disparity-distance equation was developed with the Eskilstuna River as base and with the DEM as the only parameter that was varied (cf. Brandt, 2016). Nevertheless, even though it was developed on another river, it most probably provides valuable input in the quantification of the uncertainty zone. The weakness of the equation is that the geographical area it was developed for did not have large areas of very flat slopes. Hence, the equation probably underestimates the disparity distances for coarse resolution DEMs (i.e. the 50 m in this study) (Brandt, 2016), and therefore, the uncertainty zone for the 50 m DEM is probably even bigger than this study indicates. In a flood mapping project, a good idea would be to first produce a sensitivity analysis using the regular inputs (e.g. different roughness), and then apply the disparity distance algorithm on the smallest and the largest inundation extents, respectively.

Finally, it can be concluded that it is possible to illustrate the relation between the DEM quality and the accuracy of flood map by using uncertainty zones. Furthermore, the disparity-distance algorithm (Brandt, 2016) can be used not only for existing (e.g. where extra precaution is needed) and future flood risk maps, but also on other rivers than Eskilstuna River, provided that the analyst and end users are aware that the DEM uncertainty only provides a part of the total uncertainty of the flood boundary delineation.

References

Brandt, S. A.: Resolution issues of elevation data during inundation modeling of river floods, in: Water Engineering for the Future: Choices and Challenges, edited by: Jun, B.-H., Lee, S.-I., Seo, I.-W., and Choi, G.-W., Proc. XXXI International Association of Hydraulic Engineering and Research Congress, Seoul, 11–16 September 2005, Korean Water Association, Seoul, 3573–3581, 2005.

Brandt, S. A.: Betydelse av höjdmodellers kvalitet vid endimensionell översvämningsmodellering, FoU-rapport Nr. 35, Högskolan i Gävle, 2009.

Brandt, S. A.: Modeling and visualizing uncertainties of flood boundary delineation: algorithm for slope and DEM resolution dependencies of 1D hydraulic models, Stoch. Env. Res. Risk A., doi:10.1007/s00477-016-1212-z, in press, 2016.

Brandt, S. A. and Lim, N. J.: Importance of river bank and floodplain slopes on the accuracy of flood inundation mapping, in: River Flow 2012: Volume 2, edited by: Murillo Muñoz, R. E., Proceedings of the International Conference on Fluvial Hydraulics, San José, Costa Rica, 5–7 September 2012, CRC Press/Balkema (Taylor & Francis), Leiden, The Netherlands, 1015–1020, 2012.

Casas, A., Benito, G., Thorndycraft, V. R., and Rico, M.: The topographic data source of digital terrain models as a key element in the accuracy of hydraulic flood modelling, Earth Surf. Proc. Land., 31, 444–456, 2006.

Castellarin, A., Di Baldassarre, G., Bates, P. D., and Brath, A.: Optimal cross-sectional spacing in Preissmann scheme 1D hydrodynamic models, J. Hydrol. Eng., 135, 96–105, 2009.

Cook, A. and Merwade, V.: Effect of topographic data, geometric configuration and modeling approach on flood inundation mapping, J. Hydrol., 377, 131–142, 2009.

Koivumäki, L., Alho, P., Lotsari, E., Käyhkö, J., Saari, A., and Hyyppä, H.: Uncertainties in flood risk mapping: a case study on estimating building damages for a river flood in Finland, J. Flood Risk Manag., 3, 166–183, 2010.

Lim, N. J.: Topographic data and roughness parameterisation effects on 1D flood inundation models, BSc Thesis, University of Gävle, 2009.

Lim, N. J.: Performance and uncertainty estimation of 1- and 2-dimensional flood models, MSc Thesis, University of Gävle, 2011.

Merwade, V., Olivera, F., Arabi, M., and Edleman, S.: Uncertainty in flood inundation mapping: current issues and future direction, J. Hydrol. Eng., 13, 608–620, 2008.

Pappenberger, F., Beven, K., Horritt, M., and Blazkova, S.: Uncertainty in the calibration of effective roughness parameters in HEC-RAS using inundation and downstream level observations, J. Hydrol., 302, 46–69, 2005.

Pappenberger, F., Beven, K. J., Ratto, M., and Matgen, P.: Multimethod global sensitivity analysis of flood inundation models, Adv. Water. Resour., 31, 1–14, 2008.

Raber, G. T., Jensen, J. R., Hodgson, M. E., Tullis, J. A., Davis, B. A., and Berglund, J.: Impact of Lidar nominal postspacing on DEM accuracy and flood zone delineation. Photogramm, Eng. Remote Sens., 73, 793–804, 2007.

Schumann, G., Matgen, P., Hoffmann, L., Hostache, R., Pappenberger, F., and Pfister, L.: Deriving distributed roughness values from satellite radar data for flood inundation modelling, J. Hydrol., 344, 96–111, 2007.

Werner, M. G. F.: Impact of grid size in GIS based flood extent mapping using a 1D flow model, Phys. Chem. Earth B, 26, 517–522, 2001.

Werner, M. G. F., Blazkova, S., and Petr, J.: Spatially distributed observations in constraining inundation modelling uncertainties, Hydrol. Proc., 19, 3081–3096, 2005a.

Werner, M. G. F., Hunter, N. M., and Bates, P. D.: Identifiability of distributed floodplain roughness values in flood extent estimation, J. Hydrol., 314, 139–157, 2005b.

Proc. IAHS, 373, 161–166, 2016
proc-iahs.net/373/161/2016/
doi:10.5194/piahs-373-161-2016

Effects of anthropogenic land-subsidence on inundation dynamics: the case study of Ravenna, Italy

Francesca Carisi, Alessio Domeneghetti, and Attilio Castellarin

School of Civil, Chemical, Environmental and Materials Engineering, DICAM, University of Bologna, Bologna, Italy

Correspondence to: Francesca Carisi (francesca.carisi@unibo.it)

Published: 12 May 2016

Abstract. Can differential land-subsidence significantly alter river flooding dynamics, and thus flood risk in flood prone areas? Many studies show how the lowering of the coastal areas is closely related to an increase in the flood-hazard due to more important tidal flooding and see level rise. The literature on the relationship between differential land-subsidence and possible alterations to riverine flood-hazard of inland areas is still sparse, although several geographical areas characterized by significant land-subsidence rates during the last 50 years experienced intensification in both inundation magnitude and frequency. We investigate the possible impact of a significant differential ground lowering on flood hazard over a 77 km^2 area around the city of Ravenna, in Italy. The rate of land-subsidence in the study area, naturally in the order of a few mm $year^{-1}$, dramatically increased up to 110 mm $year^{-1}$ after World War II, primarily due to groundwater pumping and gas production platforms. The result was a cumulative drop that locally exceeds 1.5 m. Using a recent digital elevation model (res. 5 m) and literature data on land-subsidence, we constructed a ground elevation model over the study area in 1897 and we characterized either the current and the historical DEM with or without road embankments and land-reclamation channels in their current configuration. We then considered these four different topographic models and a two-dimensional hydrodynamic model to simulate and compare the inundation dynamics associated with a levee failure scenario along embankment system of the river Montone, which flows eastward in the southern portion of the study area. For each topographic model, we quantified the flood hazard in terms of maximum water depth (h) and we compared the actual effects on flood-hazard dynamics of differential land-subsidence relative to those associated with other man-made topographic alterations, which resulted to be much more significant.

1 Introduction

As clearly highlighted in the recent literature, the study of hydrological processes cannot neglect the effect of anthropogenic impacts on the territory (e.g. Montanari et al., 2014 and many writings in the context of the new science of socio-hydrology, proposed by Sivapalan et al., 2012; Di Baldassarre et al., 2013 and references therein). Human and water systems are in fact closely related, so that various authors demonstrated how flood-risk evolution and the increasing of potential damages during extreme flood events are often linked to the strong land-anthropization, rather than to climate change (see e.g. Domeneghetti et al., 2015; Bouwer et al., 2010). Our study considers the human-induced land-subsidence due to the pumping of underground fluids in densely populated areas. This phenomenon has been documented, especially in the last half of the XX century, in different parts of the world, such as in Japan (see e.g. Daito and Galloway, 2015), Mexico (Toscana and Campos, 2010), Thailand (Phien-wej et al., 2006) and Bangladesh (Brown and Nicholls, 2015; Howladar and Hasan, 2014). Literature on the effects of land-subsidence in coastal areas is rich, see e.g. the effects of salt-water intrusion (see e.g. Schmidt, 2015) and the decrease of the coastal floods return period (see e.g. Yin et al., 2013), while the dynamics of hydraulic risk in rivers flood-prone areas is still poorly investigated and understood. The aim of our study is to understand whether and in to what extent the human-induced land-subsidence can change the riverine potential flooding. To investigate these aspects

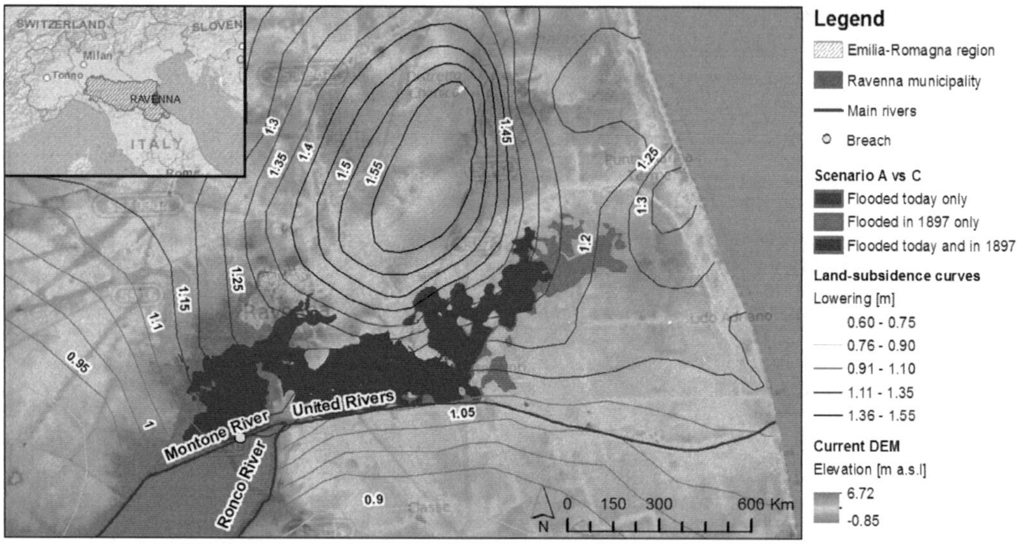

Figure 1. Study area and union of flooded area in scenarios A (current DEM without infrastructures) and C (1897's DEM without infrastructures): blue indicates areas flooded in both scenarios, red and green areas are flooded exclusively in Scenario A and C, respectively.

we focus on the most resounding case of anthropogenic land-subsidence in Italy, which is the area near the city of Ravenna (Northern Italy; see Fig. 1). Here the monitoring of ground elevation by traditional as well as more advanced techniques (see e.g. Bitelli et al., 2000) showed that the land-subsidence rate experienced a sudden acceleration in the aftermath of World War II due to an intense water and gas extraction from underground (Gambolati et al., 1991; Carminati et al., 2002). The latter produced more than 1.5 m cumulative lowering near the historical city centre of Ravenna (see Fig. 1). Ravenna is surrounded and crossed by natural streams that are characterized by artificial embankment systems protecting the city from frequent flooding. In case of extreme events or levee failures (i.e. what is usually identified as "residual flood risk"; see e.g. Castellarin et al., 2011; Di Baldassarre et al., 2009b) it would lead to higher damages, than those which would occur if rivers could expand freely in the surrounding plain. This paradox is called "levee effect" and describes the frequent phenomenon in which the flood control systems encourages urbanization in areas that are even closer to rivers (Tobin, 1995). We selected this study area because of its location (only a few kilometers from the coast) and all factors described above. We investigate the effects of land-subsidence and man-made infrastructures in the study area on the flood dynamics that is expected in case of a levee-failure in the proximity of the urban area of Ravenna. The analyses are performed by adopting fully-2-D models which consider different topographic scenarios.

2 Study area

The study area consists of a 77 km^2 area around the city of Ravenna, in the Emilia-Romagna region (Northern Italy; Fig. 1). With a municipal area of 653 km^2 and a population of 160 000 inhabitants, the city is located few kilometers away from the Adriatic coast. Ravenna is one of the oldest Italian towns, presumably founded in the eight century BC.

Although an inland city, Ravenna is directly connected to the Adriatic Sea by the Candiano Canal and is crossed by the Montone River (United Rivers after the confluence of the Ronco River). High population density, as well as a complex network of road infrastructures characterize the study area.

Like many other coastal lowlands and deltaic plains, the eastern Po plain and in particular the district of Ravenna lye on a subsiding sedimentary basin, where extremely significant changes in terms of ground elevation occurred over centuries relative to the Adriatic mean sea level. The land-subsidence rate in the area, naturally in the order of few mm $year^{-1}$, increased enormously after World War II. The main driver is believed to be the increase in the extraction of deep non-rechargeable groundwater, related to the growth of the economic activities in the Po Basin. The close relationship between groundwater pumping and land-subsidence was confirmed, among the other studies, by the lowering of subsidence rate experienced after a strong reduction of groundwater withdrawal (see Carminati et al., 2002 and references therein). Other studies identified the exploitation of several on-shore and off-shore deep gas reservoirs in the Ravenna area as an additional factor that contributed to the growing of land-subsidence rate up to some centimeters per year (see e.g. Gambolati et al., 1991). In 2005, Teatini et al. (2005) provided a detailed georeferenced map of land-subsidence in the eastern Po River plain over the period 1897–2002, based on the main levelling surveys databases available in the Ravenna area for the last century (IGM, Ravenna Reclamation Authority, Geological Service of the Ravenna Municipality, ARPA and ENI-E&P).

Although land-subsidence rate before the 1950s could be assumed to be almost constant (Teatini et al., 2005), we chose this map as the starting point to perform our study, therefore investigating the possible role of land-subsidence on flood-hazard evolution in the period 1897–2002. As shown in Fig. 1, land-subsidence drops in the last fifty years are larger than 1 m over more than one third of the study area, with peaks higher than 1.5 m over an area of 10 km^2 between the historical center and the coastline. Apart from the significant ground lowering, one must also consider the potential negative effects of differential subsidence occurred in the study area: as an example, in the northeast part of the study area the topographic lowering passes from 1.55 to 1.25 m, with a ≈ 0.30 ‰ horizontal gradient.

3 Topography of the study area: current and reconstructed conditions

The current topography of the study area is described by a 5 m Digital Elevation Model (DEM), available as a GIS Service and provided by the cartographic offices of the Emilia-Romagna for the entire region (Fig. 1).

With the aim of comparing inundation dynamics under current and historical topographic conditions, we reconstructed the ground elevations before the land-subsidence occurred during last decades. In particular, the cumulative land-subsidence contour lines between 1897 and 2002 described in Teatini et al. (2005) were used for back-warping the current 5 m DEM obtaining a historical DEM describing ground elevations in 1897. This procedure is based on the assumption that between 2002 and today no significant change in ground elevations occurred.

The area to the Northeast of the center of Ravenna suffered the greatest drop (ca. 155 cm) and for this reason it is raised the most in the back-warped DEM. The area located approximately 4 km from the city of Ravenna to the South-West experienced the opposite situation, so the ground elevation was raised about 80 cm only, when reconstructing the historical morphology.

The influence of main infrastructures on the flooding dynamics are considered by modifying the discontinuities elevation according to the real topographic characteristics of the elements: we lift the railways elevation by 1 m and we lower the greater channel elevation by 1.5 m.

4 2-D numerical model

We perform our study by means of the fully-2-D hydrodynamic model TELEMAC-2-D, which solves the 2-D shallow water Saint-Venant equations using the finite-element method within a computational mesh of triangular elements (see Galland et al., 1991; Hervouet and Van Haren, 1996). TELEMAC-2-D was adopted in previous studies and similar geographical context, proving to accurately reproduce the real flooding dynamics in a complex floodplain topography (e.g. Di Baldassarre et al., 2009a). One of the advantages of TELEMAC-2-D as finite elements model is the possibility to use structured or non-structured computational meshes. These last, in particular, provide a densification of the triangular elements at certain critical points and allow to better describe the topographical discontinuity that influences the inundation process, such as levees, road and railway embankments (Domeneghetti, 2014; Di Baldassarre et al., 2009b). For this preliminary study, we refer to a non-structured triangular mesh densified at the major discontinuities that can influence our process of flooding, such as boundaries, channels, roads and railways.

As far as what the Manning's coefficient is concerned, we rely on land-use maps available for the Emilia-Romagna region and retrieved from aerial imagery available for 2008 (AGEA-2008), classified on the base of the standardized classes aggregation adopted by the CORINE (COoRdinated INformation on the Environment) project (EEA, 2009). For each land-use class in the study area we use a different value of the Manning coefficient according to the indications provided in the literature (see e.g. Vorogushyn, 2008; Domeneghetti et al., 2013).

TELEMAC-2-D is used to simulate the inundation dynamics in the area of interest, assuming the formation of a single breach in the left embankment of the Montone River, near the confluence with the Ronco River (see Fig. 1). The breach, which is about 120 m wide and 4.5 m deep (from the embankment crest to the elevation of the ground), is assumed to develop instantaneously, triggered by a hypothetical embankment overtopping. The dimension of the breach is typical for embankment systems similar to the one of the River Montone (see e.g. breaches in the Serchio River, December 2009).

The overflowing discharge at the breach has been calculated by referring to a quasi-2-D model of the Montone-Ronco river system (see Castellarin et al., 2011 for an analogous modelling scheme) adopted for the simulation of a 30-years return period flood wave (AdB-RR, 2011). The overflowing discharges simulated at the breach by the quasi-2-D model are used in our study as boundary conditions for the 2-D model in order to simulate the inundation dynamics associated with 1897's and the current topographic conditions.

In order to assess the role of land-subsidence compared with man-made topographic alterations in river flood-hazard, independently by all other factors, the simulations are performed by considering the bathymetry evolution in the pe-

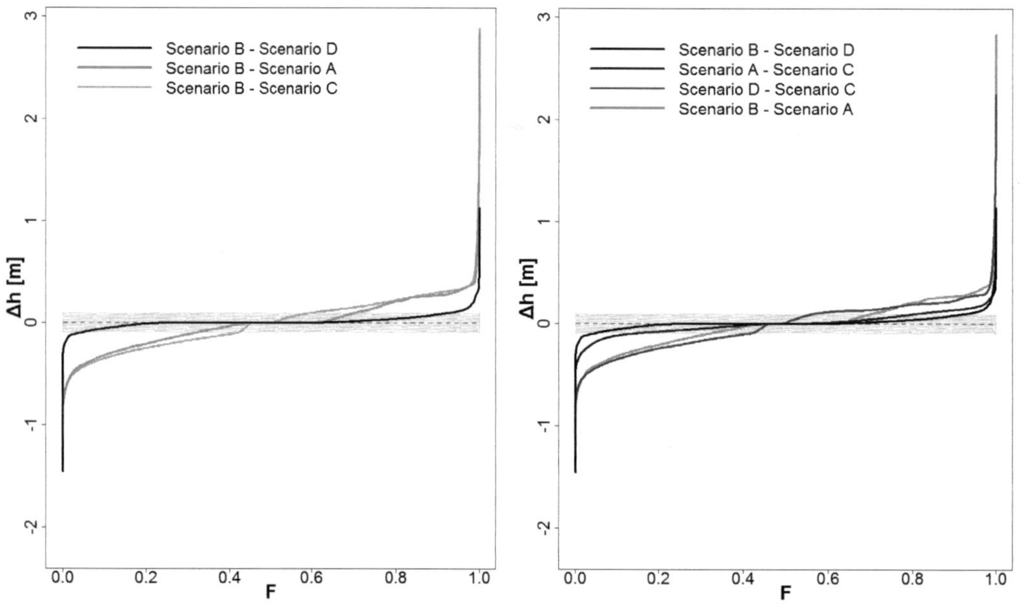

Figure 2. Cumulative distribution function of water depth differences between scenarios, Δh (i.e. differences between water depths simulated for two Scenarios X and Y); "Scenario X – Scenario Y" indicates Δh values computed as water depths simulated for Scenario X minus water depths simulated for Scenario Y; grey dashed areas highlight non significant Δh values (i.e. absolute values lower than 10 cm).

riod of interest (1897–2002). We consider four resulting topographic conditions:

- Scenario A: current morphology without infrastructures;
- Scenario B: current morphology with main infrastructures (i.e. minor channels, railways, roads, etc.)
- Scenario C: 1897 reconstructed morphology (i.e. backwarped DEM) without infrastructures
- Scenario D: 1897 reconstructed morphology with main infrastructures.

5 Results and discussion

In order to describe the flooding dynamics in the area of interest, we focus on the maximum water depth (h) resulting from the simulations for all time steps and for each scenario (A, B, C and D).

Figure 1 shows an example of the results in terms of significantly (water depths $h \geq 10$ cm) flooded areas in two different scenarios: A (current DEM) and C (1897 reconstructed DEM). The blue areas indicate the portion flooded in both topographic scenarios, while areas that are either flooded for 1897 or current are reported in green and red, respectively. Although the extent of the blue area is much larger than the green and red ones, it is evident that in the present scenario the flood-risk affects mainly the urban area of the city of Ravenna, while in 1897 the rural areas in the Eastern side were mostly impacted by inundation. The cause, as expected, is the ground lowering due to the land-subsidence, which had its peak in the historical city center.

Using as reference area the union of the significantly flooded areas in all four scenarios (representing the areas with simulated water depths $h \geq 10$ cm at least in one scenario), we computed the differences of the water depths (Δh) for all scenarios pairs. The comparison of different scenarios are shown in terms of the exceedance probability (F) of a certain difference of water depth (Δh) (see Fig. 2, left and right panels).

Very flat lines around $\Delta h = 0$ in Fig. 2 indicate that the two compared scenarios are very similar in terms of maximum simulated h. Lines that deviate from 0 indicate scenarios whose simulations provided significantly different simulated maximum h.

A first comparison, shown in the left panel of Fig. 2, is performed by considering Scenario B as reference scenario, as it represents the situation closer to reality (current DEM and schematization of major infrastructures). Water depths in any other scenarios (A, C and D) are therefore subtracted from water depths in Scenario B, in order to understand which scenario is more deviated from the real one. The black line in Fig. 2 (left panel) represents the differences between h in Scenarios B and D (1897's DEM with major infrastructures); the green line shows the comparison between Scenarios B and A (current DEM without major infrastructures); the orange line represents the differences between Scenarios B and C, the latter considering the 1897's topography and

Table 1. Comparison of different scenarios, percentage of flooded areas with significant (i.e. absolute values larger than 10 cm) Δh (i.e. differences between water depths simulated for two Scenarios X and Y); " X − Y" indicates Δh values computed as water depths simulated for Scenario X minus water depths simulated for Scenario Y.

	Scenario A	Scenario B	Scenario C	Scenario D
Scenario A		61 %	32 %	62 %
Scenario B	B − A		82 %	11 %
Scenario C	A − C	B − C		84 %
Scenario D	D − A	B − D	D − C	

the absence of major infrastructures. The results demonstrate that taking into account the land-subsidence in the study area leads to maximum water depths that are quite similar to those that result from the simulation with the current DEM (black line). Only 11 % of the reference area experiences significant Δh, i.e. larger than ±10 cm (in 4 % of the flooded extent, maximum water depths in Scenario B are lower than in Scenario D and in 7 % of the area, the opposite occurs), while in the remaining 89 % the differences can be neglected (i.e. Δh lower than ±10 cm, dashed grey areas in Fig. 2, left panel). As far as the comparison between Scenarios B and A is concerned (green line), the percentage of flooded area with negligible Δh (lower than ±10 cm) is equal to 39 %, while in the remaining 61 % of the extent the differences are more significant. On the basis of these results, it is rather evident that the effects of differential land-subsidence on flood risk in the study area are negligible if compared to the impacts of major infrastructures. The comparison between Scenarios B and C (orange line) shows 18 % of the flooded areas with negligible Δh and 82 % with significant differences in terms of maximum water depths. These values show that Scenarios B and C are the most different ones, as expected, but the cumulative distribution function of Δh in this comparison has a very similar trend to that given by the comparison between Scenarios B and A.

A second graph, shown in the right panel of Fig. 2, compares differences in terms of maximum h due to the ground drop caused by land-subsidence (starting either from a simple configuration, Scenarios A–C, and by configurations in which the infrastructure are considered: Scenarios B–D, respectively) and the differences due to the modification of major discontinuities (starting either from the current configuration, Scenarios B–A, and from the 1897's configuration – Scenarios D–C).

The results in terms of percentages of the reference area with significant Δh are presented in Table 1 and confirm that the change in elevation associated with major infrastructures is more important than land-subsidence when simulating the flooding dynamics.

6 Conclusions

Our study assesses the effects of anthropogenic land-subsidence on river flood hazard in the geographical area close to the city of Ravenna, which was affected by an important ground drop in the last century (i.e. more than 1.5 m in the historical city center). The analysis shows that large and rapid differential land-subsidence does not seem to lead to significant alterations to the flooding hazard (if we only consider the maximum water depth as local indicator).

Comparing differences arising from the comparison between simulation in the current configuration and in presence of ground lowering with those caused by the effects of major infrastructures, we can see that human-induced drivers, like construction of canals and road embankments, has an higher impact on flood-hazard than anthropogenic land-subsidence.

In addition, the study further shows the importance of an accurate identification of specific topographic data that have to be considered in the modelling exercise, which should represent the best compromise between precision, maximum expected accuracy and computational efficiency (Dottori et al., 2013).

Acknowledgements. The study is part of the research activities carried out by the working group: Anthropogenic and Climatic Controls on WateR AvailabilitY (ACCuRAcY) of Panta Rhei – Everything Flows: Change in Hydrology and Society (IAHS Scientific Decade 2013–2022).

References

AdB-RR: Piano Stralcio per il Rischio Idrogeologico – Variante al Titolo II "Assetto della rete idrografica", 2011.

Bitelli, G., Bonsignore, F., and Unguendoli, M.: Levelling and GPS networks to monitor ground subsidence in the Southern Po Valley, J. Geodynam., 30, 355–369, 2000.

Bouwer, L. M., Bubeck, P., and Aerts, J. C. J. H.: Changes in future flood risk due to climate and development in a Dutch polder area, Global Environ. Change, 20, 463–471, 2010.

Brown, S. and Nicholls, R. J,: Subsidence and human influences in mega deltas: The case of the Ganges–Brahmaputra–Meghna, Sci. Total Environ., 527–528, 362–374, 2015.

Carminati, E. and Martinelli, G.: Subsidence rates in the Po Plain, northern Italy: the relative impact of natural and anthropogenic causation, Eng. Geol., 66, 241–255, 2002.

Castellarin, A., Domeneghetti, A., and Brath, A.. Identifying robust large-scale flood risk mitigation strategies: A quasi-2D hydraulic model as a tool for the Po river, Phys. Chem. Earth, 36, 299–308, 2011.

Daito, K. and Galloway, D. L.: Preface: Prevention and mitigation of natural and anthoropogenic hazards due to land subsidence, Proc. IAHS, 372, 555–557, doi:10.5194/piahs-372-555-2015, 2015.

Di Baldassarre, G., Castellarin, A., Montanari, A., and Brath, A.: Probability weighted hazard maps for comparing different flood risk management strategies: a case study, Nat. Hazards, 50, 479–496, 2009a.

Di Baldassarre, G., Castellarin, A., and Brath, A.: Analysis of the effects of levee heightening on flood propagation: example of the River Po, Italy, Hydrol. Sci. J., 54, 1007–1017, 2009b.

Di Baldassarre, G., Kooy, M., Kemerink, J. S., and Brandimarte, L.: Towards understanding the dynamic behaviour of floodplains as human-water systems, Hydrol. Earth Syst. Sci., 17, 3235–3244, doi:10.5194/hess-17-3235-2013, 2013.

Domeneghetti, A.: Effects of minor drainage networks on flood hazard evaluation, Proc. IAHS, 364, 192–197, doi:10.5194/piahs-364-192-2014, 2014.

Domeneghetti, A., Vorogushyn, S., Castellarin, A., Merz, B., and Brath, A.: Probabilistic flood hazard mapping: effects of uncertain boundary conditions, Hydrol. Earth Syst. Sci., 17, 3127–3140, doi:10.5194/hess-17-3127-2013, 2013.

Domeneghetti A., Carisi, F., Castellarin, A., and Brath, A.: Evolution of flood risk over large areas: Quantitative assessment for the Po river, J. Hydrol., 527, 809–823, 2015.

Dottori, F., Di Baldassarre, G., and Todini, E.: Detailed data is welcome, but with a pinch of salt: Accuracy, precision, and uncertainty in flood inundation modeling, Water Resour. Res., 49, 6079–6085, 2013.

European Environment Agency (EEA): CORINE Land Cover 2006 (CLC2006) 100 m, [112/2009], Denmark, Copenhagen, 2009.

Galland, J. C., Goutal, N., and Hervouet, J. M.: TELEMAC: a new numerical model for solving shallow water equations, Adv. Water Resour., 14, 38–148, 1991.

Gambolati, G., Ricceri, G., Bertoni, W., Brighenti, G., and Vuillermin, E.: Mathematical simulation of the subsidence of Ravenna, Water Resour. Res., 27, 2899–2918, 1991.

Hervouet, J. M. and Van Haren, L.: Recent advances in numerical methods for fluid flows, in: Floodplain Processes, John Wiley & Sons, Inc., Chichester, UK, 183–214, 1996.

Howladar, M. F. and Hasan, K.: A study on the development of subsidence due to the extraction of 1203 slice with its associated factors around Barapukuria underground coal mining industrial area, Dinajpur, Bangladesh, Environ. Earth Sci., 72, 3699–3713, 2014.

Montanari, A., Ceola, S., and Baratti, E.: Panta Rhei: an evolving scientific decade with a focus on water systems, Proc. IAHS, 364, 279–284, doi:10.5194/piahs-364-279-2014, 2014.

Phien-wej, N., Giao, P. H., and Nutalaya, P.: Land subsidence in Bangkok, Thailand, Eng. Geol., 82, 187–201, 2006.

Teatini, P., Ferronato, M., Gambolati, G., Bertoni, W., and Gonella, M.: A century of land subsidence in Ravenna, Italy, Environ. Geol., 47, 831–846, 2005.

Tobin, G. A.: The levee love arrair: a stormy relationship?, J. Am. Water Resour. As., 31, 359–367, 1995.

Toscana, A. and Campos, M. M.: Environmental and social effects derived from groundwater extraction in Tláhuac and Valle-de-Chalco-Solidaridad, Metropolitan Area of Mexico City, Proc. IAHS, 414–418, 2010.

Schmidt, C. W.: Delta subsidence – An imminent threat to coastal populations, Environ. Health Persp., 123, 204–209, 2015.

Sivapalan, M., Savenjie, H. G., and Blöschl, G.: Socio-hydrology: A new science of people and water, Hydrol. Proc., 26, 1270–1276, 2012.

Vorogushyn, S.: Analysis of flood hazard under consideration of dike breaches, Dissertation, Universität Potsdam, Potsdam, Germany, 2008.

Yin J., Yu, D., Yin, Z., Wang, J., and Xu, S.: Modelling the combined impacts of sea-level rise and land subsidence on storm tides induced flooding of the Huangpu River in Shanghai, China, Clim. Change, 119, 919–932, 2013.

Proc. IAHS, 373, 167–173, 2016
proc-iahs.net/373/167/2016/
doi:10.5194/piahs-373-167-2016

Improvement of operational flood forecasting through the assimilation of satellite observations and multiple river flow data

Fabio Castelli and Giulia Ercolani

Department of Civil and Environmental Engineering, University of Florence, Florence, Italy

Correspondence to: Giulia Ercolani (giulia.ercolani@dicea.unifi.it)

Published: 12 May 2016

Abstract. Data assimilation has the potential to improve flood forecasting. However, it is rarely employed in distributed hydrologic models for operational predictions. In this study, we present variational assimilation of river flow data at multiple locations and of land surface temperature (LST) from satellite in a distributed hydrologic model that is part of the operational forecasting chain for the Arno river, in central Italy. LST is used to estimate initial condition of soil moisture through a coupled surface energy/water balance scheme. We present here several hindcast experiments to assess the performances of the assimilation system. The results show that assimilation can significantly improve flood forecasting, although in the limit of data error and model structure.

1 Introduction

The potential of data assimilation in hydrology has been demonstrated by several studies (e.g., Clark et al., 2008; Seo et al., 2009; Brocca et al., 2010; Lee et al., 2012; Laiolo et al., 2015). However, the usage of data assimilation in distributed hydrologic models for operational flood forecasts is limited by many issues: non-linear and discontinuous model structure, non-Gaussian/multiplicative errors, large dimensionality of the inverse problem, model governed by different equations, complex topology of domains such as surface drainage and river network. Moreover, the majority of studies investigates capabilities of data assimilation through synthetic experiments, while applications conducted from an operational perspective are rare, although the need for an effective transition of research advances into operational forecasting systems has been increasingly claimed in recent years (Liu et al., 2012). This work presents variational assimilation of flow data at multiple locations and of land surface temperature (LST) maps from satellite in a distributed hydrologic model that is part of the operational forecasting chain for the Arno river, in central Italy. We assess the actual gain that can be obtained in flood predictions through data assimilation in order to answer the unsolved doubts of whether the system has enough long memory, or hydrologic models are sufficiently adherent to reality, to significantly benefit from more accurate initial conditions. We show results from several hindcast experiments in the Arno basin.

2 The hydrologic model MOBIDIC

MOBIDIC (MOdello di Bilancio Idrologico DIstribuito e Continuo) is a physically-based hydrologic model (Campo et al., 2006; Castelli et al., 2009; Yang et al., 2014a, b). It is continuous in time and distributed in space, with raster-based horizontal discretization. MOBIDIC solves a coupled mass and energy balance at the surface and employs a computationally efficient representation of soil moisture dynamics, that has been recently improved in Castillo et al. (2015). It has a single layer scheme for soil, whose peculiarity is the conceptual subdivision of the layer into two non-linear reservoirs, the capillary and gravitational one. They correspond to larger pores that drains under gravity and smaller pores that hold water through capillary forces. The two components control different sets of hydrologic fluxes. In particular, available water for evapotranspiration is capillary water, that is fed by gravitational water through an absorption flux. Interactions between surface and subsurface hydrology are explicitly taken into account. Groundwater dynamics may be modeled through 2-D Dupuit approximation or as a linear reservoir. The latter method is employed in the present work. Three options are available for flow routing through the network, i.e. the lag approach, the Muskingum–Cunge method and the cascade of liner reservoirs. MOBIDIC runs operatively at the hydrologic service of Tuscany region (Servizio Idrologico Regionale, Regione Toscana) for floods forecasting and water resources management purposes.

3 The assimilation scheme

For the assimilation system the variational approach has been selected, since it requires less restrictive hypothesis than Kalman and Montecarlo filters and smoothers. The payback is the need for an adjoint model, that is challenging to derive for distributed hydrologic models. Separate adjoint models have been derived for MOBIDIC's modules of flow routing through the river network and of water and energy balance at the soil surface. The first one is devoted to the assimilation of discharge data at multiple locations, while the second one deals with LST observations from satellite.

3.1 Adjoint model of flow routing in the river network

The adjoint model of flow routing has been developed for the method of linear reservoirs in cascade, which represents the optimal compromise between complexity and representativeness of the physical process. Hence, the routing is driven by:

$$\frac{\mathrm{d}\boldsymbol{Q}}{\mathrm{d}t} = \mathbf{A}\left(\boldsymbol{q}_{\mathrm{L}} + \mathbf{U}\boldsymbol{Q} - \boldsymbol{Q}\right) = \boldsymbol{F}\left(\mathbf{A}, \boldsymbol{Q}, \boldsymbol{q}_{\mathrm{L}}\right), \tag{1}$$

where, considering a network composed by n reaches, $\boldsymbol{Q} \in \mathbb{R}^n$ are the discharges exiting each reach, $\boldsymbol{q}_{\mathrm{L}} \in \mathbb{R}^n$ are the lateral inflows (surface runoff plus groundwater flow), and $\mathbf{A} \in \mathbb{R}^{n\times n}$ is a diagonal matrix with the inverse of the characteristic time of each river on the diagonal. Lastly, $\mathbf{U} \in \mathbb{R}^{n\times n}$ is a binary matrix accounting for network topology. Following a classical variational approach (Castelli et al., 1999; Caparrini et al., 2004), the assimilation scheme provides optimal estimates of discharge initial condition and input by minimizing a penalty functional J. J contains squared errors between states predictions and observations and between current and previous values of the quantities to optimize. The physical constraint is imposed by adjoining Eq. (1) to J through a vector of Lagrange multipliers $\boldsymbol{\lambda} \in \mathbb{R}^n$.

$$\begin{aligned}
J = {} & \frac{1}{t_1 - t_0}\int_{t_0}^{t_1} \left(\boldsymbol{Q} - \boldsymbol{Q}^{\mathrm{obs}}\right)^T \frac{\mathbf{K}_Q}{2}\left(\boldsymbol{Q} - \boldsymbol{Q}^{\mathrm{obs}}\right)\mathrm{d}t \\
& + \left(\boldsymbol{Q} - \boldsymbol{Q}^{\mathrm{obs}}\right)^T \frac{\mathbf{K}_Q}{2}\left(\boldsymbol{Q} - \boldsymbol{Q}^{\mathrm{obs}}\right)\Big|_{t_1} \\
& + \left(\boldsymbol{Q}_0 - \boldsymbol{Q}_0'\right)^T \frac{\mathbf{K}_{Q_0}}{2}\left(\boldsymbol{Q}_0 - \boldsymbol{Q}_0'\right) \\
& + \frac{1}{t_1 - t_0}\int_{t_0}^{t_1} \left(\boldsymbol{q}_{\mathrm{L}} - \boldsymbol{q}_{\mathrm{L}}'\right)^T \frac{\mathbf{K}_{q_{\mathrm{L}}}}{2}\left(\boldsymbol{q}_{\mathrm{L}} - \boldsymbol{q}_{\mathrm{L}}'\right)\mathrm{d}t \\
& + \int_{t_0}^{t_1} \boldsymbol{\lambda}^T\left(\frac{\mathrm{d}\boldsymbol{Q}}{\mathrm{d}t} - \boldsymbol{F}\left(\mathbf{A}, \boldsymbol{Q}, \boldsymbol{q}_{\mathrm{L}}\right)\right)\mathrm{d}t
\end{aligned} \tag{2}$$

In Eq. (2), $\mathbf{K}_Q, \mathbf{K}_{Q_0}, \mathbf{K}_{q_{\mathrm{L}}} \in \mathbb{R}^{n\times n}$ are weighting factors applied to the various terms composing J, $\boldsymbol{Q}^{\mathrm{obs}} \in \mathbb{R}^n$ are the observed discharges available inside the assimilation window $[t_0, t_1]$, the sign $(\cdot)'$ indicates the previous value of the quantity. J is minimized if its first variation δJ vanishes, condition that, after some computations, leads to a system of ordinary differential equations that describes the time evolution of Lagrange multipliers. Furthermore, a terminal condition for the backward integration of the adjoint model and update equations for initial streamflow $\boldsymbol{Q}_0$ and lateral input $\boldsymbol{q}_{\mathrm{L}}$ that depends on $\boldsymbol{\lambda}$ are obtained. The optimal estimate of $\boldsymbol{Q}_0$ and $\boldsymbol{q}_{\mathrm{L}}$ is obtained through an iterative procedure constituted by subsequent integrations of forward and adjoint model and corresponding updates. Iterations are interrupted when updates become negligible. To note that the corrections evaluated at discharge measurement stations spread upstream thanks to the coupling between equations of flow channel routing (see Eq. 1). Since our aim is to improve both flow routing and runoff formation processes, the equations driving these latter should be included in the derivation of the adjoint model. However, it would be an extremely challenging task, mainly because of the threshold processes that characterize soil moisture dynamics. Therefore, we employ a mixed variational-Montecarlo approach. First, through the flow routing adjoint model, we estimate the optimal temporal evolution of lateral inflow q_{L}. Then, on its basis, we infer key variables that determine runoff formation through a parsimonious Montecarlo approach. As key variables, we selected initial condition of water content in capillary soil and rainfall intermittence, that is represented by the parameter f_0 (frequency of no-rainfall during a certain time step, see Castelli, 1996). An ensemble covering the extent of the assimilation window is generated for $\boldsymbol{q}_{\mathrm{L}}$ by reasonably varying initial soil moisture and f_0. Both initial capillary water and f_0 are maintained spatially homogeneous, and hence the size of the ensemble remains small (typically around 100 realizations). At each iteration of the assimilation procedure, the realization with the minimum distance from the desired trajectory of q_{L} is selected for any single reach, and the corresponding initial capillary water and f_0 are adopted for the contributing cells. Hence, a spatially distributed estimate of both quantities is obtained.

3.2 Adjoint model of soil water and energy balance

As described in Sect. 2, MOBIDIC solves a coupled water and energy balance at the surface. Hence, it reproduces the partitioning of available energy between latent and sensible heat flux that is determined by soil moisture. This characteristic allows to estimate soil moisture initial condition through the assimilation of LST observations from satellite. The advantage of assimilating LST maps instead of soil moisture products lies in their higher spatio-temporal resolution and accuracy (Campo et al., 2012).

Among MOBIDIC's driving equations, a set of three ODEs can be identified as significant for the assimilation of LST. The system is here reported in synthetic form to highlight the relevant dependencies.

$$\frac{dT_s}{dt} = F_1(T_s, T_d, H(T_s, \ldots), \mathrm{LE}(T_s, W_c, \ldots)) \tag{3}$$

$$\frac{dT_d}{dt} = F_2(T_s, T_d, \ldots) \tag{4}$$

$$\frac{dW_c}{dt} = F_3\left(W_c, W_g, \mathrm{LE}(T_s, W_c, \ldots)\right) \tag{5}$$

where H and LE are surface turbulent heat fluxes of sensible and latent heat respectively, T_s is the land surface temperature, T_d is the temperature of a deeper layer of soil, W_c and W_g are capillary and gravitational soil water content. The system can be solved pixel by pixel, being the states of a specific cell independent from those of the others. In the developed assimilation framework T_s, W_c and W_g are the analyzed states, whose optimal estimation is obtained by minimization of the penalty functional J:

$$\begin{aligned} J =& \frac{1}{t_1 - t_0}\int_{t_0}^{t_1}\left(\boldsymbol{T}_s - \boldsymbol{T}_s^{obs}\right)^T \frac{\mathbf{K}_{T_s}}{2}\left(\boldsymbol{T}_s - \boldsymbol{T}_s^{obs}\right)dt \\ &+ \left(\boldsymbol{T}_s - \boldsymbol{T}'_s\right)^T \frac{\mathbf{K}_{T_s}}{2}\left(\boldsymbol{T}_s - \boldsymbol{T}'_s\right)\Big|_{t_0} \\ &+ \left(\boldsymbol{T}_d - \boldsymbol{T}'_d\right)^T \frac{\mathbf{K}_{T_d}}{2}\left(\boldsymbol{T}_d - \boldsymbol{T}'_d\right)\Big|_{t_0} \\ &+ \left(\boldsymbol{W}_c - \boldsymbol{W}'_c\right)^T \frac{\mathbf{K}_{W_c}}{2}\left(\boldsymbol{W}_c - \boldsymbol{W}'_c\right)\Big|_{t_0} \\ &+ \frac{1}{t_1 - t_0}\int_{t_0}^{t_1}\left(\boldsymbol{W}_g - \boldsymbol{W}_g'\right)^T \frac{\mathbf{K}_{W_g}}{2}\left(\boldsymbol{W}_g - \boldsymbol{W}'_g\right)dt \\ &+ \int_{t_0}^{t_1}\left[\boldsymbol{\lambda}_1{}^T\left(\frac{d\boldsymbol{T}_s}{dt} - \boldsymbol{F}_1\right) + \boldsymbol{\lambda}_2{}^T\left(\frac{d\boldsymbol{T}_d}{dt} - \boldsymbol{F}_2\right)\right. \\ &\left. + \boldsymbol{\lambda}_1^T\left(\frac{d\boldsymbol{W}_c}{dt} - \boldsymbol{F}_3\right)\right]dt \end{aligned} \tag{6}$$

where all the state variables are vector in $\mathbb{R}^n$, being n the number of cells composing the basin. $\mathbf{K}_{T_s}$, $\mathbf{K}_{T_d}$ and $\mathbf{K}_{W_c} \in \mathbb{R}^{n \times n}$ are weighting factors applied to the various terms of the penalty functional, $\boldsymbol{T}_s^{obs} \in \mathbb{R}^n$ are measurements of LST available inside the assimilation window $[t_0, t_1]$ and the sign $(\cdot)'$ indicates the previous value of the quantity. The penalty functional is formed by quadratic errors in respect to both

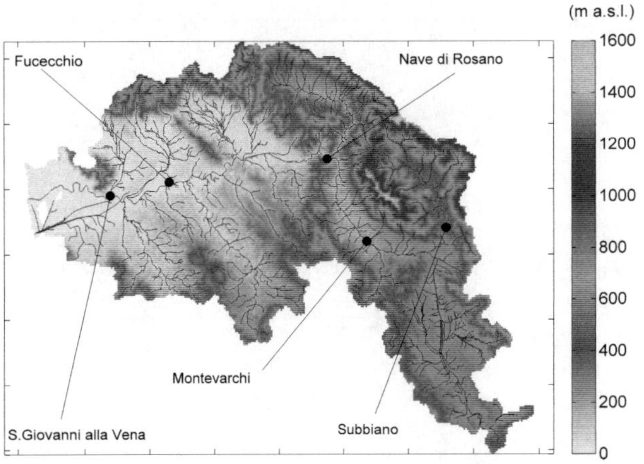

Figure 1. DEM and river network of Arno basin used in the simulations. Black circles indicate measurement stations available for experiments of flow data assimilation.

measurements and previous values of the quantity to estimate. The last term is the physical constrain adjoined through Lagrange multipliers $\boldsymbol{\lambda}_1$, $\boldsymbol{\lambda}_2$ and $\boldsymbol{\lambda}_3 \in \mathbb{R}^n$, each one corresponding to a specific state variable. J is minimized when its first variation vanishes, condition that, after some computations, leads to ODEs driving the time evolution of $\boldsymbol{\lambda}_1$, $\boldsymbol{\lambda}_2$ and $\boldsymbol{\lambda}_3$. Terminal conditions for their backward integration are also obtained, as well as update equations for initial condition of T_s, T_d, W_c and W_g that depend on Lagrange multipliers. Forward and adjoint model are solved iteratively and corresponding updates are computed. Iterations are interrupted when updates become negligible.

4 Application: flood forecasts in the Arno river basin

The assimilation scheme is tested in the Arno river basin, central Italy. The basin extends over about 8300 km^2. Figure 1 shows Digital Elevation Model, river network and flow measurement stations employed in this work. Flood forecasting is a relevant issue in Arno basin, since Arno passes through major Tuscan cities, as Florence and Pisa.

The performances of the assimilation system are assessed through several hindcast experiments. Simulations are run with the spatial and temporal resolutions that are employed operationally, i.e. 500 m and 15 min. The analysis is performed mainly in terms of flow peak prediction accuracy, since it is one of the most important skills for floods early warning.

4.1 Results from assimilation of flow data

Experiments devoted to test assimilation of multiple flow data include both high flow and false alarm (high rainfall but low flows) events that occurred in the period 2009–2014.

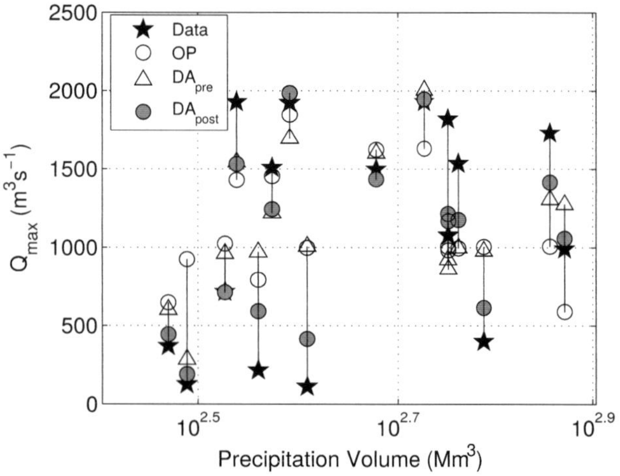

Figure 2. Flow peak versus precipitation volume at S. Giovanni alla Vena for the experiments of streamflow assimilation. Observations and open loop are black stars and white circles respectively. Values from data assimilation are white triangles/gray circles in case peaks at upstream locations are included/not included in the assimilation window.

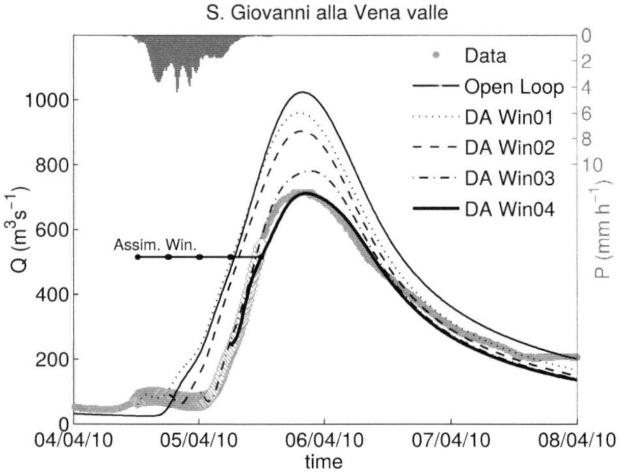

Figure 3. Hydrographs at S. Giovanni alla Vena measurement station for the false alarm event of 6 April 2010. Flow observations are gray dots, open loop simulation is the solid line, predictions using analysis obtained by assimilating flow data during the 1st, 2nd, 3rd and 4th assimilation window (Assim. Win.) are the dotted, dashed, dash-dot and solid thick line respectively.

Flow observations are available from the 5 measurement stations shown in Fig. 1, with a temporal resolution of 15 min. Sequential assimilations are realized on windows of 6 h, employing the data from all the stations simultaneously. Analysis of initial discharge in the river network and capillary water in soil, as well as the optimal estimate of the parameter f_0, are obtained for each assimilation window, and then used to run the corresponding prediction simulation. To evaluate the assimilation system we focus on results at S. Giovanni alla Vena, that is the closest station to the outlet and can be considered as representative of the overall functioning of the scheme. Figure 2 summarizes the results obtained for all the 16 examined events. Peak flows at S. Giovanni alla Vena are plotted against the corresponding total volume of rainfall. Observations and open loop are marked with black stars and white circles respectively. Forecasts from data assimilation are white triangles and gray circles. These latter correspond to the last assimilation window, that includes observed peaks at the upstream locations (Subbiano and Montevarchi, and in some cases also Nave di Rosano), while gray circles are from an antecedent window. The general behavior is, as desired, of enhanced flow peak predictions from data assimilation in respect to open loop. Gray circles show a remarkable better adherence to observations than white triangles, suggesting that the gain increases significantly when upstrem peaks are included in the assimilation window. However, in some experiments data assimilation has a slightly negative impact on predictions. This occurs when open loop is already very close to the observations.

To give further insight, hydrographs at S. Giovanni alla Vena for three events are reported in Figs. 3, 4 and 5. They

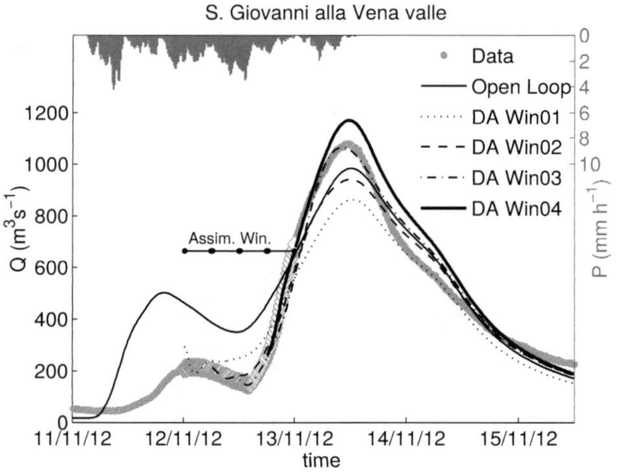

Figure 4. Hydrographs at S. Giovanni alla Vena measurement station for the flood event of 13 November 2012. Lines, symbols and acronyms maintain the same meaning of Fig. 3.

are characterized by different levels of performance. In the false alarm event of 6 April 2010 (Fig. 3), forecasts progressively improve as the assimilation window advances in time. Predictions corresponding to the last window match observations almost perfectly. Conversely, in the flood event of 13 November 2012 (Fig. 4), data assimilation initially reduces the adherence of the forecasted hydrograph to observations, especially in terms of peak flow. The reason is that assimilation attempts to lower streamflow, that in open loop is significantly overestimated during the first window. Improvements are obtained in the third step, whose window cor-

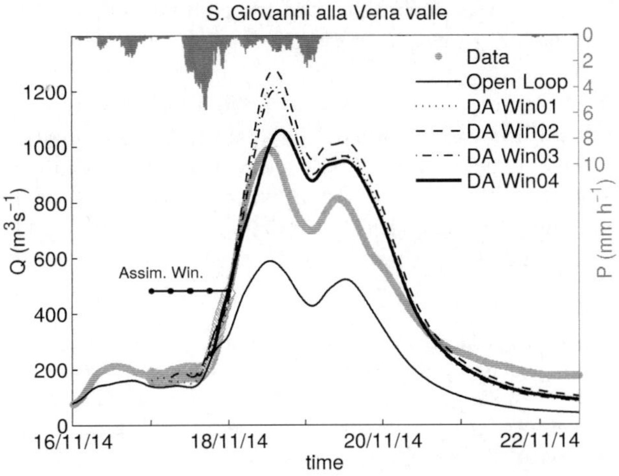

Figure 5. Hydrographs at S. Giovanni alla Vena measurement station for the flood event of 18 November 2014. Lines, symbols and acronyms maintain the same meaning of Fig. 3.

responds to the actual beginning of the event rising limb. It results into an excellent reproduction of data, that is slightly deteriorated by the subsequent assimilation. The final predicted peak is affected by an error almost equivalent to that of the open loop (about 8.4 %), although overestimation instead of underestimation. The behavior for the flood event of 18 November 2014 (Fig. 5) is intermediate. Predictions from data assimilation are always better than open loop. However, a progressive improvement as in the false alarm event of 6 April 2010 is not observed. For instance, the second assimilation slightly worsen the forecasts corresponding to the first one. The subsequent assimilation recovers the gap and predictions from the last window are definitely a significant enhancement of the open loop.

In summary, the assimilation scheme can considerably improve flood forecasts and reduce false alarms. Nevertheless, in presence of already accurate prediction, the assimilation system can not provide additional enhancement. This fact indicates that the gain obtainable through the assimilation scheme is limited by model structure, namely, model errors can not be overcome by the assimilation, as well as by possible errors in data. Conversely, an exacerbate forcing of modeled outputs toward observations can lead to a negative impact on the forecasts.

4.2 Results from assimilation of LST observations

Assimilation of LST observations is tested on a late summer event (16–18 September 2006) characterized by locally heavy rainfall (up to 250 mm) but quite low streamflow values. The employed LST observations are from MeteosatSG-SEVIRI (LandSAF product at 15 min with a spatial resolution of about 3 km). Assimilation is performed for 13 September in order to evaluate a proper initial soil moisture before rainfall starts. Figure 6 shows soil moisture at the end of 13 September for the open loop and its analysis obtained through data assimilation. Figure 7 compares streamflow observations at Nave di Rosano with hydrographs from simulations initialized with a complete dry soil, soil saturation level equal to 50 % of field capacity and soil moisture estimated from the assimilation of LST. The peak flow is well reproduced by the latter, while half-field capacity run significantly overestimates discharges and the dry run suffers underestimation. Figure 8 summarizes results for all the available flow measurement stations. Peak flows are plotted against total volume of rainfall in the drainage basin. The significant overestimation of the half-field capacity simulation that has been observed for Nave di Rosano affects also the

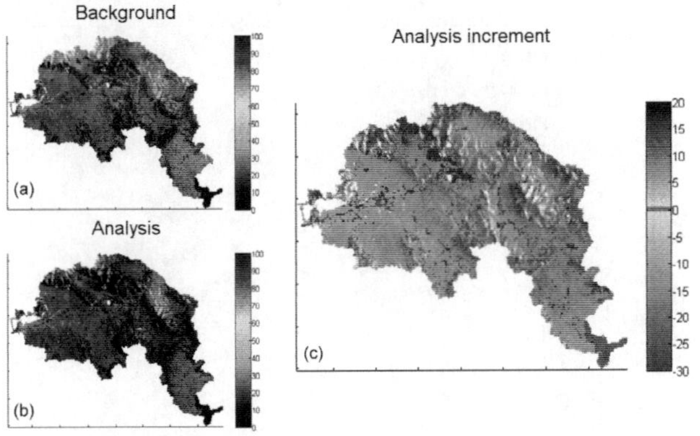

Figure 6. Soil moisture maps of background **(a)**, analysis **(b)** and increment between them **(c)** at the end of the assimilation window. Values are expressed as saturation level (%).

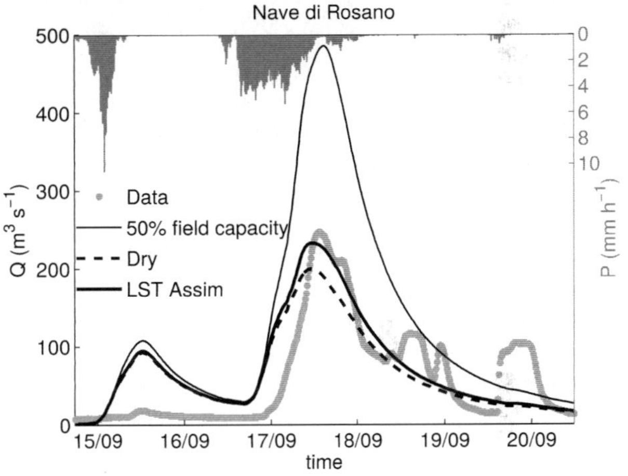

Figure 7. Hydrographs at Nave di Rosano measurement station for the test event of LST assimilation. Flow observations are gray dots, predictions initialized with a complete dry soil, soil saturation level equal to 50 % of field capacity, and soil moisture estimated from the assimilation of LST, are the solid, dashed and solid thick line respectively.

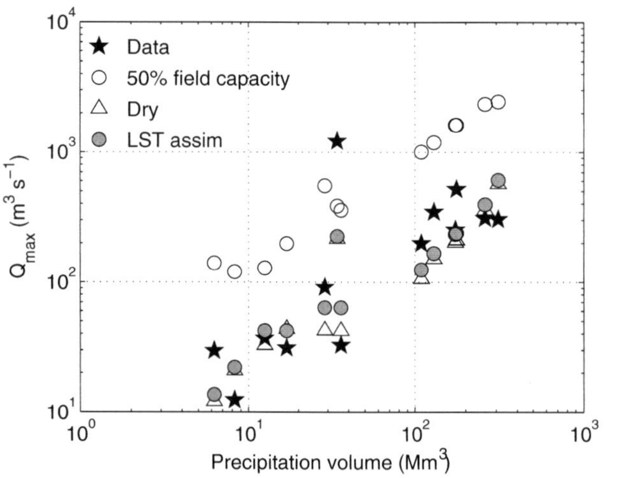

Figure 8. Flow peak versus precipitation volume at the available measurement stations for the test event of LST assimilation. Acronyms in the legend maintain the same meaning of Fig. 7.

other stations. Dry and assimilation run are closer to observations (black stars), with the latter better predicting the peak at some locations. Nevertheless, model results do not manage to reproduce the variability in the dependence of flow peak from upstream precipitation that characterizes observations. This fact reveals that the assimilation system does not add sufficiently flexibility in the model, although it improves the estimate of initial soil moisture. However, tests on several events would be necessary to properly evaluate LST assimilation.

5 Conclusions

This work presents variational assimilation of flow data at multiple locations and of LST in the distributed hydrologic model MOBIDIC, that is part of the operational forecasting chain of the Arno river in central Italy. Flow data are employed to optimally estimate initial discharge in the river network, initial capillary water in soil and rainfall intermittence parameter f_0. LST is exploited to infer initial soil moisture (capillary and gravitational). The performances of the developed assimilation system are assessed in several hindcast experiments. In particular, the scheme for assimilation of flow data is tested through 16 experiments. The scheme produces relevant improvements in flood forecasting, especially when the assimilation window includes peak flows at upstream locations. Some negative impacts of the assimilation are observed in case open loop results are already very close to measurements, suggesting that the obtainable gain is limited by the structure of the hydrologic model and flow data errors. The scheme for LST assimilation is verified on one single event with locally heavy rainfall but low streamflows. LST is assimilated 2 days in advance in respect to the event. Comparison of simulation results with observations of flow at various locations shows that the initial condition of soil moisture estimated through LST assimilation improves model performances in respect to dry soil and half-field capacity runs. However, the assimilation does not add sufficiently flexibility in the model to fully reproduce the observed variability in runoff formation process. Future work will analyze in more detail the potential of LST assimilation by performing additional hindcast experiments.

Acknowledgements. Authors acknowledge Tuscany Region for providing datasets employed in the research and funding that supported the work (research project "Attività di ricerca per la mitigazione del rischio idraulico della Regione Toscana", grant number CUP - B18C12000300002).

References

Brocca, L., Melone, F., Moramarco, T., Wagner, W., Naeimi, V., Bartalis, Z., and Hasenauer, S.: Improving runoff prediction through the assimilation of the ASCAT soil moisture product, Hydrol. Earth Syst. Sci., 14, 1881–1893, doi:10.5194/hess-14-1881-2010, 2010.

Campo, L., Caparrini, F., and Castelli, F.: Use of multi-platform, multi-temporal remote-sensing data for calibration of a distributed hydrological model: an application in the Arno basin, Italy, Hydrol. Process., 20, 2693–2712, doi:10.1002/hyp.6061, 2006.

Campo, L., Castelli, F., Caparrini, F., and Entekhabi, D.: An assimilation algorithm of satellite-derived LST observations for the operational production of soil moisture maps, in: Geoscience and Remote Sensing Symposium (IGARSS), 2012 IEEE International, Munich, Germany, 22–27 July 2012, IEEE, 1314–1317, 2012.

Caparrini, F., Castelli, F., and Entekhabi, D.: Estimation of surface turbulent fluxes through assimilation of radiometric surface temperature sequences, J. Hydrometeorol., 5, 145–159, doi:10.1175/1525-7541(2004)005<0145:EOSTFT>2.0.CO;2, 2004.

Castelli, F.: A simplified stochastic model for infiltration into a heterogeneous soil forced by random precipitation, Adv. Water Resour., 19, 133–144, doi:10.1016/0309-1708(95)00041-0, 1996.

Castelli, F., Entekhabi, D., and Caporali, E.: Estimation of surface heat flux and an index of soil moisture using adjoint-state surface energy balance, Water Resour. Res., 35, 3115–3125, doi:10.1029/1999WR900140, 1999.

Castelli, F., Menduni, G., and Mazzanti, B.: A distributed package for sustainable water management: a case study in the Arno Basin, International Association of Hydrological Sciences, IAHS Publ. 327, 52–61, 2009.

Castillo, A., Castelli, F., and Entekhabi, D.: Gravitational and capillary soil moisture dynamics for distributed hydrologic models, Hydrol. Earth Syst. Sc., 19, 1857–1869, doi:10.5194/hess-19-1857-2015, 2015.

Clark, M. P., Rupp, D. E., Woods, R. A., Zheng, X., Ibbitt, R. P., Slater, A. G., Schmidt, J., and Uddstrom, M. J.: Hydrological data assimilation with the ensemble Kalman filter: Use of streamflow observations to update states in a distributed hydrological model, Adv. Water Resour., 31, 1309–1324, doi:10.1016/j.advwatres.2008.06.005, 2008.

Laiolo, P., Gabellani, S., Campo, L., Silvestro, F., Delogu, F., Rudari, R., Pulvirenti, L., Boni, G., Fascetti, F., Pierdicca, N., Crapolicchio, R., Hasenauer, S., and Puca, S.: Impact of different satellite soil moisture products on the predictions of a continuous distributed hydrological model, Int. J. Appl. Earth Obs., 48, 131–145, doi:10.1016/j.jag.2015.06.002, 2015.

Lee, H., Seo, D.-J., Liu, Y., Koren, V., McKee, P., and Corby, R.: Variational assimilation of streamflow into operational distributed hydrologic models: effect of spatiotemporal scale of adjustment, Hydrol. Earth Syst. Sci., 16, 2233–2251, doi:10.5194/hess-16-2233-2012, 2012.

Liu, Y., Weerts, A. H., Clark, M., Hendricks Franssen, H.-J., Kumar, S., Moradkhani, H., Seo, D.-J., Schwanenberg, D., Smith, P., van Dijk, A. I. J. M., van Velzen, N., He, M., Lee, H., Noh, S. J., Rakovec, O., and Restrepo, P.: Advancing data assimilation in operational hydrologic forecasting: progresses, challenges, and emerging opportunities, Hydrol. Earth Syst. Sci., 16, 3863–3887, doi:10.5194/hess-16-3863-2012, 2012.

Seo, D.-J., Cajina, L., Corby, R., and Howieson, T.: Automatic state updating for operational streamflow forecasting 100 via variational data assimilation, J. Hydrol., 367, 255–275, doi:10.1016/j.jhydrol.2009.01.019, 2009.

Yang, J., Castelli, F., and Chen, Y.: Multiobjective sensitivity analysis and optimization of distributed hydrologic model MOBIDIC, Hydrol. Earth Syst. Sci., 18, 4101–4112, doi:10.5194/hess-18-4101-2014, 2014a.

Yang, J., Entekhabi, D., Castelli, F., and Chua, L.: Hydrologic response of a tropical watershed to urbanization, J. Hydrol., 517, 538–546, doi:10.1016/j.jhydrol.2014.05.053, 2014b.

Proc. IAHS, 373, 175–178, 2016
proc-iahs.net/373/175/2016/
doi:10.5194/piahs-373-175-2016

Open Access
PIAHS

An update on multivariate return periods in hydrology

Benedikt Gräler[1], **Andrea Petroselli**[2], **Salvatore Grimaldi**[3], **Bernard De Baets**[4], **and Niko Verhoest**[5]

[1]Institute of Hydrology, Ruhr University Bochum, Bochum, Germany
[2]Dipartimento di scienze agrarie e forestali (DAFNE Department), University of Tuscia, Tuscia, Italy
[3]Dipartimento per la innovazione nei sistemi biologici agroalimentari e forestali (DIBAF Department), University of Tuscia, Tuscia, Italy
[4]Department of Mathematical Modelling, Statistics and Bioinformatics, Ghent University, Ghent, Belgium
[5]Laboratory of Hydrology and Water Management, Ghent University, Ghent, Belgium

Correspondence to: Benedikt Gräler (benedikt.graeler@rub.de)

Published: 12 May 2016

Abstract. Many hydrological studies are devoted to the identification of events that are expected to occur on average within a certain time span. While this topic is well established in the univariate case, recent advances focus on a multivariate characterization of events based on copulas. Following a previous study, we show how the definition of the survival Kendall return period fits into the set of multivariate return periods.

Moreover, we preliminary investigate the ability of the multivariate return period definitions to select maximal events from a time series. Starting from a rich simulated data set, we show how similar the selection of events from a data set is. It can be deduced from the study and theoretically underpinned that the strength of correlation in the sample influences the differences between the selection of maximal events.

1 Introduction

Studying extremes in hydrological multivariate time series often aims at getting an estimate of the size of events to be expected in a period of 10, 50 or 100 years. This information is relevant for the construction of many hydrological structures such as dams and dykes. As most of these natural events are characterized by several variables (e.g. peak discharge, volume, duration, ...) and several locations, it is important to understand their dependence structure and which constellations result in an extreme event. Copulas allow to flexibly model the dependence between the variables and add different marginal distribution functions to build a probabilistic multivariate model. The natural ordering in univariate time series does not extend to the multivariate case calling for different tools to identify multivariate extremes.

In a previous study (Gräler et al., 2013), the practical impact of different bivariate multivariate return period definitions has been studied based on a simulated data set. Meanwhile, an additional approach, the survival Kendall return period (SKRP), has been developed (Salvadori et al., 2013). Using the same data as before, the SKRP is calculated and related to the previously studied return periods (AND, OR and Kendall return period).

Currently, multivariate maxima are often selected based on a single driving variable (e.g. peak discharge) and the associated variables (e.g. volume and duration) are studied in a multivariate setting. However, this does not a priori reflect the joint extreme characteristic that is the actual focus of such a study. Different notions of maximality can be defined following the above return period definitions. These allow to calculate the empirical joint extremeness and to select the maxima of multivariate time series.

In this paper, we will only briefly quote the key concepts. The interested reader is referred to the predecessor of this paper, Gräler et al. (2013), for further details. The following section recalls the definitions of the different multivariate return periods under study and puts them into relation. In Sect. 3, the different maxima selection regimes are presented and their effect is studied. Section 4 provides a discussion and conclusions.

2 Multivariate return periods

The driving tool underlying the multivariate return period definitions are copulas. Copulas are multivariate distribution functions defined on the unit hypercube. Based on Sklar's Theorem (Sklar, 1959), they combine marginal distribution functions $F_1, \ldots, F_d$ into multivariate distributions H via $H(x_1, \ldots, x_d) = C(F_1(x_1), \ldots, F_d(x_d))$ and solely determine the entire dependence structure. For a detailed introduction, see e.g. the book by Nelsen (2006). Going from univariate to multivariate extremes is not immediate. One major constrain is the lack of a natural ordering for problems of dimension $d \geq 2$. Typical definitions of the multivariate joint return periods include the OR case corresponding to $P(X_1 > x_1, \ldots, X_d > x_d)$ and the AND case defined for $P(X_1 > x_1 \wedge \ldots \wedge X_d > x_d)$. The Kendall return period (KRP) introduced by Salvadori et al. (2011) is an approach that shares a unique property with the univariate return periods: the critical layer separating safe from dangerous events is unique for every design return period. This is not the case for the OR and AND approaches where different regions of safe and dangerous events exist for the same return period. The basis of the Kendall return period, the *Kendall distribution function* is the distribution function of the copula's mass below its level curves.

Salvadori et al. (2013) present the *survival Kendall return period* (SKRP) to overcome limitations of the *Kendall return period* (KRP) described in Salvadori et al. (2011). The drawback of the latter is its unboundedness. The critical layer splits the region into safe and dangerous events in a way such that one of the margins might tend to infinity (even though with very small probability). This limitation is overcome by the SKRP, as the critical layer is nicely bounded, as for the OR return period, but every point on the critical layer exhibits the same return period, as in the Kendall scenario. In a way, the SKRP combines the best of both worlds. Its mathematical definition reads

$$T_{\mathrm{SKRP}} = \frac{\mu}{1 - \bar{K}(t)}$$

with $\bar{K}$ the survival Kendall distribution function given by

$$\begin{aligned}\bar{K}(t) &= P\big(\bar{F}(x_1, \ldots, x_d) \geq t\big)\\ &= P\big(\hat{C}(\bar{F}_1(x_1), \ldots, \bar{F}_d(x_d)) \geq t\big)\end{aligned}$$

and $\hat{C}$ the survival copula and $\bar{F}_i$ the marginal survival distribution functions. See Salvadori et al. (2013) for the full details.

In order to extend our previous study, we use the same data (simulated using the COSMO4SUB model, Grimaldi et al., 2012, compare Sect. 4 in Gräler et al., 2013) and adopt the same parametrization as in Gräler et al., 2013 to also calculate the SKRP for the bivariate approach. Higher dimensional approaches are out of the scope of this follow-up paper. The peak discharges Q_p are said to follow a Weibull distribution

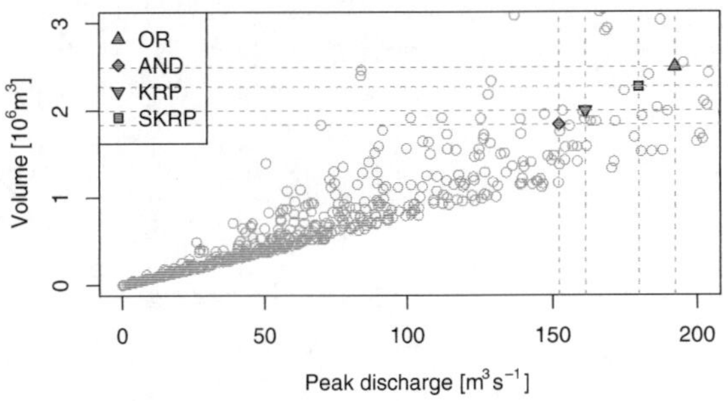

Figure 1. Comparison of different MRP definitions for a return period of 10 years (compare Figure 6 in Gräler et al., 2013).

while the associated volumes follow an exponential distribution. Recall that the selection of the annual maxima was done based on the peak discharges and the volumes are the ones corresponding to the same event, but as such not necessarily the largest one in the respective year.

Figure 1 depicts the four different multivariate return period (MRP) definitions. All points indicate the most likely bivariate event on the respective critical layers. It is important to notice that also an ensemble approach could have been taken, where a series of design events is obtained.

3 Maxima selection

In the previous section and study, the annual maxima were selected based on the maximum peak discharge and the volumes were only the corresponding, but not necessarily maximal ones. An alternate approach can be taken either based on the empirical copula or the adoption of multivariate distributions. For these, the same MRP definitions can be applied as quoted above and the largest values per year can be selected. In the following, we will follow this avenue and investigate the differences between these approaches where the copula C might be the empirical copula or an appropriate family.

For ease of notation, we will stick to bivariate events. We say that an event (x_1, x_2) is *OR-maximal*, if

$$\begin{aligned}F(x_1, x_2) = C(F_1(x_1), F_2(x_2)) &\geq C(F_1(y_1), F_2(y_2))\\ &= F(y_1, y_2)\end{aligned}$$

for all (y_1, y_2) in the same temporal window (i.e. year) as (x_1, x_2). Analogously, we say that an event (x_1, x_2) is *AND-maximal*, if

$$\begin{aligned}1 - &P(X_1 > x_1, X_2 > x_2)\\ &= F_1(x_1) + F_2(x_2) - C(F_1(x_1), F_2(x_2))\\ &\geq F_1(y_1) + F_2(y_2) - C(F_1(y_1), F_2(y_2))\end{aligned}$$

for all (y_1, y_2) in the same temporal window (i.e. year) as (x_1, x_2). Adopting also the Kendall and survival Kendall return period definitions, we say that an event is *Kendall-*

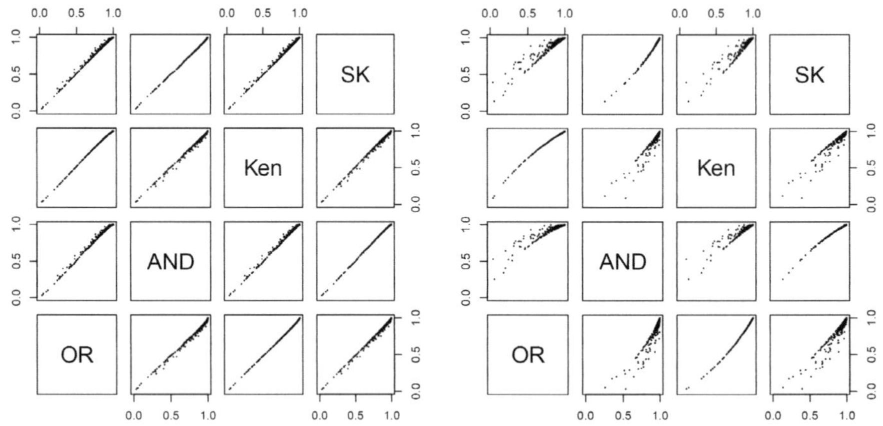

Figure 2. Comparison of the annual maximum value for the four different definitions. Left: based on 500 years of simulated rainfall data. Right: moderately correlated sample of a Gumbel copula.

maximal, if

$$K(C(F_1(x_1), F_2(x_2))) \geq K(C(F_1(y_1), F_2(y_2)))$$

for all (y_1, y_2) in the same temporal window (i.e. year) as (x_1, x_2) where K is the Kendall distribution function corresponding to C. Finally, we call an event *survival Kendall maximal*, if

$$\bar{K}(\bar{C}(1 - F_1(x_1), 1 - F_2(x_2))) \geq \bar{K}(\bar{C}(1 - F_1(y_1), 1 - F_2(y_2)))$$

for all (y_1, y_2) in the same temporal window (i.e. year) as (x_1, x_2) where $\bar{K}$ is the survival Kendall distribution function with corresponding survival copula $\bar{C}$.

To study the impact of the aforementioned definitions, we use a second run of 500 simulated years of 5.625 min resolution discharge data that were aggregated to separate rainfall events. This is different from the previous data set where only annual maxima have been used. This second data set contains 12 466 events. In order to reduce the effect of autocorrelation within this simulation, we only consider a random subset of 50 % of the data (autocorrelation plots indicate an uncorrelated time series, not shown here). We do not fit any parametric family, and solely use the empirical definitions of the above equations.

In our simulated data set, the largest event in a year often is the same for all four definitions. This is not too surprising, considering that there are on average less than 25 rainfall events in each year. What remains different, is how extreme the event is for each of the four notions. The left plot of Fig. 2 illustrates the relationship between the annual maximum value of the different definitions for the studied data set.

Identifying the single rainfall events and looking into the marginal distributions, visually reveals identical histograms for peak discharge as well as for volume for the four bivariate and respective univariate maxima selections. An overlay shows only very little variations for discharge values and volumes. Larger values of the margins tend to even better coincide.

As this data set follows a very strong correlation, we draw a sample of a Gumbel copula with a moderate Kendall's tau of 0.6, assign it to the same temporal structure as our previous data set and repeat the above analysis. The right plot of Fig. 2 is based on the copula sample and shows a larger variation in the annual maximum values for the four approaches. As in the left plot, the OR and KRP as well as the AND and SKRP approaches seem to be much more alike than the other pairwise combinations. The selected margins show a little more variability, but the histograms remain hardly distinguishable. All the variation appears in the center of the distribution.

4 Discussion and conclusion

The SKRP yields the most reliable separation into safe (subcritical) and dangerous (super-critical) events. Nevertheless, the selection of a single design event, as often required by subsequent studies, remains an open question. Here, we selected the most probable bivariate event, but any event along the critical layer separates the sub- and super-critical regions.

The differences between the four definitions of maximality were minor in the simulated rainfall time series, but this is also due to the very strong correlation. This strong dependence causes the copula to be close to the upper Fréchet-Hoeffding bound where all four definitions coincide. If all points lie close to the diagonal, there is no difference whether the critical layer follows the OR definition (enclosing the lower left rectangle), the SKRP definition (very sharply bend contour lines enclosing the lower left region), the KRP definition (very sharply bend contour lines excluding the top right region) or the AND definition (excluding the top right rectangle).

The temporal structure was not changed, only the dependence structure to investigate the effect. A less dry study area with much more rainfall events in a year will further influence the selection. However, the large extreme values appear to be extreme in each of the definitions.

Here we use the raw definitions of multivariate return periods, but an alternative would be to investigate derived measures. Requena et al. (2013) developed the *routed return period* where the water levels in a dam are used to characterize the return period of bivariate rainfall events. This idea could also be used in the same manner as presented in this paper to initially select the largest events from the original time series.

The investigated data set features a very long time series. Shorter time series might be more sensitive to changes of the maximum selection regime applied, as few events might have a strong influence on the selection of the marginal distributions. The influence of these outer properties needs to be further investigated. An avenue of future research is to consider the joint extremeness for the selection of extremes to be fed into a peak over threshold approach.

References

Gräler, B., van den Berg, M. J., Vandenberghe, S., Petroselli, A., Grimaldi, S., De Baets, B., and Verhoest, N. E. C.: Multivariate return periods in hydrology: a critical and practical review focusing on synthetic design hydrograph estimation, Hydrol. Earth Syst. Sci., 17, 1281–1296, doi:10.5194/hess-17-1281-2013, 2013.

Grimaldi, S., Petroselli, A., and Serinaldi, F.: A continuous simulation model for design-hydrograph estimation in small and ungauged watersheds, Hydrolog. Sci. J., 57, 1035–1051, doi:10.1080/02626667.2012.702214, 2012.

Nelsen, R. B.: An Introduction to Copulas, Springer Science+Buisness, New York, second edn., 272 pp., ISBN: 978-0-387-28659-4, 2006.

Requena, A. I., Mediero, L., and Garrote, L.: A bivariate return period based on copulas for hydrologic dam design: accounting for reservoir routing in risk estimation, Hydrol. Earth Syst. Sci., 17, 3023–3038, doi:10.5194/hess-17-3023-2013, 2013.

Salvadori, G., De Michele, C., and Durante, F.: On the return period and design in a multivariate framework, Hydrol. Earth Syst. Sci., 15, 3293–3305, doi:10.5194/hess-15-3293-2011, 2011.

Salvadori, G., Durante, F., and Michele, C.: Multivariate return period calculation via survival functions, Water Resour. Res., 49, 2308–2311, 2013.

Sklar, A.: Fonctions de répartition á *n* dimensions et leurs marges, Publ. Inst. Statist. Univ. Paris, 8, 229–231, 1959.

Proc. IAHS, 373, 179–182, 2016
proc-iahs.net/373/179/2016/
doi:10.5194/piahs-373-179-2016

Open Access
PIAHS

Up-scaling of multi-variable flood loss models from objects to land use units at the meso-scale

Heidi Kreibich, Kai Schröter, and Bruno Merz

German Research Centre for Geosciences GFZ, Section 5.4 Hydrology, Potsdam, Germany

Correspondence to: Heidi Kreibich (heidi.kreibich@gfz-potsdam.de)

Published: 12 May 2016

Abstract. Flood risk management increasingly relies on risk analyses, including loss modelling. Most of the flood loss models usually applied in standard practice have in common that complex damaging processes are described by simple approaches like stage-damage functions. Novel multi-variable models significantly improve loss estimation on the micro-scale and may also be advantageous for large-scale applications. However, more input parameters also reveal additional uncertainty, even more in upscaling procedures for meso-scale applications, where the parameters need to be estimated on a regional area-wide basis.

To gain more knowledge about challenges associated with the up-scaling of multi-variable flood loss models the following approach is applied: Single- and multi-variable micro-scale flood loss models are up-scaled and applied on the meso-scale, namely on basis of ATKIS land-use units. Application and validation is undertaken in 19 municipalities, which were affected during the 2002 flood by the River Mulde in Saxony, Germany by comparison to official loss data provided by the Saxon Relief Bank (SAB).

In the meso-scale case study based model validation, most multi-variable models show smaller errors than the uni-variable stage-damage functions. The results show the suitability of the up-scaling approach, and, in accordance with micro-scale validation studies, that multi-variable models are an improvement in flood loss modelling also on the meso-scale. However, uncertainties remain high, stressing the importance of uncertainty quantification. Thus, the development of probabilistic loss models, like BT-FLEMO used in this study, which inherently provide uncertainty information are the way forward.

1 Introduction

Losses from natural disasters have dramatically increased during the last few decades and floods have generated the largest economic losses, also in Germany (Kreibich et al., 2014). Flood risk analyses are gaining more and more attention in the fields of flood design, prevention, and risk-management (EU flood risk directive 2007/60/EC). Flood risk analyses are performed on different spatial scales (Meyer and Messner, 2005; de Moel et al., 2015): At the micro-scale the assessment is based on single elements at risk. For instance, in order to estimate the loss to a community in case of a certain flood scenario, loss is calculated for each affected object (e.g. building). At the meso-scale the assessment is based on spatial aggregations. Typical aggregation units are land use units, e.g. residential areas. At the macro-scale large-scale spatial units are the basis for loss estimation. Typically, administrative units are used, e.g. municipalities, regions, countries. The classification in micro-, meso- and macro-scale is, on the one hand, related to the spatial extent of the loss assessment. On the other hand, there is a methodological distinction: Meso- and macro-scale approaches differ from micro-scale approaches in their need for aggregation. Loss is assessed for aggregated objects, e.g. land use units. Commonly a bottom-up approach is used, which starts with a detailed analysis and modelling of single elements at risk (micro-scale) and develops an up-scaling procedure for application on basis of land-use units (e.g. Kreibch et al., 2010).

The objective of this study is to gain additional knowledge about challenges associated with the up-scaling of flood loss models, particularly multi-variable loss models. Single- and multi-variable flood loss models are up-scaled to be applied on the meso-scale in a test area. Validation is undertaken via comparisons with official loss data.

2 Loss models

The following seven loss models, which estimate direct economic loss to residential buildings, are up-scaled to be used at the meso-scale.

The probabilistic, multi-variable model BT-FLEMO developed by Merz et al. (2013) is an ensemble of 200 regression trees, which use the following predictors: water depth, return period, contamination indicator, inundation duration, flow velocity indicator, floor space of building, building value and precautionary measures indicator.

The rule based FLEMOps+r model (Elmer et al., 2010) calculates flood loss using five different classes of water depth, three individual building types, two classes of building quality and three classes of return period (Fig. 1).

The regression tree loss model (RT2) developed by Merz et al. (2013) has eight leaves using the predictors water depth, floor space of building, return period, monthly net income of the household.

The stage-damage function of MURL (2000) calculates the loss ratio (rloss [–]) of residential buildings by the equation $\text{rloss} = 0.02\,\text{wst}$, where wst is the water depth [m]. For water levels of more than 5 m the loss ratio is set to 0.1 (Fig. 1).

The stage-damage function of ICPR (2001) estimates the loss ratios of residential buildings by the relation $\text{rloss} = (2\,\text{wst}^2 + 2\,\text{wst})/100$. Estimated loss ratios >1 are set to 1, i.e. total loss (Fig. 1).

The stage-damage function of HYDROTEC (2001, 2002) uses the root function $\text{rloss} = (27\sqrt{\text{wst}})/100$. Estimated loss ratios >1 are set to 1 (Fig. 1).

The stage-damage function sd-f is taken from Merz et al. (2013) and uses the equation $\text{rloss} = 0.0142 + 0.0127 \times \sqrt{(\text{wst} \times 100)}$ (Fig. 1).

3 Up-scaling approach

The up-scaling approach of Kreibich et al. (2010) is followed, i.e. the model structures are not changed, but the input variables of the micro-scale loss models are estimated area-wide for spatially aggregated meso-scale units. The following data is used to estimate the model-input variables for the test area, i.e. 19 municipalities that were affected by the 2002 flood at the river Mulde in Saxony, Germany.

The inundation patterns including water depths distribution of the 2002 flood in the 19 case study municipalities are taken from Grabbert (2006) and Apel et al. (2007). Return periods are taken from Elmer et al. (2010). Contamination indicator, inundation duration, flow velocity indicator, precautionary measures indicator and monthly net income are estimated on basis of empirical flood damage data collected via computer aided telephone interviews with households affected by the 2002 flood in the case study area (Thieken et al., 2005). The average floor space of residential buildings as well as the average building value per municipality are taken from the Germany-wide exposure dataset of Kleist et al. (2006). The residential building type composition and the mean residential building quality per municipality are derived following the approach of Thieken et al. (2008).

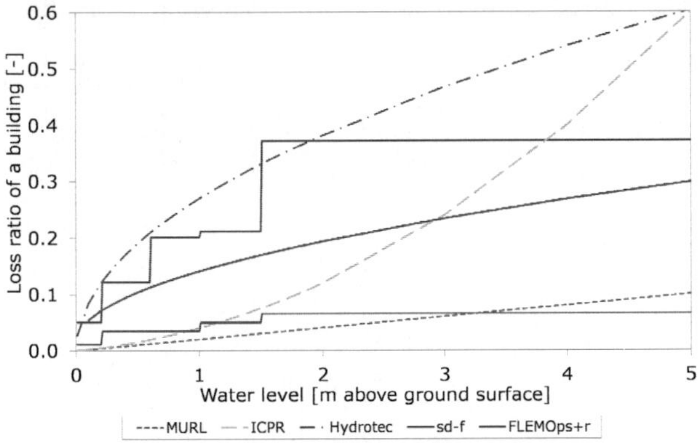

Figure 1. Stage-damage functions as well as upper and lower bounds of the rule based multi-variable model FLEMOps+r used in this up-scaling study.

These meso-scale input variables are estimated on the municipal level, except for water depth and return period, which are given, in a more spatially differentiated format. Water depth is modeled area-wide with 10 m grid resolution. Return periods are estimated on sub-catchment level. All input variables are processed to be available as raster data sets with a cell size of 10 m × 10 m. For each grid cell, the loss ratio is estimated by applying the seven loss models on basis of the meso-scale input variables. These loss ratios are then multiplied by the specific building value assigned to the corresponding grid cell. Finally, the loss estimates are aggregated per municipality. The specific building values were extracted from Wünsch et al. (2009).

4 Application and Validation

Meso-scale model application and validation are conducted in 19 municipalities located at the river Mulde in Saxony, Germany. The area was strongly affected by the 2002 flood. Flood loss was well documented by the Saxon Relief Bank, which was in charge of the loss adjustment and management in Saxony after the flood in 2002.

The above listed seven loss models are used to estimate direct economic losses of residential buildings for the 19 municipalities of the case study area. The modelled aggregated absolute loss to residential buildings per municipality is compared to the official loss information provided by the Saxon Relief Bank (2005).

Table 1. Error statistics of meso-scale model performance (MBE: mean bias error, MAE: mean absolute error).

	MURL	ICPR	HYDROTEC	sd-f	FLEMOps+r	RT2	BT-FLEMO
MBE [Mill. €]	−10.4	−4.2	8.4	−2.0	0.9	−2.6	−0.1
MAE [mill. €]	11.5	10.5	12.1	9.8	9.0	10.1	9.2

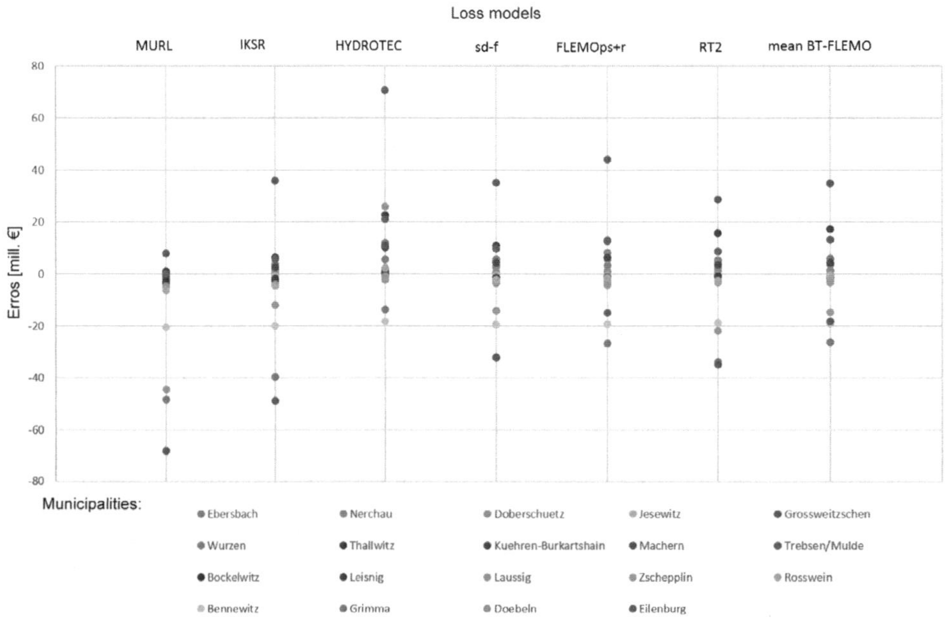

Figure 2. Errors, i.e. estimated loss minus officially reported loss [million euros] for all tested models and municipalities.

5 Results and Discussion

The two best performing models in terms of mean absolute error are the multi-variable models FLEMOps+r and BT-FLEMO (Table 1). BT-FLEMO additionally provides a nearly unbiased prediction. The models which were developed on basis of empirical loss data from recent flood events in 2002, 2005 and 2006, namely sd-f, FLEMOps+r, RT2 and BT-FLEMO, perform better than the other models MURL, ICPR, and HYDROTEC which were developed on basis of expert judgement and loss data from floods between 1978 and 1994.

Multi-variable models outperform stage-damage functions and are as such an improvement in flood loss modelling also on the meso-scale. However, the model sd-f with one input variable shows in comparison still relatively good error statistics (Table 1), and is as such also suitable for meso-scale loss estimation. Uncertainties of loss estimation remain high, which underlines the importance of uncertainty quantification. The probabilistic loss model BT-FLEMO is as such a significant advancement.

The error statistics are strongly influenced by five municipalities, which appear particularly problematic for loss estimation (Fig. 2): Bennewitz, Eilenburg, Grimma and Döbeln where most models underestimate the loss as well as Grossweitzschen where all models overestimate the loss. The different loss models provide coherent results in terms of underestimation in some municipalities and overestimation in others. The within model variation of loss predictions are smaller than within municipalities (Fig. 2). Seifert et al. (2010) reported similar patterns for loss estimation for the commercial sector. They relate large errors in loss estimation to high uncertainties in the exposure estimation, particularly in municipalities with a small fraction of affected companies. This is in accordance with the presented results: smaller errors in loss estimation are obtained for municipalities which incurred larger total loss in comparison to municipalities which incurred a total loss below about EUR 3 million officially reported loss (exception is Bennewitz).

6 Conclusions

Multi-variable models outperform stage-damage functions and are as such an improvement in flood loss modelling also on the meso-scale. However, more input variables also reveal additional uncertainty, even more in up-scaling procedures, where the model input variables are estimated on a regional area-wide basis. Hence, a suitable compromise between model performance and number of variables should be aimed for. Further research should develop more spatially differentiated estimation methods for key input variables used in flood loss estimation at the meso-scale. The use of remote sensing data is a way forward in this respect (Gerl et al., 2014). This study shows that uncertainties of loss estimation remain high. Therefore, thorough validations and uncertainty analyses are necessary for the development of reliable loss models as a basis for application in flood risk studies. The development of probabilistic loss models, possibly with less input variables for an easier meso-scale application, is the way forward.

In municipalities where all loss models significantly over- or underestimate the official loss report it is likely that errors and uncertainty result from other sources along the loss estimation chain, e.g. from inundation modelling or exposure estimation. A further source is the uncertainty of the official loss data. These other sources of uncertainty are not addressed in this study. However, comprehensive uncertainty analyses including all components of the flood risk model chain, as for instance undertaken by Apel et al. (2009), should be further developed.

References

Apel, H., Aronica, G. T., Kreibich, H., and Thieken, A. H.: Evaluation of different modelling strategies for flood risk assessment in urban areas, in: Proceedings of the 32nd Congress of IAHR July 1–6 2007, Venice, Italy, International Association of Hydraulic Engineering & Research, 2007.

Apel, H., Aronica, G. T., Kreibich, H., and Thieken, A. H.: Flood risk assessments – How detailed do we need to be?, Nat. Hazards, 49, 79–98, 2009.

de Moel, H., Jongman, B., Kreibich, H., Merz, B., Penning-Rowsell, E., and Ward, P. J.: Flood risk assessments at different spatial scales, Mitigation and Adaptation Strategies for Global Change, 20, 865–890, 2015.

Elmer, F., Thieken, A. H., Pech, I., and Kreibich, H.: Influence of flood frequency on residential building losses, Nat. Hazards Earth Syst. Sci., 10, 2145–2159, doi:10.5194/nhess-10-2145-2010, 2010.

Gerl, T. M., Bochow, M., and Kreibich, H.: Flood Damage Modeling on the Basis of Urban Structure Mapping Using High-Resolution Remote Sensing Data, Water, 6, 2367–2393, 2014.

Grabbert, J.-H.: Analyse der schadensbeeinflussenden Faktoren des Hochwassers 2002 und Ableitung eines mesoskaligen Abschätzungsmodells für Wohngebädeschäden [Analysis of parameters influencing loss during the flood event 2002 and development of a meso-scale estimation model for residential building damages], Unpublished thesis, University of Potsdam, 2006.

HYDROTEC: Hochwasser-Aktionsplan Angerbach [Flood action plan for the river Angerbach]. Teil I: Berichte und Anlagen, National Environmental Agency of the City of Düsseldorf, Aachen, Germany, 2001.

HYDROTEC: Hochwasser-Aktionsplan Lippe [Flood action plan for the river Lippe]. Teil I: Berichte und Anlagen, National Environmental Agency of the City of Lippstadt, Aachen, Germany, 2002.

ICPR (International Commission for the Protection of the Rhine): Rhine-Atlas, ICPR, Koblenz, available at: http://www.rheinatlas.de, 2001.

Kleist, L., Thieken, A. H., Köhler, P., Müller, M., Seifert, I., Borst, D., and Werner, U.: Estimation of the regional stock of residential buildings as a basis for a comparative risk assessment in Germany, Nat. Hazards Earth Syst. Sci., 6, 541–552, doi:10.5194/nhess-6-541-2006, 2006.

Kreibich, H., Seifert, I., Merz, B., and Thieken, A. H.: Development of FLEMOcs – A new model for the estimation of flood losses in companies, Hydrolog. Sci. J., 55, 1302–1314, 2010.

Kreibich, H., Bubeck, P., Kunz, M., Mahlke, H., Parolai, S., Khazai, B., Daniell, J., Lakes, T., and Schröter, K.: A review of multiple natural hazards and risks in Germany, Nat. Hazards, 74, 2279–2304, 2014.

Merz, B., Kreibich, H., and Lall, U.: Multi-variate flood damage assessment: a tree-based data-mining approach, Nat. Hazards Earth Syst. Sci., 13, 53–64, doi:10.5194/nhess-13-53-2013, 2013.

Meyer, V. and Messner, F.: National Flood Damage Evaluation Methods: A Review of Applied Methods in England, the Netherlands, the Czech Republic and Germany, UFZ-Discussion Papers, www.floodsite.net, 2005.

MURL (Ministerium für Umwelt, Raumordnung und Landwirtschaft des Landes Nordrhein-Westfalen): Potentielle Hochwasserschäden am Rhein in Nordrhein-Westfalen, Düsseldorf, 2000.

Saxon Relief Bank (SAB): Schäden durch das Hochwasser 2002, Angaben aus der Fördermitteldatenbank [Damages of the flood 2002. Data from the Compensation Data Bank], February, 2005.

Seifert, I., Kreibich, H., Merz, B., and Thieken, A. H.: Application and validation of FLEMOcs – a flood loss estimation model for the commercial sector, Hydrolog. Sci. J., 55, 1315–1324, 2010.

Thieken, A. H., Müller, M., Kreibich, H., and Merz, B.: Flood damage and influencing factors: New insights from the August 2002 flood in Germany, Water Resour. Res., 41, W12430, doi:10.1029/2005WR004177, 2005.

Thieken, A. H., Olschewski, A., Kreibich, H., Kobsch, S., and Merz, B.: Development and evaluation of FLEMOps – a new Flood Loss Estimation MOdel for the private sector, in: Flood Recovery, Innovation and Response, edited by: Proverbs, D., Brebbia, C. A., and Penning-Rowsell, E., WIT Press, 315–324, 2008.

Wünsch, A., Herrmann, U., Kreibich, H., and Thieken, A. H.: The Role of Disaggregation of Asset Values in Flood Loss Estimation: A Comparison of Different Modeling Approaches at the Mulde River, Germany, Environ. Manage., 44, 524–541, 2009.

Proc. IAHS, 373, 183–187, 2016
proc-iahs.net/373/183/2016/
doi:10.5194/piahs-373-183-2016

A framework of integrated hydrological and hydrodynamic models using synthetic rainfall for flash flood hazard mapping of ungauged catchments in tropical zones

Worapong Lohpaisankrit[1], **Günter Meon**[1], **and Tawatchai Tingsanchali**[2]

[1]Department of Hydrology, Water Management and Water Protection, University of Braunschweig, Braunschweig, 38106, Germany
[2]School of Engineering and Technology, Asian Institute of Technology, Pathumthani, 12120, Thailand

Correspondence to: Worapong Lohpaisankrit (w.lohpaisankrit@gmail.com)

Published: 12 May 2016

Abstract. Flash flood hazard maps provide a scientific support to mitigate flash flood risk. The present study develops a practical framework with the help of integrated hydrological and hydrodynamic modelling in order to estimate the potential flash floods. We selected a small pilot catchment which has already suffered from flash floods in the past. This catchment is located in the Nan River basin, northern Thailand. Reliable meteorological and hydrometric data are missing in the catchment. Consequently, the entire upper basin of the main river was modelled with the help of the hydrological modelling system PANTA RHEI. In this basin, three monitoring stations are located along the main river. PANTA RHEI was calibrated and validated with the extreme flood events in June 2011 and July 2008, respectively. The results show a good agreement with the observed discharge data. In order to create potential flash flood scenarios, synthetic rainfall series were derived from temporal rainfall patterns based on the radar-rainfall observation and different rainfall depths from regional rainfall frequency analysis. The temporal rainfall patterns were characterized by catchment-averaged rainfall series selected from 13 rainstorms in 2008 and 2011 within the region. For regional rainfall frequency analysis, the well-known L-moments approach and related criteria were used to examine extremely climatic homogeneity of the region. According to the L-moments approach, Generalized Pareto distribution was recognized as the regional frequency distribution. The synthetic rainfall series were fed into the PANTA RHEI model. The simulated results from PANTA RHEI were provided to a 2-D hydrodynamic model (MEADFLOW), and various simulations were performed. Results from the integrated modelling framework are used in the ongoing study to regionalize and map the spatial distribution of flash flood hazards with four levels of flood severities. As an overall outcome, the presented framework can be applied in areas with inadequate runoff records.

1 Introduction

Flash floods are distinguished from river floods by high flow-velocities and rapid rises of water levels within a short period. Due to their typical characteristics, flash floods are difficult to forecast and usually result in the limitation of time available for warning to be prepared. Thus, they are the most dangerous events (Collier, 2007). Based on statistical analysis of Jonkman and Vrijling (2008), the average mortality for flash floods is highest compared to drainage and river floods. Therefore, flash flood hazard maps are urgent necessity.

Flash flood hazard maps are an essential tool for flood risk mitigation and management. However, flash flood hazard mapping is complicated and faces many uncertainties because flash floods often occur in ungauged basins. Thus, flash flood studies are challenging and still poorly understood for many countries, especially tropical regions. One of the major challenges is the scarcity of reliable runoff data and high

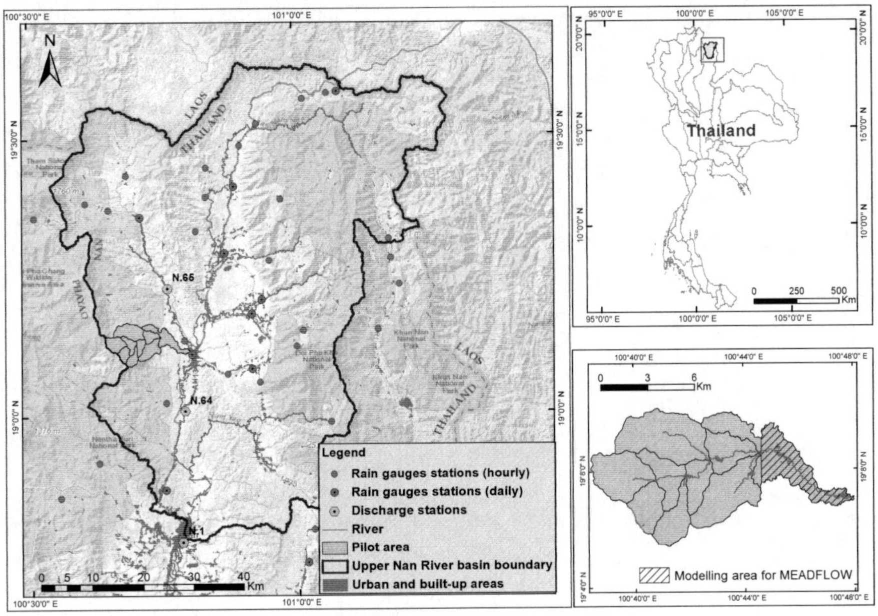

Figure 1. Study area: upper Nan River basin (left), a pilot catchment (right bottom).

resolution rainfall data in both spatial and temporal scales (Marchi et al., 2010). As will be seen, this is one of the main challenges this study aims to solve.

Physically based distributed hydrological models, rainfall-runoff models with high spatial and temporal resolutions, help to understand mechanisms of flash floods. They can be used to deal with the lack of quantitative information in ungauged basins as well. The advantage of the distributed hydrological models is that their input parameters are often related to the physical attributes of the catchments (Ogden et al., 2000). Thus, hydrological models have been applied in flash flood studies by numerous researchers (Rulli and Rosso, 2002; Reed et al., 2007; Braud et al., 2010). Even though many advanced hydrological models are available, the lack of temporal high resolution of rainfall deprives researchers of accurately estimating potential of flash floods.

Sarhadi et al. (2012) mapped floodplains at ungauged rivers based on flood frequency analysis of annual maximum streamflow data and multivariate linear regression models using catchment characteristics. Flood hazard maps were created with the help of a one dimensional hydraulic model; notwithstanding, rainfall which is the most important factor in flash floods was omitted. In contrast to Sarhadi et al. (2012), this study applied regional frequency analysis to rainfall data since flash floods are often triggered by heavy rainfall.

The aim of the present study is to propose a framework for flash flood hazard mapping. We selected a small catchment with 76 km^2 as a pilot study area. The innovative aspect of this study is that multidisciplinary approaches were applied:

- Synthetic rainfall series were derived from radar-rainfall observation to estimate flash flood potential.
- Regional rainfall frequency analysis was introduced to test homogeneity of the region.
- Finally, integrated hydrological and hydrodynamic modelling was used to delineate flash flood hazard maps.

2 Study region and data

The Nam Rim basin, an ungauged subbasin of the upper Nan River basin, was selected as a pilot catchment covering 76 km^2 (Fig. 1). The Nam Rim stream flows directly into the main river of the Nan River. Elevations in the catchment range from 1455 m to 222 m a.m.s.l. The mean catchment slope is 38 %. In this catchment, flash floods which include debris flow and mudslides have caused flooding on agricultural lands and settlement areas, as seen for example during tropical depression Haima in 2011 (DMR, 2011).

Rainfall data recorded every 24 h with a standard rain gauge. The rainfall data is available from 1993 to 2012. Another type of rain gauges automatically measures rainfall every 15 min. Since 2006, automatic rain gauges have started to operate. Unfortunately, many of them had been operated for only three or four years and contained too few heavy rainfall events for statistical analysis. Other important meteorological data, namely, wind speed, sunshine duration and temperature, were obtained for the purpose of hydrological modelling. In addition, Radar-rainfall observations are available

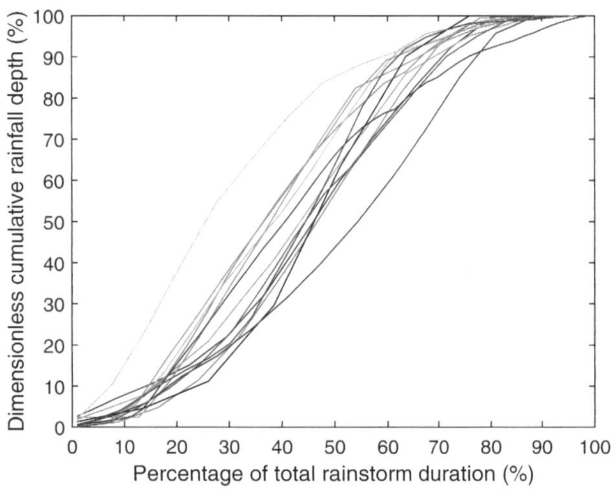

Figure 2. Dimensionless mass curve of 13 rainstorms selected in June, July and August of 2008 and 2011.

for this study area. The radar station with a 240 km observation range covering the study area is located at Chiang Rai (19°57′41″ N, 99°52′53″ E).

In addition to the temporal data required for integrated hydrological and hydrodynamic modelling, spatial data are essential. The spatial data available for this study are digital elevation model (DEM) with 10 m resolution, and land use and soil maps at 1 : 25 000.

3 Methodology

3.1 Development of synthetic rainfall series

The development of synthetic rainfall series comprises two main procedures: temporal rainfall patterns and regional frequency analysis.

The analysis of temporal rainfall patterns was based on radar-rainfall observations. Radar-rainfall images are available every one hour. The temporal patterns of rainfall were characterized by catchment-averaged rainfall series selected from 13 rainstorms of different rainfall durations in June, July and August of 2008 and 2011. In this study, the temporal patterns of rainfall were described in terms of a dimensionless mass curve. The dimensionless mass curve is a plot between dimensionless cumulative rainfall depth and percentage of total rainstorm duration (Fig. 2). To select the representative temporal pattern of rainfall for this region, the 13 temporal patterns of rainfall were averaged.

Regional frequency analysis (RFA) has become necessary when high-resolution rainfall time series are too short for analysis. Thus, annual maximum 24 h (AM24H) rainfall data at 11 rain gauge stations were selected to construct the frequency characteristics of the region. The AM24H rainfall data at each station were screened by Grubbs's test (Grubbs, 1969) for outliers. The outliers which depart significantly from the remaining data were excluded from the analysis because they may distort the statistical properties of data (Chow et al., 1988). In order to test the homogeneity of the selected rain gauges and choose the regional distribution function, L-moment techniques such as discordance and heterogeneity measures were applied (Hosking and Wallis, 1997).

As a result of the previous described steps, synthetic rainfall series are the combination of the temporal rainfall pattern and distribution functions. These rainfall series are a key input for hydrological modelling in order to estimate potential flash floods.

3.2 Hydrological modelling

The objective of the hydrological modelling in this study is to estimate runoff in ungauged basins. The upper Nan River basin was modelled by using PANTA RHEI. The physically based PANTA RHEI model is a semi-distributed hydrological modelling system at the watershed scale (Förster et al., 2014). The watershed is divided into subbasins and into hydrological response units (HRUs).

PANTA RHEI is designed with a modular structure which allows users to choose hydrological approaches based on their research objectives and available data. The major hydrological processes of the model include runoff formation, runoff concentration and channel runoff routing.

To improve the model performance, daily rainfall data were disaggregated with hourly rainfall data observed at neighbouring stations and radar-rainfall images.

To calibrate the model, simulated hydrographs from PANTA RHEI were statistically compared with observed hydrographs from the three river discharge stations: N.65, N.64 and N.1 as shown in Fig. 1. The calibrated parameters were carried out to simulate flood hydrographs in the pilot catchment.

3.3 Flash flood hazard mapping

MEADFLOW, a 2-D hydrodynamic model, was applied to route simulated discharge from PANTA RHEI and calculate the flood depths and velocities at downstream areas of the pilot catchment as shown in Fig. 1. The MEADFLOW model requires as input data:

- the drainage network extracted from the digital elevation model,
- the Manning-Strickler roughness coefficients which can be estimated from suggested values provided by Chow (1959),
- the upstream and downstream boundaries which were assigned from the simulated discharge and bankfull elevation, respectively.

As mentioned earlier, the pilot catchment is an ungauged catchment. Therefore, maximum flood marks, which oc-

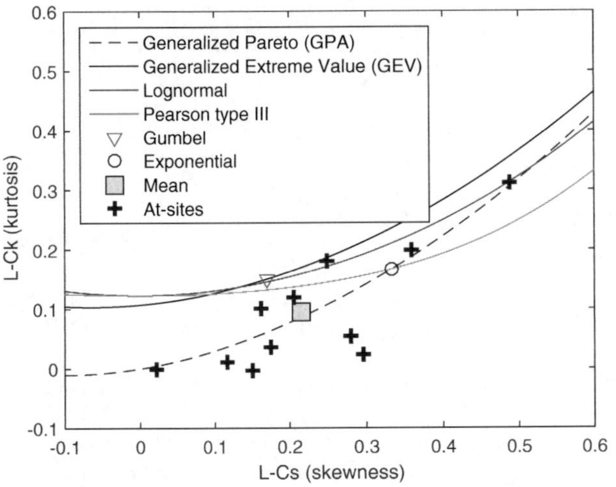

Figure 3. L-moment ratio diagram of 11 rain gauges (at-sites) and regional rainfall data (mean) for selecting the best-fit frequency distribution.

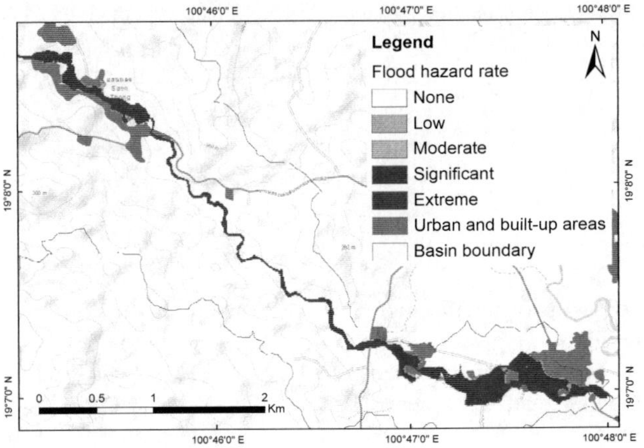

Figure 4. Flood hazard map for downstream areas of the pilot catchment based on the scenario of 15Tr12H.

curred in 2011, and information from eyewitness reports were used to approximately calibrate the model.

To produce a flash flood severity grid, DEFRA (2006) determined the flood hazard rating (FHR) based on the combination of flood depths and velocities. The FHR is computed as

$$\mathrm{FHR} = D(V + 0.5) + \mathrm{DF}, \qquad (1)$$

where FHR is the flood hazard rating (–); D is the depth of flooding (m); V is the flow velocity (m s^{-1}); and DF is the debris factor (–). The debris factor is equal to 1 for $D > 0.25$ m and equal to zero otherwise.

4 Results and discussion

4.1 Regional frequency analysis

Based on the heterogeneity test of the AM24H rainfall data from 11 stations, the region is fairly homogeneous in extremely climatic conditions. The L-moment ratio diagram (Fig. 3) was applied to visually compare the relation of L-Cs and L-Ck for the candidate distributions with corresponding relations obtained from the observed rainfall stations (at-site) and the average of the regional data. The square symbol defined by the regional average L-Cs and L-Ck is close to the GPA distribution. Thus, the GPA distribution is identified as the best fit frequency distribution for this region.

4.2 Hydrological model performance

In order to characterize the flash floods, PANTA RHEI was tested during short periods of extreme flood events. The extreme flood events triggered the upper Nan River basin in July–August 2008 and June 2011. Thus, the period of 25–29 June 2011 were selected as the calibration period of the model. The validation period was between 17 July and 11 August 2008. The results of the model simulation were evaluated by visual comparison with the observed hydrographs and by the help of statistical indices, namely, efficiency index (EI), coefficient of determination (R^2) and relative peak error (RPE) in Table 1.

According to the analysis of synthetic rainfall series using PANTA RHEI, 15 year return period of rainfall occurring within 12 h (15Tr12H) has potential to cause flash floods in the catchment. The simulated flash flood compares to the flooding in 2011.

4.3 Flood hazard maps

The flood hazard map is based on the scenario of 15Tr12H which is in similar manner to 2011 flood occurring at downstream areas of the pilot catchment as shown in Fig. 4. The flood hazard ratings were classified into four levels of flood severities: low, moderate, significant and extreme. The four levels of classification were established based on potential flash flood hazard to people as a function of water depth and velocity (DEFRA, 2006). The key parameters needed to calculate the flood hazard rating was done by using MEADFLOW.

5 Conclusions

This paper presents the framework for flash flood hazard mapping in a small ungauged catchment as an example. In tropical countries, such as those in Southeast Asia, intensive guidelines for mapping flash flood hazard are needed. For the creation of flash flood scenarios, our analysis is based on regional frequency analysis. In the regional frequency analysis of rainfall events, L-moments approach is a practical and ro-

Table 1. Statistical indices indicating performance of the PANTA RHEI model

Discharge stations	Calibration (26–30 June 2011)			Validation (17 July–11 August 2008)		
	EI	R^2	RPE (%)	EI	R^2	RPE (%)
N.65	0.78	0.91	0.18	0.77	0.82	−4.71
N.64	0.95	0.96	0.04	0.83	0.87	6.60
N.1	0.80	0.84	−0.95	0.86	0.87	3.41

bust tool for identifying the best frequency distribution of a region.

PANTA RHEI was employed to the parent watershed of the ungauged catchment. The capability of PANTA RHEI for producing extreme flood events reasonably performs when subdaily rainfall series are input.

To produce flash flood hazard maps at the ungauged catchment, simulated results from PANTA RHEI were integrated with MEADFLOW. Outcomes from MEADFLOW may vary with response to roughness coefficients. Thus, high flood marks and information from eyewitness are necessary for estimating the coefficients.

Using the presented framework, various flash flood hazard maps based on different rainfall scenarios can be evaluated. This information is considered to be useful in the context of flood protection and management.

Acknowledgements. The present research is financially supported by German Academic Exchange Service (DAAD) (funding programme number 57076385). The authors would like to acknowledge the Leichtweiss Institute for Hydraulic Engineering and Water Resources (LWI) and the Ingenieure für Wasser, Umwelt und Datenverarbeitung GmbH (IWUD) for technical support. The authors would like to express their appreciation to the Royal Irrigation Department (RID), Thai Meteorological Department (TMD), Department of Water Resources (DWR) and Land Development Department (LDD) for providing information and datasets.

References

Braud, I., Roux, H., Anquetin, S., Maubourguet, M.-M., Manus, C., Viallet, P., and Dartus, D.: The use of distributed hydrological models for the Gard 2002 flash flood event: Analysis of associated hydrological processes, J. Hydrol., 394, 162–181, doi:10.1016/j.jhydrol.2010.03.033, 2010.

Chow, V. T.: Open-channel hydraulics, McGraw-Hill Kogakusha Ltd, Tokyo, 1959.

Chow, V. T., Maidment, D. R., and Mays, L. W.: Applied hydrology, McGraw-Hill, New York, 1988.

Collier, C. G.: Flash flood forecasting: What are the limits of predictability?, Q. J. Roy. Meteor. Soc., 133, 3–23, doi:10.1002/qj.29, 2007.

DEFRA: Flood Risks to People: Phase 2, FD2321/TR2 Guidance Document, DEFRA/Environment Agency, London, 2006.

DMR: Landslides risk maps at community levels, Nan provice (in Thai), Department of Mineral Resources (DMR), Bangkok, 2011.

Förster, K., Meon, G., Marke, T., and Strasser, U.: Effect of meteorological forcing and snow model complexity on hydrological simulations in the Sieber catchment (Harz Mountains, Germany), Hydrol. Earth Syst. Sci., 18, 4703–4720, doi:10.5194/hess-18-4703-2014, 2014.

Grubbs, F. E.: Procedures for Detecting Outlying Observations in Samples, Technometrics, 11, 1–22, 1969.

Hosking, J. and Wallis, J. R.: Regional frequency analysis: An approach based on L-moments, Cambridge University Press, Cambridge and New York, 1997.

Jonkman, S. N. and Vrijling, J. K.: Loss of life due to floods, Journal of Flood Risk Management, 1, 43–56, doi:10.1111/j.1753-318X.2008.00006.x, 2008.

Marchi, L., Borga, M., Preciso, E., and Gaume, E.: Characterisation of selected extreme flash floods in Europe and implications for flood risk management, J. Hydrol., 394, 118–133, doi:10.1016/j.jhydrol.2010.07.017, 2010.

Ogden, F. L., Sharif, H. O., Senarath, S., Smith, J. A., Baeck, M. L., and Richardson, J. R.: Hydrologic analysis of the Fort Collins, Colorado, flash flood of 1997, J. Hydrol., 228, 82–100, doi:10.1016/S0022-1694(00)00146-3, 2000.

Reed, S., Schaake, J., and Zhang, Z.: A distributed hydrologic model and threshold frequency-based method for flash flood forecasting at ungauged locations, J. Hydrol., 337, 402–420, doi:10.1016/j.jhydrol.2007.02.015, 2007.

Rulli, M. C. and Rosso, R.: An integrated simulation method for flash-flood risk assessment: 1. Frequency predictions in the Bisagno River by combining stochastic and deterministic methods, Hydrol. Earth Syst. Sci., 6, 267–284, doi:10.5194/hess-6-267-2002, 2002.

Sarhadi, A., Soltani, S., and Modarres, R.: Probabilistic flood inundation mapping of ungauged rivers: Linking GIS techniques and frequency analysis, J. Hydrol., 458-459, 68–86, doi:10.1016/j.jhydrol.2012.06.039, 2012.

Proc. IAHS, 373, 189–192, 2016
proc-iahs.net/373/189/2016/
doi:10.5194/piahs-373-189-2016

Open Access
PIAHS

Long term prediction of flood occurrence

Cristina Aguilar[1], Alberto Montanari[2], and María José Polo[3]

[1]Fluvial dynamics and hydrology research group, Andalusian Institute of Earth System Research, University of Granada, Granada, Spain
[2]Department DICAM, University of Bologna, Bologna, Italy
[3]Fluvial dynamics and hydrology research group, Andalusian Institute of Earth System Research, University of Cordoba, Cordoba, Spain

Correspondence to: Cristina Aguilar (caguilar@ugr.es)

Published: 12 May 2016

Abstract. How long a river remembers its past is still an open question. Perturbations occurring in large catchments may impact the flow regime for several weeks and months, therefore providing a physical explanation for the occasional tendency of floods to occur in clusters. The research question explored in this paper may be stated as follows: can higher than usual river discharges in the low flow season be associated to a higher probability of floods in the subsequent high flow season? The physical explanation for such association may be related to the presence of higher soil moisture storage at the beginning of the high flow season, which may induce lower infiltration rates and therefore higher river runoff. Another possible explanation is persistence of climate, due to presence of long-term properties in atmospheric circulation. We focus on the Po River at Pontelagoscuro, whose catchment area amounts to 71 000 km^2. We look at the stochastic connection between average river flows in the pre-flood season and the peak flows in the flood season by using a bivariate probability distribution. We found that the shape of the flood frequency distribution is significantly impacted by the river flow regime in the low flow season. The proposed technique, which can be classified as a data assimilation approach, may allow one to reduce the uncertainty associated to the estimation of the flood probability.

1 Introduction

Perturbations occurring in large catchments may impact the flow regime for several weeks and months, therefore providing a physical explanation for the occasional tendency of floods to occur in clusters (Montanari, 2012). In the Po river, for instance, it has been observed that some flood events have been preceded by long lasting average flows. The physical explanation for such association may be related to the presence of higher than usual soil moisture storage, which may induce lower infiltration rates and therefore higher river runoff. Another possible explanation is persistence of climate, due to presence of long-term properties in atmospheric circulation.

It is well known that river flows are affected by forms of persistence that are not fully understood yet (O'Connell et al., 2015). These are referred to as the "Hurst Phenomenon", or the "Hurst Effect". The Hurst Effect has been physically explained as an implication of the principle of maximum entropy (Koutsoyiannis et al., 2011; Koutsoyiannis, 2014) and implies the presence of long-term cycles over a multitude of time scales. Therefore, the presence of long memory is connected to the possible occurrence of long-term cycles that imply the persistence of high and extreme flows.

With the idea that extreme floods may be induced by long term stress, rather than a short sequence of extreme rainfall, this paper explores the following research question: can higher than usual river discharges in the low flow season be associated to a higher probability of floods in the subsequent high flow season? An application in the Po River is carried out in order to set up a methodology to update the uncertainty associated to the estimation of flood occurrence probability.

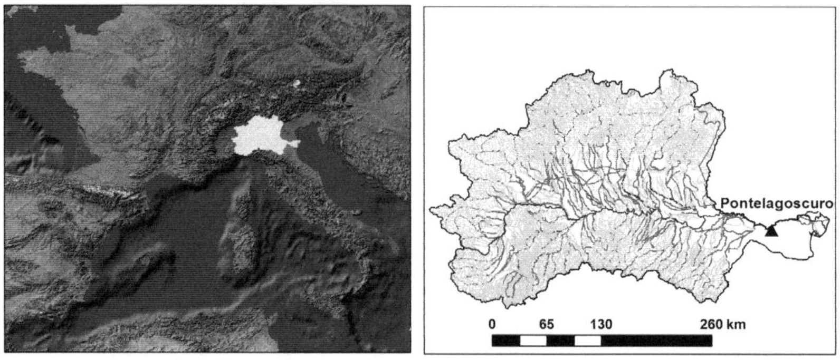

Figure 1. Po River Basin. Location (left), drainage network and closure section at Pontelagoscuro (right).

2 Study site and data sources

The Po River whose catchment has an area of about 71 000 km^2 is the longest river entirely flowing in the Italian Peninsula (Fig. 1). The average annual precipitation in the catchment is 78 km^3, of which 60 % reaches the closure river cross-section at Pontelagoscuro where the mean annual flow is about 1470 $m^3 s^{-1}$. An intense exploitation of water resources for irrigation, hydro-power production, civil and industrial use is found in the catchment. Even though the situation is currently sustainable on average, it might be problematic during drought periods (Montanari, 2012). The hydrological behavior of the Po River is described in detail in recent studies (Zanchettin et al., 2008; Montanari, 2012; Zampieri et al., 2015).

Daily discharge time series for the Po River Basin in Pontelagoscuro were analyzed in this study. The observation period of the complete series was 1920–2009. The discharge pattern shows a typical pluvial regime and thus a strong seasonality with two flood seasons in spring and autumn (Fig. 2).

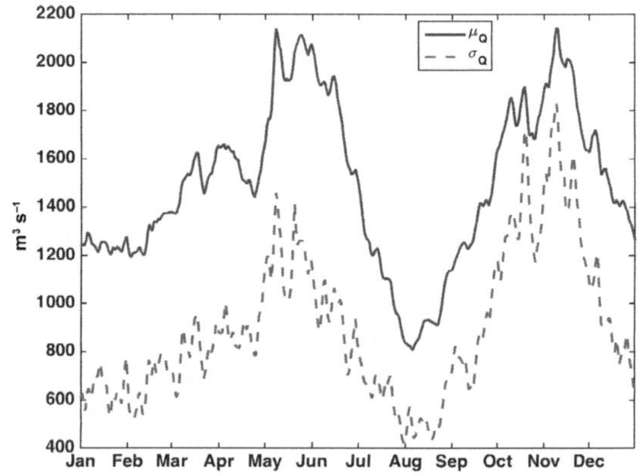

Figure 2. Daily mean value μ_Q ($m^3 s^{-1}$) and daily standard deviation σ_Q ($m^3 s^{-1}$) of the daily flows in the observation period (1920–2009) at Pontelagoscuro.

3 Bivariate probability distribution fitting

In order to look at the stochastic connection between the average river flows in the pre-flood season and the peak flows in the flood season a bivariate probability distribution function is fitted to observed data sets. In what follows, random variables and their outcomes are identified with bold and unbold characters, respectively. The yearly random variables included in the analysis were:

- Monthly mean flow in the pre-flood season, Q_m.
- Peak flow in the flood season, Q_p.

First, the time series $Q_m(t)$ and $Q_p(t)$ with sample size n, where n is the number of years in the observation period, are extracted from the observed data sets. Then, the Normal Quantile Transform (NQT) is applied in order to make their marginal probability distributions Gaussian, therefore obtaining the normalized observations $NQ_m(t)$ and $NQ_p(t)$. A detailed description of the application of the NQT in hydrological studies can be found in the literature (e.g. Moran, 1970; Montanari and Brath, 2004; Montanari, 2005; Montanari and Grossi, 2008; Bogner et al., 2012).

Finally, a bivariate Gaussian distribution function between both canonical Gaussian random variables is fitted. The parameters of the distribution are the mean $\mu(NQ_m) = 0$ and $\mu(NQ_p) = 0$ and the standard deviation $\sigma(NQ_m) = 1$ and $\sigma(NQ_p) = 1$ of the normalized series, and the Pearson's cross correlation coefficient between both normalized series, $\rho(NQ_m, NQ_p)$. In the presence of dependence between NQ_m and NQ_p, the correlation coefficient will be significantly different from zero. The bivariate Gaussian distribution implies that, for an arbitrary (observed) $NQ_m(t)$, the probability distribution of NQ_p is Gaussian, with parameters (Eqs. 1, 2):

$$\mu\left(NQ_p\right) = \rho\left(NQ_m, NQ_p\right) \cdot NQ_m(t), \quad (1)$$

$$\sigma\left(NQ_p\right) = \left(1 - \rho^2\left(NQ_m, NQ_p\right)\right)^{0.5}. \quad (2)$$

Then, by taking the inverse of the NQT one can infer the updated probability distribution of Q_p conditioned to the ob-

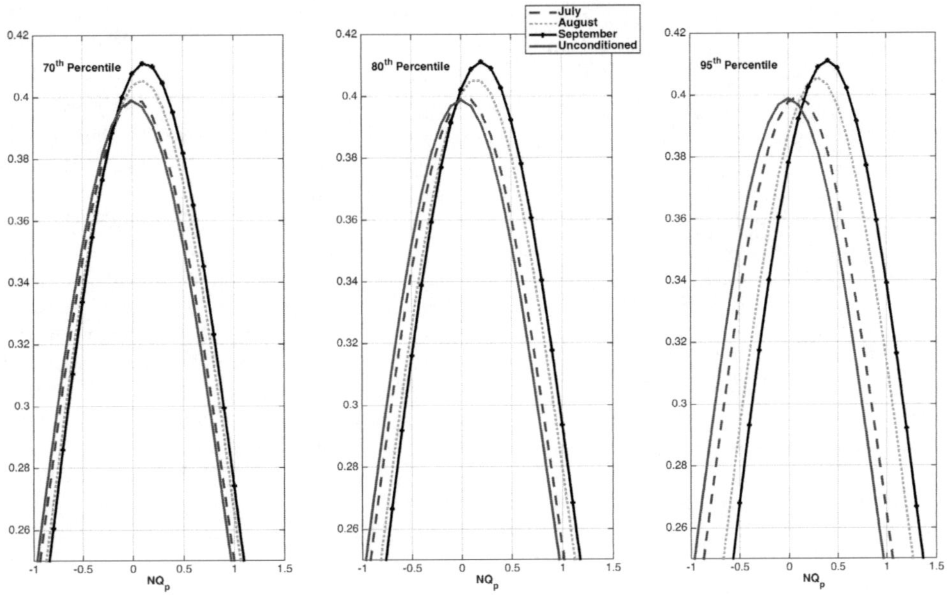

Figure 3. Probability distribution functions of the normalized dependent variable (NQ_p) conditioned to the occurrence of the 70th, 80th and 95th percentiles of the normalized variables in the pre-flood season.

served outcome $Q_m(t)$. Therefore, once the parameters of the distribution are computed, the probability distribution of the peak flow can be updated after observing the average flow in the considered low flow season.

The following two main assumptions are applied in this study. (1) The peak flows season covers the months of October and November in the Po River. Thus, the low flow season is assessed in the previous months to the peak flows season (July–September). Nevertheless, the methodology allows the user to select the seasons arbitrarily so that it can be applied to any other study site or hydrological regime. (2) For the sake of comparison, peak flows can be adequately modeled through the EV1 distribution.

In order to infer the actual impact of the dependence between peak flows and average flow in the low flow season, the unconditioned flood frequency distribution and the updated distributions inferred for several levels higher-than-average values of mean flow (e.g. 70, 80, and 95 % quantiles) in the pre-flood season were compared.

4 Results

The correlation coefficient between NQ_p and NQ_m was calculated by considering different observation periods for Q_m. In detail, we assumed that Q_m is given by the monthly mean flow in each of the 9 months preceding the high flow season (from September to January). Table 1 shows the decrease in the correlation coefficient as the considered low flow period moves backward, as one would expect.

Table 1. Correlation coefficient between NQ_p and NQ_m for varying low flow season in the abscissa.

Month	$\rho(NQ_m, NQ_p)$
September	0.24
August	0.18
July	0.06
June	0.02
May	−0.06
April	−0.13
March	−0.18
February	−0.04
January	−0.07

The effect of the identified dependence on peak flow estimation, for an assigned return period, is shown in Fig. 3 for three different levels of mean flow (70, 80, and 95 % quantiles) in the considered pre-flood season. The probability distribution functions (pdf) of the normalized observed variable, NQ_p, with mean zero and standard deviation 1 is also displayed for the sake of comparison and denoted as unconditioned in Fig. 3. We can appreciate that the higher the cross correlation value, the lower the variability in the distribution of the normalized dependent variable and the higher the mean value. For example, when estimating the probability distribution of NQ_p conditioned to the occurrence of the 95th quantile value in the normalized mean flow in September, the pdf is centered around a mean vaue of 0.4 and presents a standard deviation of 0.97. In contrast, if one attempts to estimate the probability distribution of NQ_p conditioned to the occurrence

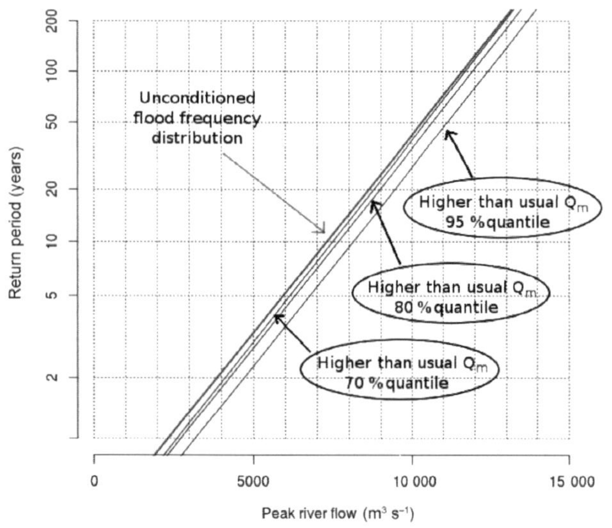

Figure 4. Peak flows in the flood season (October–November) vs return period modeled through the EV1 distribution function.

of the 95th quantile of the normalized mean flow in June, no significant change is found in the estimate with respect to the unconditioned distribution. In fact, the resulting probability density function (pdf) for NQ_p is centered around a mean value of 0.03 with a standard deviation of 0.998. In what follows, September was selected as the pre-flood season in the study site.

Once a pre-flood season was identified it is possible to update the flood frequency distribution after the observation of $Q_m(t)$. Figure 4 shows the comparison between the unconditioned flood frequency distribution and the simulated updated distributions when the flow in September is higher than usual (70, 80, and 95 % quantile). For example, the unconditioned expected flood for a return period of 200 years, 12 507 $m^3 s^{-1}$, increases up to 13 790 $m^3 s^{-1}$ when the mean flow in September corresponds to its 95 % quantile).

5 Conclusions

We found that the peak flow of the Po River is dependent on the average flow of the pre-flood season. Thus, we conclude that it is possible to update the flood frequency distribution basing on discharge observations during the low flow season. To this end, we use a bivariate Gaussian distribution function to model the above dependence. The methodology herein proposed can be applied to any other study site once the flood season is identified and the parameters of the bivariate distribution confirm the presence of the above stochastic dependence.

Several possible physical explanation can be postulated for the sensitivity of the peak flow to the mean discharge in the preceding low flow season, such as the impact of the catchment storage or soil moisture, which in turn impact the formation of net rainfall, and the existence of memory in the weather. Current research is focusing on gaining a better understanding of the processes leading to the formation of the flood flows and in particular the related weather dynamics. Furthermore, we are carrying out experiments on several other rivers in the attempt to relate the above dependence to catchment properties.

Acknowledgements. Cristina Aguilar acknowledges funding by the Jose Castillejo Programme (Grant number CAS14/00432) and Juan de la Cierva Fellowship Programme (Grant number JCI-2012-12802) of the Spanish Ministry of Economy and Competitiveness. The present work was (partially) developed within the framework of the Panta Rhei Research Initiative of the International Association of Hydrological Sciences (IAHS).

References

Bogner, K., Pappenberger, F., and Cloke, H. L.: Technical Note: The normal quantile transformation and its application in a flood forecasting system, Hydrol. Earth Syst. Sci., 16, 1085–1094, doi:10.5194/hess-16-1085-2012, 2012.

Koutsoyiannis, D.: Entropy: from thermodynamics to hydrology, Entropy, 16, 1287–1314, doi:10.3390/e16031287, 2014.

Koutsoyiannis, D., Paschalis, A., and Theodoratos, N.: Two-dimensional Hurst–Kolmogorov process and its application to rainfall fields, J. Hydrol., 398, 91–100, doi:10.1016/j.jhydrol.2010.12.012, 2011.

Montanari, A.: Deseasonalisation of hydrological time series through the normal quantile transform, J. Hydrol., 313, 274–282, doi:10.1016/j.jhydrol.2005.03.008, 2005.

Montanari, A.: Hydrology of the Po River: looking for changing patterns in river discharge, Hydrol. Earth Syst. Sci., 16, 3739–3747, doi:10.5194/hess-16-3739-2012, 2012.

Montanari, A. and Brath, A.: A stochastic approach for assessing the uncertainty of rainfall-runoff simulations, Water Resour. Res., 40, W01106, doi:10.1029/2003WR002540, 2004.

Montanari, A. and Grossi, G.: Estimating the uncertainty of hydrological forecasts: A statistical approach, Water Resour. Res., 44, W00B08, doi:10.1029/2008WR006897, 2008.

Moran, P.: Simulation and Evaluation of Complex Water Systems Operations, Water Resour. Res., 6, 1737–1742, doi:10.1029/WR006i006p01737, 1970.

O'Connell, P. E., Koutsoyiannis, D., Lins, H. F., Markonis, Y., Montanari, A., and Cohn, T.: The scientific legacy of Harold Edwin Hurst (1880–1978), Hydrol. Sci. J., doi:10.1080/02626667.2015.1125998, online first, 2015.

Zampieri, M., Scoccimarro, E., Gualdi, S., and Navarra, A.: Observed shift towards earlier spring discharge in the main Alpine rivers, Sci. Total Environ., 503–504, 222–232, doi:10.1016/j.scitotenv.2014.06.036, 2015.

Zanchettin, D., Traverso, P., and Tomasino, M.: Po River discharge: a preliminary analysis of a 200-year time series, Climatic Change, 88, 411–433, doi:10.1007/s10584-008-9395-z, 2008.

Proc. IAHS, 373, 193–199, 2016
proc-iahs.net/373/193/2016/
doi:10.5194/piahs-373-193-2016

An analysis of changes in flood quantiles at the gauge Neu Darchau (Elbe River) from 1875 to 2013

Christoph Mudersbach[1,2], Jens Bender[2,3], and Fabian Netzel[1]

[1]Bochum University of Applied Sciences, Institute for Water and Environment Lennershofstr. 140, 44801 Bochum, Germany
[2]wbu consulting Ingenieurgesellschaft mbH, Schelderberg 16a, 57072 Siegen, Germany
[3]University of Siegen, Research Institute for Water and Environment, Paul-Bonatz-Str. 9–11, 57076 Siegen, Germany

Correspondence to: Christoph Mudersbach (christoph.mudersbach@hs-bochum.de) and Jens Bender (bender@wbu-consulting.de)

Published: 12 May 2016

Abstract. Within this investigation, we focus on a detailed analysis of the discharge data of the gauge Neu Darchau (Elbe River). The Elbe River inflows onto the North Sea. The gauge Neu Darchau is the most downstream discharge gauge of the Elbe River before it becomes an estuary. We follow the questions, whether the discharge characteristics of the Elbe River have changed over the last decades and how much common flood quantiles (i.e. 100-year flood) are affected by the latest extreme events in 2002, 2006, 2011, and 2013. Hence, we conduct (i) trend and seasonality analysis and (ii) an assessment of time-dependencies of flood quantiles by using quasi non-stationary extreme value statistics with both block maxima and peak-over-threshold approaches. The (iii) significance of the changes found in flood quantiles are assessed by using a stochastic approach based on autoregressive models and Monte Carlo simulations. The results of the trend analyses do show no clear evidences for any significant trends in daily mean discharges and increasing flood frequencies. With respect to the extreme events in 2002, 2006, 2011, and 2013 our results reveal, that those events do not lead to extraordinary changes in the 100-year floods. Nevertheless, in the majority an increase in the 100-year floods over the recent decades can be stated. Although these changes are not significant, for many time series of the 100-year flood quantiles there is a clear tendency towards the upper confidence band.

1 Introduction

During the last decades several severe floods occurred in different river basins in Germany (e.g. 1993 and 1995 Rhine; 1997 Odra; 1999, 2001, 2006 and 2013 Danube, 2002, 2006, 2011 and 2013 Elbe) (Petrow and Merz, 2009). In a public perception, there seems to be a trend to more frequent hydrological extreme events resulting in a rising flood hazard.

Thus, the questions arise whether the flood frequencies of the Elbe River have changed over the last decades and how much the design discharges (i.e. the 100-year flood) are affected?

Many recent studies focused on trend estimations in flood magnitude, frequency and seasonality of European or German rivers (e.g. Mudelsee et al., 2003, 2004; Petrow et al., 2009; Petrow and Merz, 2009; Beurton and Thieken, 2009; Stahl et al., 2010; Bormann et al., 2011). Mudelsee et al. (2003) investigated discharge data sets from the Elbe and Odra River for the past 80 to 150 years. For the Elbe River (gauge Dresden), they found a decreasing trend in winter floods, while summer floods do not show any significant trend.

Interestingly, although the extreme values are of major importance for river training, flood risk management and design purposes there is a lack of publications covering the aspect of changes in extreme flood quantiles (e.g. 100-year flood). In this paper we would like to make a contribution to close this gap.

In order to gain the above mentioned objective we analyse the development of the 100-year flood quantile based on different flood indicators over time. Since we use the terms *flood indicator*, *flood frequency indicator* and *flood quantile* in this paper we first define them as follows:

A *flood indicator* is a time series of observed peak discharges of independent flood events. For gaining flood indicators several methods are commonly used, such as the block maxima approach or the peak-over-threshold approach. In this paper we use six different flood indicators. In addition, we calculate one *flood frequency indicator*, which is defined as the number of flood events per year exceeding a certain threshold. Separated from the flood indicator a *flood quantile* is defined as a statistically calculated flood event – by means of extreme value theory – with a certain exceedance probability (e.g. 100-year flood). The calculation of flood quantiles is based on the flood indicators.

In this paper we perform the following steps (applied to the gauge Neu Darchau):

- trend and seasonality analysis of daily mean discharges
- trend analysis of the flood frequency indicator
- assessment of time-dependent changes in flood quantiles (100-year flood) based on six different flood indicators by using extreme value statistics with both, block maxima (Generalized Extreme Value distribution, GEV) and peak-over-threshold (Generalized Pareto distribution, GPD) approach
- an estimation of confidence levels for the time-dependent flood quantiles by means of Monte-Carlo simulations using autoregressive models.

More details of this investigation can be found in Mudersbach et al. (2015).

2 Data

2.1 Daily mean discharge data

Within this investigation, we do not perform spatial analyses of different discharge data sets over Europe or Germany, but focus on detailed analyses of the discharge data of the gauge Neu Darchau (Elbe River). The gauge Neu Darchau is the most downstream discharge gauge of the Elbe River before it becomes an estuary. The gauge is considered as a benchmark for the total discharge of the Elbe River (WSA Lauenburg, 2012). Thus, the discharge statistics of this gauge is of special importance for all designing purposes downstream of this location. Particular demands to design approaches are existent, e.g. due to nuclear power plants, the tidal weir at Geesthacht and the flood protection of Hamburg. The main threat of extreme floods downstream of the weir Geesthacht in the Elbe River evolves from storm surges from the North Sea. However, the river discharge is also one important parameter in design approaches, especially if a storm surge coincides with a high river discharge.

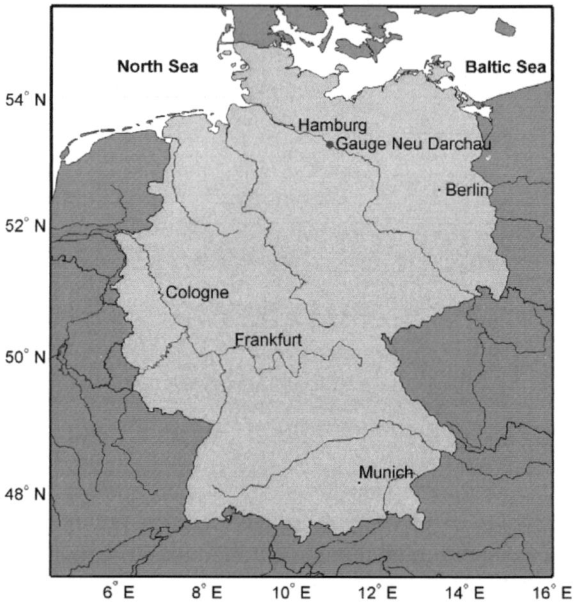

Figure 1. Location of gauge Neu Darchau at the Elbe River.

The discharge data from the gauge Neu Darchau at Elbe location 536.4 km (Fig. 1) were obtained from the Water and Shipping Office Lauenburg (WSA Lauenburg), which is the official gauge operator. The records comprise daily mean discharge values from 1 November 1874 to 31 October 2013 (hydrological year in Germany: 1 November to 31 October), resulting in a time series covering 139 years without any gaps. The discharge measurements operate regularly since 1874 without any discontinuities. Since 1 November 1997 data with a resolution in time of 15 min are available. Prior to that, several measurements per day (not equally distributed) are the basis for the daily mean discharge data. Since the catchment size amounts to 131 950 km^2, these daily measurements are also suitable for deriving flood quantiles (Bender et al., 2015).

The most extreme flood was recorded on 25 March 1888 with HHQ $= 4400\,m^3\,s^{-1}$. Since it was a severe winter flood, it is known that during this event an ice jam significantly influenced the flood stage measurement at gauge Neu Darchau (WSA Lauenburg, 2012). For this flood event the WSA Lauenburg provides a corrected peak discharge of $2310\,m^3\,s^{-1}$ on 24 March 1888 instead of $4400\,m^3\,s^{-1}$ on 25 March 1888 (N. Rölver, personal communication, 2012). We decided to use this corrected value instead of the original value, since the original – and obviously incorrect – value would significantly affect the extreme value statistics. Figure 2 illustrates the corrected daily mean discharge series at gauge Neu Darchau from 1875 to 2013.

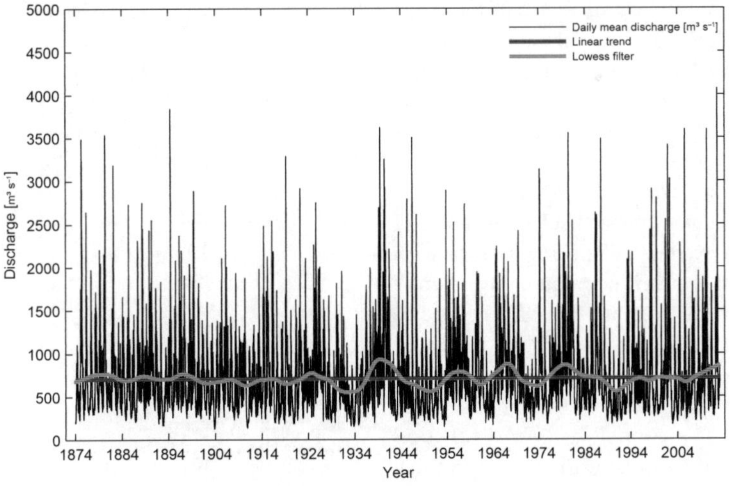

Figure 2. Daily mean discharge of gauge Neu Darchau from 1875 to 2013 with linear and non-linear trend (LOWESS, span = 10 years).

The severe flood events on 23 August 2002 ($Q = 3410\,m^3\,s^{-1}$), 9 April 2006 ($Q = 3590\,m^3\,s^{-1}$), 23 January 2011 ($Q = 3593\,m^3\,s^{-1}$), and 11 June 2013 ($Q = 4070\,m^3\,s^{-1}$) may lead to the assumption, that extreme events occurred more frequently within the last 15 years. While the floods in 2006 and 2011 occurred within the typical flood period (December–May), the severe floods in 2002 and 2013 occurred during summer. It is worthwhile to mention, that the 2013 event is a new record, i.e. this value is higher than all values observed before. Even if the latest extreme events in 2002, 2006, 2011 are very noticeable, the time series reveals that extreme floods with a comparable magnitude already occurred in former years.

2.2 Flood indicators and flood frequency indicators

Six flood indicators are analysed in this study using both, the block maxima and the peak-over-threshold (POT) approach (e.g. Leadbetter, 1991; Bayliss and Jones, 1993; Coles, 2001). The most common flood indicator in flood trend studies is the annual maximum discharge, i.e. the largest daily mean discharge that occurs in each hydrological year. This flood indicator is labelled as AMF. The annual maxima approach has extensively been used in the past (e.g. Acero et al., 2011). However, it can be a wasteful method if further data of extremes are available (Coles, 2001). Conversely, if no extreme flood occurs within a year, the maximum value will still be selected. To overcome these shortcomings, some alternative approaches came up in hydrological statistics.

In the r-largest approach (e.g. Smith, 1986; Coles, 2001), not only the annual maxima ($r = 1$) are considered in the sample, but e.g. the two ($r = 2$) or three ($r = 3$) largest annual values. The advantages and disadvantages of this method are obvious. Given a year with several extreme floods, using the r-largest method extends the data basis by including more of the available information concerning extreme discharge events. In contrast, if a year has no major floods, using the r-largest approach still considers the r-largest events of this year within the sample.

The POT approach (also known as partial duration series) provides a more flexible representation of floods compared to the AMF approach, since it accounts for stochastically and unequally distributed occurrences of floods. A POT sample is created using all independent values exceeding a predefined threshold u. The main advantage of the POT approach is therefore the consideration of all severe floods within a flood intensive year, while years with no extreme events are neglected. Thus, a POT time series captures more information concerning the entire flood characteristics of a river than using AMF. The key challenge of the POT approach, however, is the threshold selection, since statistical methods (e.g. extreme value distribution) may react very sensitive to different thresholds. Selecting suitable thresholds is therefore a complex task representing the main difficulty associated with the POT approach.

A common threshold selection criteria is to use a standard frequency factor f, so that the threshold u can be estimated from the daily mean discharge series Q by:

$$u = \mu_Q + f \cdot \sigma_Q \tag{1}$$

where μ_Q and σ_Q are the mean and standard deviation of the daily mean discharge series Q, respectively. Rosbjerg and Madsen (1992) prefer to use a standard frequency factor of $f = 3$, but take care for the condition $N \geq 2$.

We use the standard frequency factors $f = 0.78$, $f = 1.23$ and $f = 2.18$, leading to a mean number of floods per year of $N = 3$, $N = 2$, and $N = 1$, respectively. Although the factor $f = 2.18$ ($N = 1$) violates the condition $N \geq 2$, we also consider this factor for comparing with the AMF data set. Hence, we name the partial series as POT-0.78, POT-1.23, and POT-

2.18. For flood frequency analyses we make use of the flood frequency indicator POT-0.78F which counts the number of floods per year exceeding the threshold $u = \mu_Q + 0.78 \cdot \sigma_Q$.

3 Methods

For trend estimation of the daily mean discharge data and the flood frequency indicator we used linear regressions with an ordinary least square estimation technique. The significance of the trends was tested using the Mann-Kendall test (Mann, 1945; Kendall, 1975) on the 95 % confidence level. In addition, we fitted a non-linear function to the daily mean discharge time series, using locally weighted scatterplot smoothing (LOWESS). The applied LOWESS function fits simple models to localized subsets of the data using linear least-squares fitting and a first-degree polynomial (Cleveland, 1981).

Nowadays, the Generalized Extreme Value distribution (GEV) and the Generalized Pareto distribution (GPD) have been established as the main distribution functions for extreme value statistics (e.g. Coles, 2001).

We use the maximum likelihood estimation (MLE) method to estimate the distribution parameters (Smith, 1986; Hosking and Wallis, 1997; Davison and Smith, 1990).

As introduced in Sect. 1 we try to clarify, how the flood quantiles have changed over the last decades. Time-dependent changes in flood quantiles can be investigated using non-stationary statistical extreme value approaches (e.g. Mendez et al., 2007; Mudersbach and Jensen, 2010; Bender et al., 2014). However, in order to reflect the real situation in statistical or engineering practice, where design values are mainly determined by use of stationary extreme value approaches, here we use a quasi non-stationary extreme value approach in order to analyse the influence of new extreme events on the flood quantiles. The quasi non-stationary approach is based on the above mentioned stationary extreme value distributions (i.e. GEV and GPD) and a stepwise analysis of different time series lengths by what the uncertainties over time can be assessed. This approach is supported by Serinaldi and Kilsby (2015), who recommend a suitable uncertainty assessment accounting to temporal persistence when using stationary extreme value models. For our approach, firstly, flood quantiles using stationary extreme value statistics are computed for a time period from 1875 to 1950. Afterwards, the time series is extended incrementally by one year until the entire time series from 1875 to 2013 is analysed. In this paper we only focus on the $Q_{0.99}$ flood quantile (i.e. 100-year flood).

As a result of the quasi non-stationary method we derive time series of the 100-year flood based on the different flood indicators. With only these information one can assess whether any time-dependent behaviour exists. However, the question which cannot be answered is, whether the observed changes are significant or not. To overcome this shortfall, one need appropriate confidence bands for the 100-year flood time series. This confidence bands can be calculated by applying the following approach: Having a large set of independent synthetic daily mean discharge data of the gauge Neu Darchau, one may compare the calculated 100-year floods from the original series with quantiles of the 100-year flood values calculated from the large synthetic data set. Synthetic data sets can be simulated using autoregressive models, which have been used widely in hydrological time series analysis (e.g. Salas, 1980; Hosking, 1984; Maidment, 1992; Modarres, 2007; Lohani et al., 2012; Malamud and Turcotte, 2012). Thus, we simulate a large data set of synthetic daily mean discharges by using autoregressive (AR) models following Maidment (1993).

Using an AR3-model we can simulate a set of synthetic time series. Here, we simulate 200 synthetic daily discharge series with a length of 139 years. For all synthetic time series the same analyses as for the original discharge time series are repeated. For calculating the confidence bands we chose the 90 % quantile of the resulting 100-year flood time series from the synthetic data sets. Thus, the calculation of the confidence bands is independent from the calculation of the 100-year flood quantiles from the observed data. This procedure enables one to assess, whether the flood quantiles of the observed data show any significant changes in comparison to an (assumed) homogeneous hydrological system, which is represented by the synthetic time series.

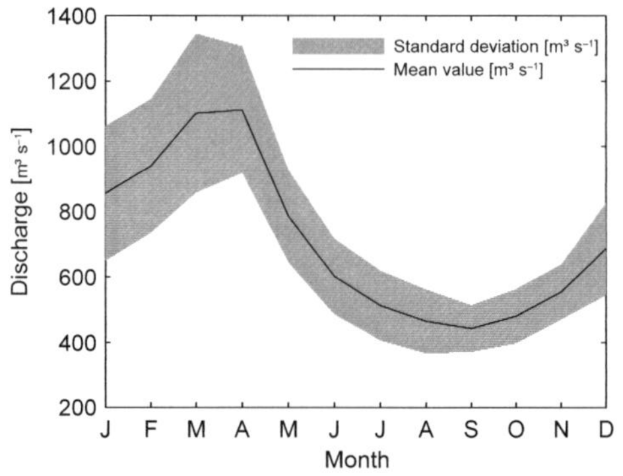

Figure 3. Seasonal cycle of monthly mean discharge data (black line) with corresponding standard deviation of gauge Neu Darchau from 1875 to 2013.

4 Results

4.1 Trends in daily mean discharge and seasonal cycle

The overall mean of the daily mean discharge from 1875 to 2013 (hydrological years) amounts to $MQ = 710\,m^3\,s^{-1}$ with a standard deviation of $s = 446\,m^3\,s^{-1}$ (Fig. 2). The linear trend is non-significant at the 95 % confidence level. The LOWESS filter with a filter span of 10 years illustrate, that the series can be intersected into two periods. A first one

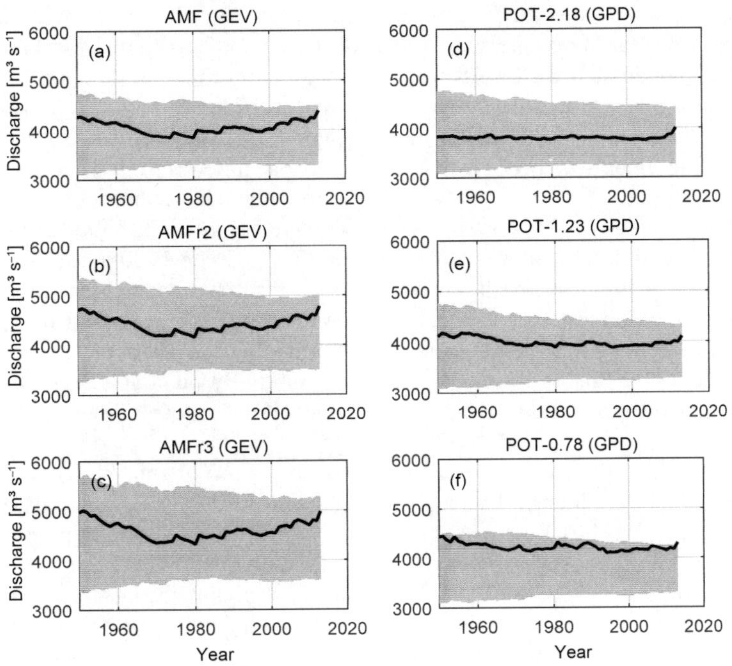

Figure 4. Time-dependent 100-year floods (thick black line) from 1950 to 2013 derived by extreme value statistics and the different flood indicators. The grey area refer to the 90 % confidence levels.

from 1875 to about 1920, where a rather small variance exists, followed by a period up to the end with higher magnitudes. Interannual or decadal variations in discharges may be explained to a certain extend by the North Atlantic Oscillation (NAO) (Villarini et al., 2011; Ionita et al., 2011).

As the annual cycle contributes to the largest of all variations in the discharges, the annual cycle is analysed more in detail in Fig. 3, which shows the mean values (thick black line) of each month from 1875 to 2013. The grey line represents the associated standard deviation. Maximum discharge values occur in March and April, with mean values of $MQ_{\text{March}} = 1101\,\text{m}^3\,\text{s}^{-1}$ and $MQ_{\text{April}} = 1112\,\text{m}^3\,\text{s}^{-1}$. The minimum value appears in September with $MQ_{\text{September}} = 443\,\text{m}^3\,\text{s}^{-1}$. The standard deviation is also largest in March and April. The winter season is therefore the most important season for the genesis of floods. Trend estimates with standard errors are performed for each month and the significance is tested at a 95 % confidence level using the Mann-Kendall test (not shown). The results indicate that there is no evidence for general changes with time. Except for the month May, no significant trends can be found at the 95 % confidence level. Since the trend in May is also rather small, there seems to be no physical reason. It is worthwhile to mention, that there is a tendency for a positive trend in the summer season and a negative trend in the winter season.

The overall results of analysing the daily mean (not the extreme) discharge data indicate, that there are considerable annual to decadal variations existent, but no clear evidence for any long term trend in the discharge characteristics is given. As a conclusion, there is no need to reject the Neu Darchau data set for any ongoing statistical analyses, although many river regulations works may affect the time series.

4.2 Changes in flood quantiles

As introduced in Sect. 1 we try to clarify how flood quantiles have changed over the last decades at gauge Neu Darchau and whether these changes are significant or not. The basis for calculating the flood quantiles are the six flood indicators described in Sect. 2.2. A typical problem in stationary extreme value statistics is to find a trade-off between the extrapolation bias and the non-stationarity bias. The shorter a time series is, the more valid is the assumption of stationarity, but the uncertainty with extrapolating to high quantiles will rise significantly. Vice versa, the longer a time series is, the smaller the uncertainty with extrapolating to high quantiles, but the less valid is the assumption for stationarity. This problem was solved by two conditions: (1) the first time period analysed (1875 to 1950) has a length of 76 years, which is a reliable basis for determining the 100-year flood. (2) Each period is linearly detrended and is therefore considered as a stationary sample.

Figure 4 illustrates the results of the quasi non-stationary extreme value analysis for the six flood indicators POT-0.78, POT-1.23, POT-2.18, AMF, AMFr2, and AMFr3. The black lines represent the time-dependent development of the 100-

year floods for each time span. This means, the first value plotted at the year 1950 is the 100-year flood resulting from 1875 to 1950. The next value, plotted at 1951, results from the time series 1875 to 1951 and so on.

The grey area refers to the 90 % confidence bands calculated by means of autoregressive models and Monte-Carlo simulations.

Results from the AMF time series (Fig. 4a) shows a decreasing trend from 1950 to around 1980, followed by an increase until 2013. It can also be stated, that for the recent years a rather sharp increase can be observed. As the confidence bands depend on the underlying time series length, the confidence bands get smaller as the time series record increases. From 1980 to 2013 the 100-year flood time series steadily move towards the upper confidence band. In 2013, the 100-year flood series is still within the confidence bands but remains very close to the upper confidence band. The general characteristics of the AMFr2 (Fig. 4b) and AMFr3 (Fig. 4c) time series is comparable to the AMF time series. A remarkable difference can be seen in the absolute values of the 100-year floods. For example, for the year 2013 the 100-year floods vary from 4384 m^3 (AMF) to 4979 m^3 (AMFr3), which is mainly a result of the different shape parameters of the GEV.

Results from analysing the POT-2.18 time series with the GPD (Fig. 4d) show a rather stationary behaviour from 1950 to 2009. From 2009 until 2013 there is a sharp increase of the 100-year floods from approximately 3780 to 4005 m^3. The confidence bands also get smaller with increasing time series length. In the POT-2.18 case the 100-year flood values move more or less in the mean range of the confidence bands. The results for the POT-1.23 time series (Fig. 4e) show not that stationary behaviour than for the POT-2.18 series. From 1950 to approximately 2000 there is a slight decrease of the 100-year floods and a moderate increase until 2013. Results from analysing the POT-0.78 time series (Fig. 4f) with the GPD reveal a decreasing development of the 100-year floods from 1950 to about 2000 and a slight increase until 2013. However, for the whole time series there is a convergence to the upper confidence band, which is exceeded in 2013.

5 Discussion and conclusions

The main objective of this paper is to answer the questions, whether the flood frequencies of the Elbe River have changed over the last decades and how much the design discharges (i.e. the 100-year flood) are affected. Therefore, we use a quasi non-stationary extreme value approach flanked by Monte Carlo simulations in order to assess the significance of the changes found.

The results of the trend analyses reveal no clear evidences for increasing flood frequencies. These results point out, that the time series of the gauge Neu Darchau can be assumed as stationary. Of course, many river trainings works were conducted in the Elbe River over the last decades, which may lead to an inhomogeneous data set. Due to the fact, that the major part of these river training works took place in the upper part of the Elbe River, the effect on the downstream located gauge Neu Darchau is rather small. This basically confirms the findings from (Mudelsee et al., 2004) who analysed extreme floods in central Europe. A large number of dams in the upper part of the Elbe River catchment started their operation between 1900 and 1960. Since reservoirs damp the flood peak level due to retention effects, a dependency between the found reduction in flood quantiles and seasonal amplitude until 1980 and the reservoirs is very likely. Furthermore, Busch et al. (2012) found that the peak discharge of flood events can be reduced due to dam regulation up to 359 to 757 $m^3\,s^{-1}$ at gauge Dresden and 183 to 616 $m^3\,s^{-1}$ at the gauge Wittenberge (approximately 80 km upstream of gauge Neu Darchau). As we did not separate anthropogenic (e.g. river engineering/training) and climate change effects on the discharge data in our analyses, no quantitative statements regarding the interactions of these effects can be made. In addition, our results found for the gauge Neu Darchau are not directly transferrable to more upstream located gauges.

With respect to the extreme events in 2002, 2006, 2011, and 2013 the results from the quasi non-stationary extreme value approach reveal, that those events do not lead to extraordinary changes in the 100-year floods. Nevertheless, in the majority of the flood indicators an increase in the 100-year floods over the last decades can be stated. Comparing the absolute values one can see that the more data a flood indicator contains (e.g. POT-0.78 and AMFr3) the higher the 100-year floods are.

From a more general point of view it is important to mention, that the stability of a flood with a given return period strongly depends on the analysed record length. The shorter the analysed time series, the larger the variability and the uncertainty of return discharge estimates. Merz et al. (2011) discussed this topic with respect to the return period of the 2002 flood at the Elbe River gauge Dresden. This phenomenon is also present in this study. To avoid large uncertainties in our analyses due to short time series we use as shortest time series a length of 76 years (1875 to 1950).

References

Acero, F. J., Garcia, J. A., and Cruz Gallego, M.: Peaks-over threshold study of trends in extreme rainfall over the Iberian, Peninsula, J. Climate, 24, 1089–1105, 2011.

Bayliss, A. C. and Jones, R. C.: Peaks-over-threshold flood database: Summary statistics and seasonality, Report No. 121, Institute of Hydrology, Wallingford, 1993.

Bender, J., Wahl, T., and Jensen, J.: Multivariate design in the presence of non-stationarity, J. Hydrol., 514, 123–130, 2014.

Bender, J., Mudersbach, C., and Jensen, J.: Verwendung diskreter Abflusszeitreihen für die Ermittlung von Bemessungshochwasserabflüssen, Tagungsband 38, Dresdner Wasserbaukolloquium, Dresdner Wasserbauliche Mitteilungen, 2015.

Beurton, S. and Thieken, A. H.: Seasonality of floods in Germany, Hydrolog. Sci. J., 54, 62–76, 2009.

Bormann, H., Pinter, N., and Elfert, S.: Hydrological signatures of flood trends on German rivers: Flood frequencies, flood heights and specific stages, J. Hydrol., 404, 50–66, 2011.

Busch, N., Balvin, P., Hatz, M., and Krejci, J.: Bewertung von Einflüssen tschechischer und thüringer Talsperren auf Hochwasser an Moldau und Elbe in Tschechien und Deutschland mittels Einsatz mathematischer Modelle, Technischer Bericht BfG-1725, Bundesanstalt für Gewässerkunde, Koblenz, 2012 (in German).

Cleveland, W. S.: Lowess – a Program for Smoothing Scatterplots by Robust Locally Weighted Regression, Am. Stat., 35, 54–54, 1981.

Coles, S.: An Introduction to Statistical Modeling of Extreme Values, Springer-Verlag, New-York, 2001.

Davison, A. C. and Smith, R. L.: Models for exceedances over high thresholds (with discussion), J. Roy. Stat. Soc. B, 52, 393–442, 1990.

Hosking, J. R. M.: Modeling persistence in hydrological time series using fractional differencing, Water Resour. Res., 20, 1898–1908, 1984

Hosking, J. R. M. and Wallis, J. R.: Regional frequency analysis. Cambridge University-Press. Cambridge, 1997.

Ionita, M., Rimbu, N., and Lohmann, G.: Decadal variability of the Elbe River streamflow, Int. J. Climatol., 31, 22–30, 2011.

Kendall, M. G.: Rank Correlation Methods, Griffin, London, 1975.

Leadbetter, M. R.: On a basis for "Peaks over Threshold modeling", Stat. Probabil. Lett., 12, 357–362, 1991.

Lohani, A. K., Kumar, R., and Singh, R. D.: Hydrological time series modeling: A comparison between adaptive neuro-fuzzy, neural network and autoregressive techniques, J. Hydrol., 442–443, 23–35, 2012.

Malamud, B. D. and Turcotte, D. L.: Time Series: Analysis and Modelling, in: Environmental modelling. Finding simplicity in complexity, edited by: John Wainwright, J. and Mulligan, M., 2nd Edn., Chichester, West Sussex, Hoboken, NJ, Wiley, 2012.

Mann, H. B.: Nonparametric tests against trend, Econometrica, 13, 245–259, 1945.

Mendez, F., Menendez, M., Luceno, A., and Losada, I. J.: Analyzing monthly extreme sea levels with a time-dependent GEV Model, J. Atmos. Ocean. Tech., 24, 894–911, 2007.

Merz, B., Bittner, R., Grünewald, U., and Piroth, K.: Management von Hochwasserrisiken. Mit Beiträgen aus den RIMAX-Forschungsprojekten, Schweizerbart, Stuttgart, 2011 (in German).

Modarres, R.: Streamflow drought time series forecasting, Stoch. Env. Res. Ris. A., 21, 223–233, 2007.

Mudelsee, M., Börngen, M., Tetzlaff, G., and Grünewald, U.: No upward trends in the occurrence of extreme floods in central Europe, Nature, 425, 166–169, 2003.

Mudelsee, M., Börngen, M., Tetzlaff, G., and Grünewald, U.: Extreme floods in central Europe over the past 500 years: Role of cyclone pathway "Zugstrasse Vb", J. Geophys. Res., 109, D23101, doi:10.1029/2004JD005034, 2004.

Mudersbach, C. and Jensen, J.: Non-stationary extreme value analysis of annual maximum water levels for designing coastal structures at the German North Sea coastline, Journal of Flood Risk Management, 3, 52–62, 2010.

Mudersbach, C., Bender, J., and Netzel, F.: An analysis of changes in flood quantiles at the gauge Neu Darchau (Elbe River) from 1875 to 2013, Stoch. Env. Res. Ris. A., doi:10.1007/s00477-015-1173-7, online first, 2015.

Petrow, T. and Merz, B.: Trends in flood magnitude, frequency and seasonality in Germany in the period 1951–2002, J. Hydrol., 371, 129–141, 2009.

Petrow, T., Zimmer, J., and Merz, B.: Changes in the flood hazard in Germany through changing frequency and persistence of circulation patterns, Nat. Hazards Earth Syst. Sci., 9, 1409–1423, doi:10.5194/nhess-9-1409-2009, 2009.

Rosbjerg, D. and Madsen, H.: On the choice of threshold level in partial durations series. XVII Nordic Hydrological Conference, Alta, Norway, NHP Rep no. 30, 604–615, 1992.

Salas, J. D.: Applied modeling of hydrologic time series, Water Resources Publication, Littleton, 1980.

Serinaldi, F. and Kilsby, C. G.: Stationarity is undead: Uncertainty dominates the distribution of extremes, Adv. Water Resour., 77, 17–36, 2015.

Smith, R. L.: Extreme Value Theory Based On The Largest Annual Events, J. Hydrol., 86, 27–43, 1986.

Stahl, K., Hisdal, H., Hannaford, J., Tallaksen, L. M., van Lanen, H. A. J., Sauquet, E., Demuth, S., Fendekova, M., and Jódar, J.: Streamflow trends in Europe: evidence from a dataset of near-natural catchments, Hydrol. Earth Syst. Sci., 14, 2367–2382, doi:10.5194/hess-14-2367-2010, 2010.

Villarini, G., Smith, J. A., Serinaldi, F., and Ntelekos, A. A.: Analyses of seasonal and annual maximum daily discharge records for central Europe, J. Hydrol., 399, 299–312, 2011.

WSA Lauenburg: Official website of the gauge operator, available at: http://www.wsa-lauenburg.wsv.de (last access: 13 November 2013), 2012.

Proc. IAHS, 373, 201–208, 2016
proc-iahs.net/373/201/2016/
doi:10.5194/piahs-373-201-2016

Open Access
PIAHS

Small-scale (flash) flood early warning in the light of operational requirements: opportunities and limits with regard to user demands, driving data, and hydrologic modeling techniques

Andy Philipp[1], Florian Kerl[1], Uwe Büttner[1], Christine Metzkes[1,2], Thomas Singer[2], Michael Wagner[2], and Niels Schütze[2]

[1]Saxon State Office for Environment, Agriculture and Geology, Water, Soil, and Waste, 01109 Dresden, Germany
[2]Institute of Hydrology and Meteorology, Dresden University of Technology, 01069 Dresden, Germany

Correspondence to: Andy Philipp (andy.philipp@smul.sachsen.de)

Published: 12 May 2016

Abstract. In recent years, the Free State of Saxony (Eastern Germany) was repeatedly hit by both extensive riverine flooding, as well as flash flood events, emerging foremost from convective heavy rainfall. Especially after a couple of small-scale, yet disastrous events in 2010, preconditions, drivers, and methods for deriving flash flood related early warning products are investigated. This is to clarify the feasibility and the limits of envisaged early warning procedures for small catchments, hit by flashy heavy rain events. Early warning about potentially flash flood prone situations (i.e., with a suitable lead time with regard to required reaction-time needs of the stakeholders involved in flood risk management) needs to take into account not only hydrological, but also meteorological, as well as communication issues. Therefore, we propose a threefold methodology to identify potential benefits and limitations in a real-world warning/reaction context. First, the user demands (with respect to desired/required warning products, preparation times, etc.) are investigated. Second, focusing on small catchments of some hundred square kilometers, two quantitative precipitation forecasts are verified. Third, considering the user needs, as well as the input parameter uncertainty (i.e., foremost emerging from an uncertain QPF), a feasible, yet robust hydrological modeling approach is proposed on the basis of pilot studies, employing deterministic, data-driven, and simple scoring methods.

1 Introduction

For Saxony, considering the last two decades, the hydrologically most intense and most disastrous events occurred in August 2002, August/September 2010, as well as June 2013 (LfULG, 2015). Total damage for the aforementioned events sums up to 9 billion Euros (ca. 6.1 in 2002, 0.85 in 2010 and 2.0 in 2013). Especially in August/September 2010, flashy events caused large parts of total damages. In this light, the Saxon State Government mandated an independent commission to make suggestions for improving flood risk management actions (Jeschke et al., 2010). One of the commission's recommendations was to line out the potentials and limits of small-scale flash flood early warning approaches (i.e., based on hydrological forecasts).

As the authority responsible for operational flood forecasting and warning, the Saxon Flood Center drafted a corresponding project with a preferably holistic view on flood risk management procedures, especially, when it comes to small-scale and flashy events. Therefore, a threefold approach is proposed, aiming at (1) the assessment of the demands and requirements of potential users of early warning products; (2) the verification of driving meteorological data for the targeted spatio-temporal scales; (3) checking the usefulness of a preferably broad range of modeling approaches with regard to model skill, robustness, and regional applicability, for

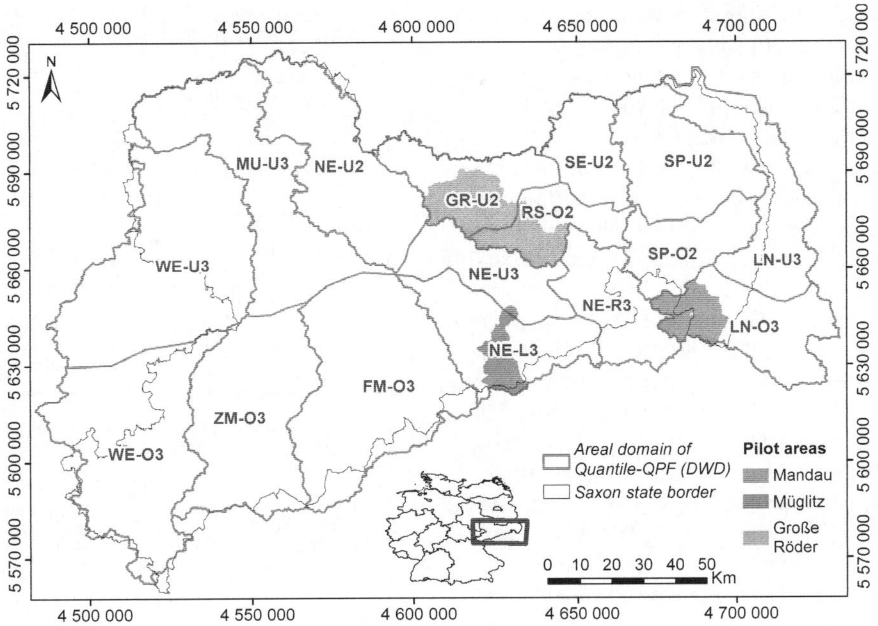

Figure 1. Overview map indicating the areal domain of the Quantile-QPF for Saxony (16 regions; e.g., "FM-O3" indicates the parts of the Freiberger Mulde catchment above 300 m a.s.l.). The area of the regions ranges between approximately 600 and 2700 km^2. Furthermore, the hydrological pilot areas (cf. Sect. 2.3) are shown. Gauss conformal projection with reference at 12° E (Zone 4). The Thumbnail map is showing the location of Saxony within Germany.

small, potentially ungauged basins. The paper at hand provides a short overview of the current state of work and illustrates a way towards an operational early warning system for small catchments in Saxony.

2 Methods

2.1 User survey

To investigate the needs and demands of potential users of an envisaged flood early warning system, a quantitative survey was carried out, based on an online questionnaire. The questionnaire comprised 15 questions, with 12 multiple-choice questions, two questions with gradually-scaled answers, and one question for the submission of verbal comments. Strictly speaking, the survey comprised quantitative and qualitative elements. For the sake of brevity, the full questionnaire is not presented herein but can be found in Philipp et al. (2015).

The surveyed sample was selected systematically (i.e., not randomly) and included all legal users (i.e., according to the Saxon Flood Alarm Bylaw; HWMO, 2014) of Flood Center products ($n = 578$) who were reachable via email to be invited for participating in the online survey ($n = 491$). The interviewee affiliation spanned administration/authorities at local/district/state level, fire departments and civil protection agencies, as well as the private sector. It has to be stated that the interviewees do not represent lay people since they participate in the official flood management procedures on a legal and regular basis.

The survey results were evaluated using descriptive statistics and subgroup analyses by means of contingency tables. Therefore, given answers were investigated in an user-group specific manner, i.e., more than one variable is considered at a time (multivariate approach). A question to address was whether specific user groups answered differently or not. Such an effect can be induced by strongly differing sizes of sub-samples or can indicate a truly diverse response behavior. The literature suggests χ^2-based dependency measures to clarify such questions (Sachs, 1999). For the present study, Cramér's V and χ^2-based p values were used.

2.2 Verification of QPFs

The verification of meteorological data comprised two Quantitative Precipitation Forecasts (QPFs) which are operationally used by the Saxon Flood Center: the deterministic numerical weather prediction COSMO-DE product (Baldauf et al., 2011) and the probabilistic "Quantile Forecast" (QF) for 16 specific areas in Saxony (cf. Fig. 1), issued by German Met Service's Regional Center in Leipzig. The two QPFs are compared against a Quantitative Precipitation Estimate (QPE), emerging from rain gauge data, which was spatially interpolated (Ordinary Kriging) to derive areal precipitation estimates. Additionally, weather radar data (Met Service's RADOLAN-RW product; Sacher et al., 2011) was employed as another QPE reference. A comprehensive overview of the herein considered QPFs and QPEs is given in Table 1.

The Quantile Forecast represents a probabilistic, qualitative expert estimate of areal precipitation for the next 36 h and consists of three values/quantiles per forecasting time step. Since the forecast is issued for 16 specific areas in Saxony (i.e., river catchments with topographic partitioning according to elevation), verification was based on the comparison of areal rainfall for the mentioned 16 regions, and spanned a period from April 2011 to June 2014.

The comparison of areal rainfall was based on consecutive 6 h sums, starting from 06:00 and 18:00 UTC. 6 h sums were chosen to accommodate the coarsest temporal resolution of the investigated products, given by the Quantile-QPF. The product features areal rainfall totals (for the 16 forecasting regions) with 0.9, 0.5, and 0.1 exceedance probability for two consecutive 6 h and two further 12 h intervals. The forecast is updated twice a day (at 06:00 and 18:00 UTC). However, a main task of the herein presented verification was to evaluate the quality of this product against highly resolved numerical weather prediction output (i.e., COSMO-DE).

A QPF/QPE comparison typically employs a number of tools and methods (Jolliffe and Stephenson, 2012), ranging from simple diagnostic (e.g., time series and totals comparisons, residual and bias analyses, scatter and frequency plots) to integral, quantitative methods. Analyses are often based on threshold-oriented contingency table evaluation and deliver typical verification/skill scores, e.g., False Alarm Rate, Probability Of Detection (FAR, POD) or combined products, e.g., Receiver Operating Characteristic curves (ROC curves; Fawcett, 2006). A prototypical work flow of threshold-oriented skill assessment is shown in Fig. 2. More detailed information on the herein employed QPF/QPE verification methodology (as well as concerning the results) can be obtained from Kerl and Philipp (2015).

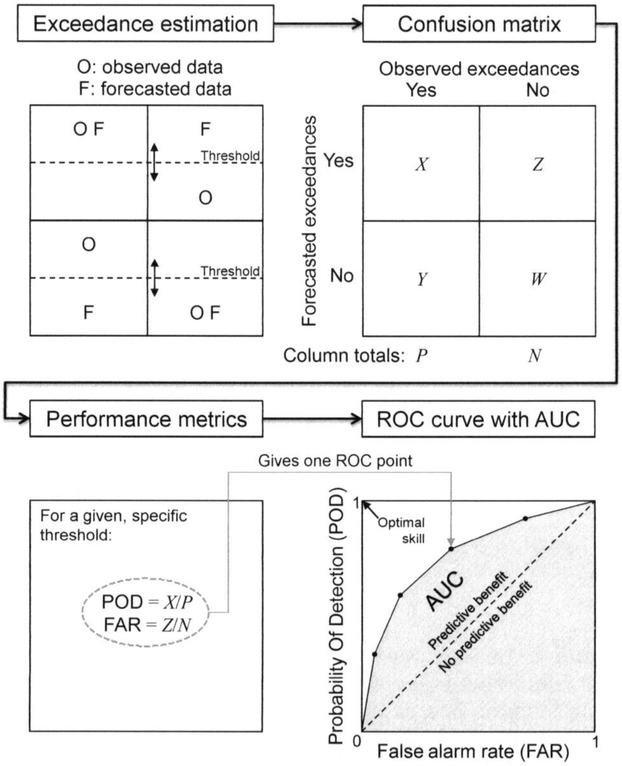

Figure 2. Typical work flow for deriving threshold-exceedance based skill scores, e.g., False Alarm Rate, Probability Of Detection (FAR, POD) or combined products, e.g., Receiver Operating Characteristic (ROC) curves and Area Under Curve (AUC) values.

2.3 Hydrological modeling approaches

Three different hydrological modeling techniques were implemented and applied for three pilot areas in Saxony (cf. Fig. 1): first, a semi-distributed deterministic model (DeHM), second, a data-driven, neural-network model (DaHM) and, third, a simple classification model, based on the scoring of flood-relevant parameters (ScoHM). Subsequently, the modeling concepts and their application (with regard to calibration, data assimilation, etc.) are briefly described. Only snow-free conditions were regarded for model development and application.

2.3.1 Deterministic hydrological model (DeHM)

DeHM model's topology is based on a nodal representation of sub-catchments. Runoff generation is portrayed by the SCS Curve Number method. Runoff concentration is either modeled via an arbitrarily long cascade of linear reservoirs or via response-function convolution. Channel routing is described with either a time-lag function, a cascade of linear reservoirs, Muskingum method, or a translation-diffusion model. Since there are a number of multi-purpose and flood-retention reservoirs in the pilot areas, flood control was specifically included in the model.

Model calibration was based on event-specifically masked hydrograph data and employed a mixed performance criterion after Li et al. (2015). Data assimilation/state updating was realized with a simplified Kalman filter with error variances, following Blöschl et al. (2014). More details on the DeHM model and its application can be found in Schwarze et al. (2015).

2.3.2 Data-driven hydrological model (DaHM)

DaHM is an artificial neural network model, employing a feed forward two-layer perceptron (Hagan et al., 2002). The input vector features flow, rainfall, and cumulative rainfall data with the general 15-element form I : $\left[Q_{t-[0\ldots3]};\ P_{t-[0\ldots3]};\ P^{\mathrm{c}}_{t-[0\ldots6]}\right]$ (with hourly values of flow Q, rainfall P, and cumulative rainfall P^{c}). Adding to that, and depending on the considered lead time in the forecast-

Table 1. Overview of the considered QPF and QPE products.

Product	Provider	QPF/QPE	Type	Temporal resolution	Spatial resolution	Lead time	Update cycle
COSMO-DE	German Met Service (DWD)	QPF	Deterministic numerical weather prediction (gridded)	1 h	2.8×2.8 km	21/27 h[a]	3 h
Quantile Forecast (QF)	DWD-RWB-LZ[b]	QPF	Probabilistic forecast of mean areal precipitation	6/12 h[c]	Forecast regions from ca. 600 to 2700 km^2	36 h	12 h
Interpolated rain gauge data	DWD	QPE	89 stations for the area of Saxony plus 25 km buffer	1 h	1×1 km[d]	–	1 h
RADOLAN-RW	DWD	QPE	Rain gauge adjusted weather radar estimate (gridded)	1 h	1×1 km	–	1 h

[a] 27 h since February 2014. [b] DWD's Regional Service Center (Regionale Wetterberatung) in Leipzig. [c] Product comprises two consecutive 6 h and two further 12 h intervals. [d] Data gridded via Ordinary Kriging.

ing case, inputs for the rainfall forecast were included, e.g., for forecasting Q_{t+6}, the input P_{t+6} is added, for Q_{t+12}, $P_{t+6;\,t+12}$, respectively, whereas the P_{t+x} values represent specific QPF lead times.

The Levenberg-Marquardt algorithm was applied for network training, whilst allowing the number of hidden neurons range from 3 to 13. Event-wise masked hydrograph data and hourly areal rainfall were used for training. 15 training runs were evaluated for each specific hidden-neuron configuration and the best network was selected. Schwarze et al. (2015) give more details on the training and validation of the DaHM model.

2.3.3 Scoring model (ScoHM)

The basic concept of scoring models is – in contrast to deterministic and data-driven concepts – not to simulate or reproduce the development of process variables (e.g., flow) but to empirically determine the current and/or expected further state of a variable by means of a simple classification-based, additive assessment of influencing parameters (i.e., scoring). The employed scoring model resembles the Flooding Susceptibility Assessment approach proposed by Collier and Fox (2003). The method is twofold; first, a baseline susceptibility is derived, based on morphological features, e.g., slope, land cover, etc. Second, a time-variant, dynamic susceptibility is calculated, incorporating the Standardized Precipitation Index (SPI; Edwards and McKee, 1997), cumulative precipitation measures, and the response of a linear reservoir being charged with hourly precipitation.

The scoring is carried out according to Table 2; baseline sub-scores and the SPI sub-score are mapped linearly, according to the range of each respective feature. For the remaining dynamic susceptibility sub-scores, frequency analyses were applied to deliver specific percentiles that are in turn connected to specific sub-score values, e.g., P-sums within the 75th–90th percentile-range of the data result in a sub-score of 1, etc. The method requires only one effective parameter, namely the recession constant of the incorporated linear reservoir, which was manually adjusted to a global value of 8 h.

In contrast to the DeHM and DaHM models, the ScoHM approach does not rely on observed flow data at all; neither in the sense of directly including auto-correlative signals, as applies for the data-driven DaHM model (in form of the Q_{t-x} inputs), nor indirectly via data assimilation/state updating, as applies for the deterministic DeHM model. Therefore, the ScoHM approach might offer a robustly transferable methodology for hydrological prediction in small, ungauged basins.

3 Results

3.1 User survey

Subsequently, the most important results of the user survey (cf. Sect. 2.1) are presented in a concise manner; a more detailed report can be found in Philipp et al. (2015). The response rate was 76 % ($n = 373$), which is extraordinarily high and is mainly a result of the systematic sampling. For 11 out of 15 questions, user-group specific replies were not distinguishable in a statistical sense. The outcomes of the statistical analysis of the survey data can be summarized as follows:

Information and pathways: (1) The interviewees request selective, event-related information or inform themselves on an event-related basis (rather than on a regular basis). (2) 37 % of all users trust that a more regular and more frequent distribution of warning products will provide increased security for their management decisions, even if the meteorological and hydrological trend remains unchanged. (3) All

Table 2. ScoHM scoring system.

	Parameter description	Upper parameter limits	Sub-score range
Baseline susceptibility	Mean catchment slope	0.02/0.08/0.14/0.20/∞	0 to 4
	Catchment shape factor[a]	0.20/0.40/0.60/0.80/1.00	0 to 4
	Degree of surface sealing	0.05/0.20/0.35/0.50/1.00	0 to 4
	Proportion of fast runoff components[b]	0.10/0.23/0.37/0.50/1.00	0 to 4
Dynamic susceptibility[c]	SPI over the last 30 days[d]	−3/−2/−1/0/1/2/∞	−3 to 3
	Precipitation sum over the last 7 days	Sub-score percentiles[e]	0 to 4
	Precipitation sum over 12/24/48 h[f]	based on actual data from	0 to 4
	Linear reservoir outflow[g]	01/2010 to 09/2015	0 to 4
Total susceptibility score	−3 to 31		

[a] Catchment being more circular for values near unity. [b] According to Peschke et al. (1999). [c] In contrast to Collier and Fox (2003), snow-specific dynamic sub-scores were not considered. [d] SPI values rounded to integers. [e] Percentiles: 75th/90th/95th/99th/100th. [f] Only highest sub-score is considered. [g] Linear reservoir being charged with hourly precipitation.

groups, except the group "private persons", attach greatest importance to the internet in contrast to other communication channels (e.g., fax, video text, voice mail). The official flood warnings issued by fax or email are also used for information by a majority of users. (4) A high availability of warning services and products is deemed important by a vast majority of users, especially in case of flooding.

Flood warning products: (1) A short-termed, but more precise warning is preferred over a long-term estimation, carrying presumably more uncertainty. (2) The majority of users (> 65 %) are interested in receiving a possibly reliable forecast of the peak water level. 45 % of users would appreciate being informed about the peak timing. (3) Most popular products for fulfilling early warning purposes are forecasted hydrographs with uncertainty bands (about 50 % of all persons interviewed), as well as catchment-oriented classification products ("traffic light", approximately 40 % of all persons interviewed).

Lead time and miscellaneous: (1) The minimum required lead times amount to ≤ 3 (9 % of users), ≤ 6 (27 %), ≤ 12 (50 %), ≤ 24 (83 %), ≤ 72 h (98 %). (2) A lead time of ≤ 12 h is deemed to be adequate by a slim majority of users in small catchments (< 200 km^2). (2) The interviewed user groups vary significantly in terms of the replies given when being asked for the requested updating frequency of flood warnings and their communication via email or fax. (3) Furthermore, the interviewees of various user groups specifically replied to the questions concerning the quality of current products and the quality of the work of the Saxon Flood Center. (4) Moreover, no significant differences in the response behavior of the various user groups could be identified by statistical means.

3.2 Verification of QPFs

The investigated QPFs (COSMO-DE and Quantile Forecast) were compared against areal precipitation estimates, based on gridded rain gauge data, and, additionally, a radar-based QPE (RADOLAN-RW product). First, threshold exceedance frequencies were derived from the QPEs and QPFs for threshold values from 10 to 30 mm/6 h (cf. Fig. 3). COSMO-DE delivers exceedance frequencies which are close to the ones obtained from rain gauge data. RADOLAN slightly underestimates the threshold exceedance frequencies from rain gauge data, whereas the chance of underestimation is higher at lower thresholds, and vice versa. Threshold exceedances drawn from the Quantile Forecast's 50th and 10th percentiles are generally more frequent than the observed ones (i.e., from rain gauge data), whereas the 90th percentile underestimates observed frequencies.

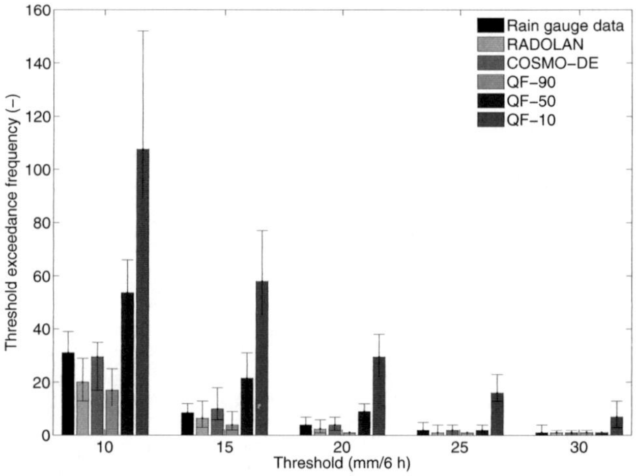

Figure 3. Threshold exceedance frequencies of 6-hourly areal precipitation sums for QPEs (gridded rain gauge data, RADOLAN-RW) and QPFs (COSMO-DE, Quantile Forecast) from April 2011 to June 2014. The bars show the median of exceedance frequencies for the respective precipitation products for the 16 forecast areas (cf. Fig. 1). The whiskers illustrate the minimum and maximum values.

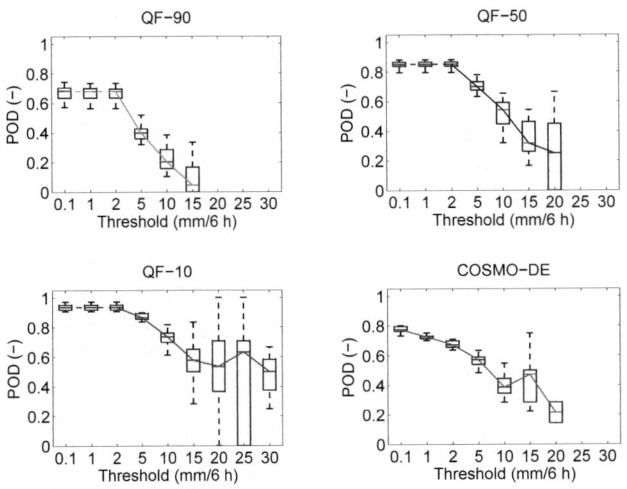

Figure 4. Probability Of Detection (POD) according to thresholds of areal precipitation sums ranging from 0.1 to 30 mm/6 h for the Quantile Forecast and COSMO-DE from April 2011 to June 2014. The box plots indicate the spread of POD over the 16 forecast areas.

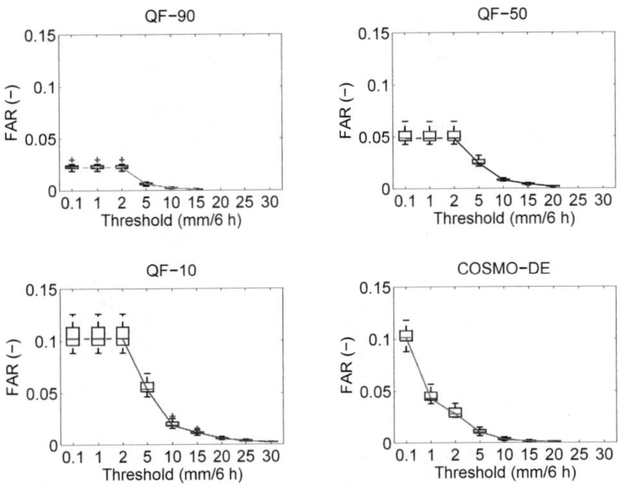

Figure 5. False Alarm Rate (FAR) according to thresholds of areal precipitation sums ranging from 0.1 to 30 mm/6 h for the Quantile Forecast and COSMO-DE from April 2011 to June 2014. The box plots indicate the spread of FAR over the 16 forecast areas.

Second, for a more in-depth view at the regarded QPFs, the contingency-based measures POD and FAR were evaluated down to thresholds of 0.1 mm/6 h (Figs. 4 and 5). Due to product-specific conventions of the Quantile Forecast (areal precipitation sum < 4.5 mm/6 h is set to zero), the results are constant for thresholds < 4.5 mm. Following Winterrath et al. (2012), a minimum of 10 observed or predicted threshold exceedances should be required for the calculation of skill scores. Therefore, POD and FAR were not always evaluated for higher thresholds. Generally, higher precipitation thresholds are connected with lower POD and lower FAR values, and vice versa. Furthermore, for POD, the skill variance amongst the forecast areas increases with increasing precipitation thresholds. POD = FAR indicates a boundary for which the considered QPF has no predictive benefit anymore. This boundary is not reached for both QPFs, concerning the investigated thresholds. Finally, for the regarded QPFs, COSMO-DE exhibits the highest performance with regard to POD/FAR relations and skill variance amongst the forecast areas.

3.3 Hydrological model validation

The three presented models (DeHM, DaHM, and ScoHM) were applied for the three aforementioned pilot areas (cf. Fig. 1). The herein investigated QPEs (gridded rain gauge data and RADOLAN data) and QPFs (COSMO-DE and Quantile Forecast; cf. Sects. 2.2 and 3.2) were used as meteorological drivers (for the current state of work, on the QPF side, ScoHM was charged with the Quantile Forecast only). Validation for the DeHM and DaHM models is straightforward since modeled hydrographs are simply compared against observed ones. Model evaluation is a bit more delicate for the ScoHM results, since the ScoHM output (i.e., dimensionless scores) does only qualitatively correlate with observed flow values. Therefore, a quantile-mapping procedure (Piani et al., 2009) was applied to relate thresholds of Q with corresponding total-score values.

Model performance was evaluated on the basis of threshold-oriented contingency table analyses, i.e., it is checked if modeled output matches/exceeds a certain observed flow level or not. More specifically, the variation of threshold values delivers a set of corresponding skill scores, e.g., POD values with corresponding FARs. These POD/FAR tuples were used to establish catchment-specific Receiver Operating Characteristic curves (Fawcett, 2006). The curves were finally integrated to deliver the Area Under Curve (AUC), with values near unity for a near-perfect model prediction and near 0.5 for no predictive skill (cf. Fig. 2 and Sect. 2.2). For brevity, results are presented and discussed for the Mandau catchment only, featuring four river gauges.

Generally, different combinations of lead times and update cycles (i.e., the time after which a new forecast is processed) were investigated; herein, results for an update cycle length of 12 h are presented. Event-specifically masked, hourly hydrograph data and hourly rainfall observations were used during model validation. Data which were employed in model calibration/training were not used for validation purposes. Calibration/training data originated from the period of 2006 to 2011 (11 events), validation data from 2010 to 2015 (10 events). Figure 6 comprehensively shows the validation results for the Mandau pilot area. For the Quantile Forecast, results for the 50th percentile are exemplarily shown.

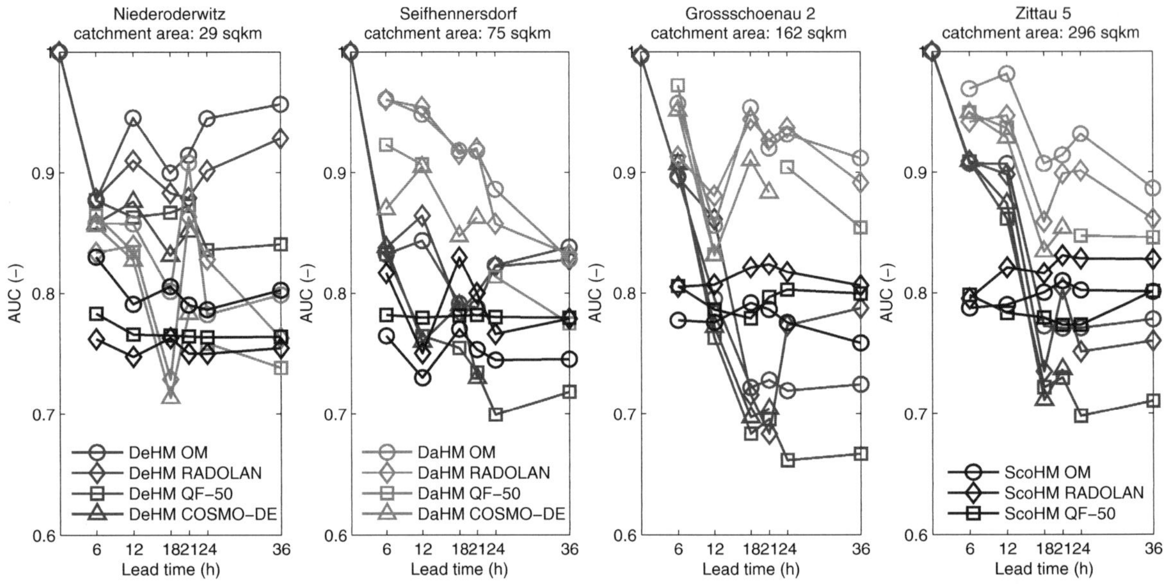

Figure 6. Results of hourly, threshold-oriented evaluation for DeHM, DaHM and ScoHM output in the Mandau pilot area, based on Area Under Curve values. Lead times range from 6 to 36 h, update cycle is 12 h. OM: ombrometer data (i.e., gridded rain gauge data); RADOLAN: QPE from weather radar scans; QF-50: 50th percentile of Quantile Forecast; COSMO-DE: numerical weather prediction output. Skill for DeHM at a lead time of zero is based on true model output after assimilation/updating and can be slightly smaller than unity (e.g., apparent for Großschönau 2).

For the smallest sub-catchment, Niederoderwitz (29 km^2), DeHM performs best; for the three larger sub-catchments, DaHM features the highest Area Under Curve values. However, DeHM and DaHM performance trends to decrease with increasing lead time; ScoHM features a quite constant/robust skill development. The reason for this might be that for shorter lead times (6 h) the auto-correlative Q_{t-x} signal, included directly or indirectly in the DeHM and DaHM model (cf. Sect. 2.3), leads to improved performance. This does not apply for the ScoHM results, since the model is not dependent on observed flow data. Generally, ScoHM exhibits Area Under Curve values around 0.8 which indicates a good overall predictive skill, foremost, when keeping in mind the generality and straightforwardness of the model approach.

It can be further seen from Fig. 6 that QPE data delivers highest predictive skill with a tendency of RADOLAN outperforming the rain gauge data. Predictive skill under QPF data (Quantile Forecast and COSMO-DE) is mostly lower. For different QPFs as drivers, resulting skills do not differ greatly. Apparently, the observed differences in QPF quality (cf. Sect. 3.2) do not systematically impact hydrological model skill. Furthermore, it is important to say that validation was carried out on the basis of hourly values; a more general evaluation, e.g., comparing only the highest values within a specific temporal window (e.g., 6 h), would yield considerably higher skill scores.

Finally, it should be stated that the results for the other investigated pilot areas are consistent with the herein presented findings for the Mandau pilot region when focusing on catchments with areas of up to 200 km^2. For larger scales, when wave translation and diffusion impact flood expression, the deterministic and data-driven models outperform the scoring approach since it does not account for such processes.

4 Conclusions and outlook

In this study, user demands, driving data, and hydrologic modeling techniques were evaluated within a real-word application context in order to illustrate a way towards a flash flood early warning strategy for (sub-)mesoscale catchments in Saxony. First, results suggest that the majority of potential users of flood warnings would be satisfied with forecasting lead times of up to 24 h and that users are foremost interested in predicted peak water/alarm levels (rather than peak timing). Second, on the basis of meteorological verification results, highly resolved numerical weather prediction data seem to provide the best predictive skill, compared to more general, areally integrated products. Third, differences in the quality of meteorological driving data do not greatly influence hydrological model skill. Fourth, a clear statement on the superiority of one hydrological model over another cannot be made.

In fact, if simple classification models would be sufficient to satisfy warning needs (e.g., providing the information whether or not a specific threshold is likely to be exceeded in the next forecasting interval), results show that such a modeling approach (i.e., ScoHM) performs with favorable skill, compared to more sophisticated modeling techniques, and without introducing cumbersome parameter estimation problems and limited (DeHM) or even non-existent (DaHM) regional transferability. However, overall forecasting skill always decreases with increasing randomness of driving events and conditions, i.e., the more rare/focused/intense the flood-causing processes and/or the longer the lead time, the smaller the chance of correct detection/warning.

Further research is currently carried out regarding the statewide implementation and comparative evaluation of the herein considered approaches to gain more insight into the dependencies of meteorological drivers, hydrological models, spatio-temporal scaling effects, and regional transferability. Meteorological verification will be carried out for smaller spatio-temporal scales and with a temporally extended data set. Additionally, the set of QPFs will be extended to German Met Service's 21-member ensemble product, COSMO-DE-EPS. Thus, allowing a statewide, comprehensive probabilistic verification and validation of the presented hydrological models.

Acknowledgements. The meteorological driving data were provided by courtesy of German Meteorological Service (Deutscher Wetterdienst). The authors would like to thank one anonymous referee for thoroughly reviewing the manuscript. The manuscript was funded by Saxon State Ministry of the Environment and Agriculture Grant/project number: 45-8904.20/1.

References

Baldauf, M., Förstner, J., Klink, S., Reinhardt, T., Schraff, C., Seifert, A., and Stephan, K.: Short description COSMO-DE (LMK) and its data bases on DWD's data server, Technical Report, DWD, 2011 (in German).

Blöschl, G., Nester, T., Parajka, J., and Komma, J.: Flood forecasting on the Austrian Danube and data assimilation, Hydrol. Wasserbewirts., 58, 64–72, 2014 (in German).

Collier, C. G. and Fox, N. I.: Assessing the flooding susceptibility of river catchments to extreme rainfall in the United Kingdom, International Journal of River Basin Management, 1, 225–235, 2003.

Edwards, D. and McKee, T.: Characteristics of 20th century drought in the United States at multiple time scales, Atmospheric Science Paper, 634, 1–155, 1997.

Fawcett, T.: An introduction to ROC analysis, Pattern Recogn. Lett., 27, 861–874, 2006.

Hagan, M. T., Demuth, H. B., Beale, N., and De Jesus, O.: Neural Network Design, Published by Martin Hagan, 2nd Edn., 2002.

HWMO: Verwaltungsvorschrift des Sächsischen Staatsministeriums für Umwelt und Landwirtschaft zum Hochwassernachrichten- und Alarmdienst im Freistaat Sachsen (Saxon Flood Alarm Bylaw), Saxon State Government, 2014 (in German).

Jeschke, K., Greiff, B., Kolf, R., Burk, H.-P., Merker, H., Bogatsch, C., Fritzsche, C., and Vogel, M.: Commission's report on the assessment of the flood information and warning procedures in Saxony during the August 2010 flood event, Technical Report, Saxon State Government, 2010 (in German).

Jolliffe, I. T. and Stephenson, D. B. (Eds.): Forecast Verification: A Practitioners Guide for Atmospheric Science, Wiley, 2012.

Kerl, F. and Philipp, A.: Verification of operationally available QPF and QPE products for the area of Saxony (Germany) from 04/2011 to 06/2014 (in German), Technical Report, Saxon State Office for Environment, Agriculture and Geology, 2015.

LfULG: Event analysis of the 2013 Flood (in German), Technical Report, Saxon State Office for Environment, Agriculture and Geology, 2015.

Li, Y., Ryu, D., Western, A. W., and Wang, Q. J.: Assimilation of stream discharge for flood forecasting: updating a semi-distributed model with an integrated data assimilation scheme, Water Resour. Res., 51, 3238–3258, 2015.

Peschke, G., Etzenberg, C., Müller, G., Töpfer, J., and Zimmermann, S.: The expert system FLAB: a tool for the delineation of landscape units with similar runoff generation mechanisms, Technical Report, IHI Zittau, 1999 (in German).

Philipp, A., Kerl, F., and Müller, U.: Demands by potential users for a flood early warning system for Saxony, Hydrol. Wasserbewirts., 1, 4–22, 2015 (in German).

Piani, C., Haerter, J. O., and Coppola, E.: Statistical bias correction for daily precipitation in regional climate models over Europe, Theor. Appl. Climatol., 99, 187–192, 2009.

Sacher, D., Weigl, E., Podlasly, C., and Winterrath, T.: RADOLAN/RADVOR-OP: description of the composite format, Technical Report, DWD, 2011 (in German).

Sachs, L.: Applied Statistics, Springer Berlin Heidelberg, 9th Edn., 1999.

Schwarze, R., Singer, T., Stange, P., Wagner, M., and Schütze, N.: Development and implementation of deterministic and data-driven methods for hydrological forecasting in small, partly ungauged basins for the purpose of deriving flood early warnings, Technical Report, Dresden University of Technology, 2015 (in German).

Winterrath, T., Weigl, E., Reich, T., Rosenow, W., and Stephan, K.: RADVOR-OP: radar-based, real-time precipitation nowcasting for operational purposes, Technical Report, DWD, 2012 (in German).

Proc. IAHS, 373, 209–214, 2016
proc-iahs.net/373/209/2016/
doi:10.5194/piahs-373-209-2016

Open Access
PIAHS

Inflow forecasting using Artificial Neural Networks for reservoir operation

Chuthamat Chiamsathit, Adebayo J. Adeloye, and Soundharajan Bankaru-Swamy

Institute for Infrastructure and Environment, Heriot-Watt University, Edinburgh, EH14 4AS, UK

Correspondence to: Adebayo J. Adeloye (a.j.adeloye@hw.ac.uk)

Published: 12 May 2016

Abstract. In this study, multi-layer perceptron (MLP) artificial neural networks have been applied to forecast one-month-ahead inflow for the Ubonratana reservoir, Thailand. To assess how well the forecast inflows have performed in the operation of the reservoir, simulations were carried out guided by the systems rule curves. As basis of comparison, four inflow situations were considered: (1) inflow known and assumed to be the historic (Type A); (2) inflow known and assumed to be the forecast (Type F); (3) inflow known and assumed to be the historic mean for month (Type M); and (4) inflow is unknown with release decision only conditioned on the starting reservoir storage (Type N). Reservoir performance was summarised in terms of reliability, resilience, vulnerability and sustainability. It was found that Type F inflow situation produced the best performance while Type N was the worst performing. This clearly demonstrates the importance of good inflow information for effective reservoir operation.

1 Introduction

The planning of reservoirs for various purposes including flood and drought control relies on the historic inflow data at the reservoir site. Due to natural variability and other factors (e.g. climate and land-use changes), however, the inflow situation when the reservoir is being operated will be different. It is therefore important that reservoirs are properly operated so that they continue to perform satisfactorily during changing hydro-climatology.

Reservoir operation concerns taking decisions on water release from a reservoir based on the amount of water available vis-à-vis the demand placed on the system. The available water is the sum of starting period storage and the inflow expected during the period. Consequently, effective reservoir operation relies on reliable forecast of the inflow into the reservoir. Traditional forecasting methods using hydrologic, hydraulic and time-series models require specification of the functional relationship of the model which can be problematic (Zhang et al., 1998), which is why focus has recently shifted to the use of data-driven techniques that do not require knowledge of this functional relationship. In particular, artificial neural networks (ANN) have been widely used to forecast reservoir inflows (see e.g. Edossa and Babel, 2012; Mohammadi et al., 2005) due to their effectiveness and flexibility and have been proven to be superior to other approaches such as regression-based and time series models.

The aim of this study is to apply multi-layer perceptron (MLP)-ANN for the one-month-ahead inflow forecasting for the Ubonratana reservoir, Thailand. To investigate the effect of the forecasts on reservoir operation performance, four situations were considered for the one-month-ahead inflow: (1) inflow is known and assumed to be the historic (Type A); (2) inflow is known and assumed to be the ANN forecast (Type F); (3) inflow is known and assumed to be the historic average for the given month (Type M); and (4) inflow is not known and the release decision is conditioned only on the starting reservoir storage (Type N). Simulations of the Ubonratana reservoir were then carried out with these alternative inflow scenarios and the resulting reservoir performance was summarised in terms of reliability, resilience, vulnerability and sustainability.

In the next section, further details about the methodology will be given. This is then followed by the presentation of the case study. Next the results are presented and discussed and finally, the main conclusions are given.

2 Methodology

2.1 Artificial neural networks modelling

The theory and mathematical basis of ANN have been described excellently by Shamseldin (1997). Essentially, the structure of ANN comprises an input layer, an output layer and one or more hidden layers as illustrated in Fig. 1. The schematic in Fig. 1 has a single hidden layer which is generally sufficient to approximate any complex, non-linear function (Mulia et al., 2015). The layers contain nodes or neurons which are connected by weights. Determining optimal values for these weights and other parameters of the network is the purpose of the ANN training exercise.

For a given problem, the number of nodes in the output layer is fixed by the problem, e.g. in the current work, it is the 1-month ahead inflow forecast. The input nodes must be determined by the factors known to affect the output variable and this has been achieved through an examination of the cross-correlation matrix (see Adeloye and De Munari, 2006). The number of neurons in the hidden layer is much more difficult to arrive at and is normally determined as part of the training by trial and error as described by Adeloye and De Munari (2006).

Training is often improved through the use of early-stop-rule (ESR) that helps to avoid over-fitting. In ESR, the available data are divided into three parts: (i) a training set, used to determine the network weights and biases, (ii) a validation set, used to estimate the network performance and decide when the training should be stopped, and (iii) a test set, used to verify the effectiveness of the stopping criterion and to estimate the expected performance in the future.

The tested ANN architectures (in trying to arrive at the best value for the number of hidden neurons) were compared using the correlation coefficient (R) criterion, i.e.:

$$R=\frac{\sum y_{\text{sim}}y_{\text{obs}}-\frac{\sum y_{\text{sim}}\sum y_{\text{obs}}}{N}}{\sqrt{\left(\sum y_{\text{sim}}-\frac{(\sum y_{\text{sim}})^2}{N}\right)\left(\sum y_{\text{obs}}^2-\frac{(\sum y_{\text{obs}})^2}{N}\right)}}\qquad(1)$$

where y_{sim} and y_{obs} are respectively the simulated and observed values of the output variable and N is the number of exemplars used.

2.2 Reservoir performance simulation

Reservoir behaviour simulation employed the mass balance equation (McMahon and Adeloye, 2005):

$$S_{t+1}=S_t+Q_t-D'_t-E_t\qquad(2)$$

subject to the operational policy for the reservoir, where S_t and S_{t+1} are respectively storage at the beginning and end of time t; Q_t is the inflow to the reservoir during t; E_t is the net evaporation (evaporation minus direct rainfall) in period t; D'_t is the total water release towards meeting the target demand of D_t during t.

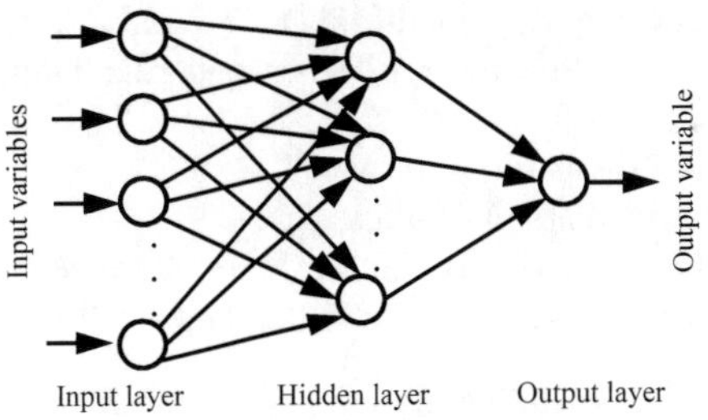

Figure 1. Schematic of artificial neural network.

As noted previously, the water available for allocation during t, WA_t, is:

$$\text{WA}_t=S_t+Q_t\qquad(3)$$

and assumes that the inflow is known at the start of the month when making the release decision. In practice, however, this is not the case and assumptions about the size of the anticipated inflow must be made. If the actual inflow turns out to be exactly the same as the assumed inflow, then the end of period storage will be exactly as given by Eq. (2). If, however, there is a discrepancy, the actual end of period storage will be different from Eq. (2).

Let the actual end-of-period storage be $S_{\text{end},t}$, the relationships between this and S_{t+1} for each of the assumed inflow knowledge assumptions become:

1. Type A: $\text{WA}_t=S_t+Q_t$ and $S_{\text{end},t}=S_{t+1}$
2. Type F: $\text{WA}_t=S_t+Q'_t$ and $S_{\text{end},t}=S_{t+1}+Q_t-Q'_t$
3. Type M: $\text{WA}_t=S_t+\overline{Q}_t$ and $S_{\text{end},t}=S_{t+1}+Q_t-\overline{Q}_t$
4. Type N: $\text{WA}_t=S_t$ and $S_{\text{end},t}=S_{t+1}+Q_t$

where Q_t is the observed (correct) inflow during time t, Q'_t is the corresponding forecast inflow, $\overline{Q}_t$ is the historic mean flow for the month of time t, and $S_{\text{end},t}$ is the adjusted end-of-period storage.

With the available water determined, release then takes place guided by the rule curves as follows:

Case 1: For $\text{WA}_t\geq\text{URC}_m$ this is the excess operation case, i.e., $D'_t\geq D_t$

$$D'_t=S_t+Q_t-E_t-\text{URC}_m\qquad(4)$$

$$Y_t=D'_t-D_t\qquad(5)$$

Case 2: For $\text{LRC}_m<\text{WA}_t<\text{URC}_m$ this is the normal operation case, i.e., $D'_t\leq D_t$

$$Y_t=0\qquad(6)$$

$$\text{If }\text{WA}_t-D_t\geq\text{LRC}_m,\ D'_t=D_t\qquad(7)$$

$$\text{If }\text{WA}_t-D_t<\text{LRC}_m,\ D'_t=\text{WA}_t-\text{LRC}_m\qquad(8)$$

Case 3: For $\text{WA}_t\leq\text{LRC}_m$ this is the deficit operation case, i.e., $D'_t=0$ (No water released), where URC_m is the upper rule curve during month $m\,(=1, 2, 3, \ldots, 12)$ of the year; LRC_m is the lower rule curve during month m; Y_t is the excess water released during period t. In general, $t = 12(y-1)+m$ for years $y = 1, 2, 3, \ldots, n$, where n is the number of years in the data record.

Once the simulation is complete, performance indices are then evaluated as follows (McMahon and Adeloye, 2005):

i. Time-based Reliability (R_t): $R_t = N_s/N$, where N_s is the total number of intervals out of N that the demand was met.

ii. Volume-based Reliability (R_v): $R_\mathrm{v} = \sum_{t=1}^{N} D'_t / \sum_{t=1}^{N} D_t$, $\forall D'_t \le D_t$

iii. Resilience: $\phi = 1/\left(\frac{f_\mathrm{d}}{f_\mathrm{s}}\right) = \frac{f_\mathrm{s}}{f_\mathrm{d}}$; $0 < \phi \le 1$, where ϕ is resilience, f_s is number of continuous sequences of failure periods and f_d is the total duration of the failures, i.e. $f_\mathrm{d} = N - N_s$.

iv. Vulnerability: $\eta = \frac{\sum_{k=1}^{f_\mathrm{s}} \left(\frac{\max(sh_k)}{D_k}\right)}{f_\mathrm{s}}$, where $\max(sh_k)$ is the maximum water shortage in failure sequence k and D_k corresponding demand.

v. Sustainability index (Sandoval-Solis et al., 2011): $\lambda = (R_t\phi(1-\eta))^{1/3}$, where there are multiple users or sectors, each of the above indices will be evaluated for each sector and these can later be combined to determine a weighted group (or global) index. This was done for the sustainability index λ using:

$$\lambda_\mathrm{G} = \sum_{j=1}^{M} w_j \lambda_j \tag{9}$$

where w_j is a weight, given by (Sandoval-Soils et al., 2011):

$$w_j = \frac{\mathrm{DS}^j}{\sum_{j=1}^{M} \mathrm{DS}^j} \tag{10}$$

and λ_G is the group sustainability; λ_j is the sustainability for users category j; w_j is the weighting for user j; M is the total number of users sectors and DS^j is the average annual water demand for users sector j.

3 Study area and data

The Ubonratana reservoir is the largest, single multi-purpose reservoir in the upper Chi River Basin in north-eastern Thailand. The dam provides water for consumptive uses (domestic, industrial, irrigation), Pong River in-stream flow augmentation as well as flood control (EGAT, 2002). However, the water deliveries first pass through turbines for power generation (installed capacity = 25.2 MW) before being allocated to the other uses. The release is prioritised in the order of public (i.e. domestic and industrial), instream flow augmentation and irrigation. The maximum storage capacity of the reservoir is 2431 Mm^3 at elevation of 182 m above mean sea level (m a.m.s.l). Minimum water level of the reservoir is 175 m a.m.s.l. or 581.67 Mm^3 which has been prescribed for the purpose of hydropower generation.

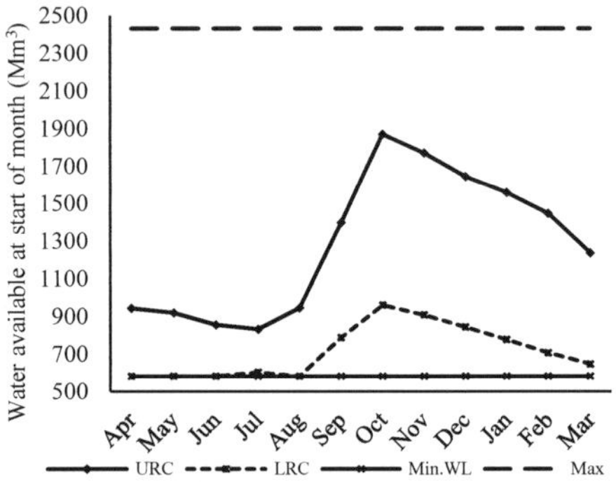

Figure 2. Rule curves for Ubonratana reservoir.

Data collected for the study included daily reservoir inflows, evaporation, area-height-storage relationship, weekly and monthly water requirements and operating rule curves for the reservoir. The observed monthly inflow from April 1970 to March 2012 and rainfall from April 1981 to March 2012 were provided by the Electricity Generating Authority of Thailand (EGAT) and the Royal Irrigation Department (RID). The analysis, however, used the overlapping period of April 1982 to March 2012 (i.e. 360 months) for which the rainfall and runoff data were complete. Data on historical water releases to the various sectors were also provided by the RID. The gross water requirements for the analysis period were 28 952 Mm^3, i.e. average monthly of: 0.98 Mm^3 for public (municipal and industrial) demands; 18.83 Mm^3 for downstream requirements; and 60.6 Mm^3 for irrigation. The original rule curves were also provided by the EGAT; the improved versions of these (see Fig. 2) developed by Chiamsathit et al. (2014) were used in the current study.

4 Results and discussion

4.1 ANN inflow forecasts

Based on extensive testing involving the examination of the auto-correlation function (acf – Fig. 3a), partial-autocorrelation function (pcf – Fig. 3b) and cross-correlation function (ccf – Fig. 3c), six input variables (i.e. current month historic mean inflow, lagged inflows ($t-1$, $t-2$, $t-3$), and lagged rainfall ($t-1$, $t-2$)) were used for the ANN modelling. The acf (Fig. 3a) shows infinite attenuation with only the first three lags of inflow being significant. Additionally,

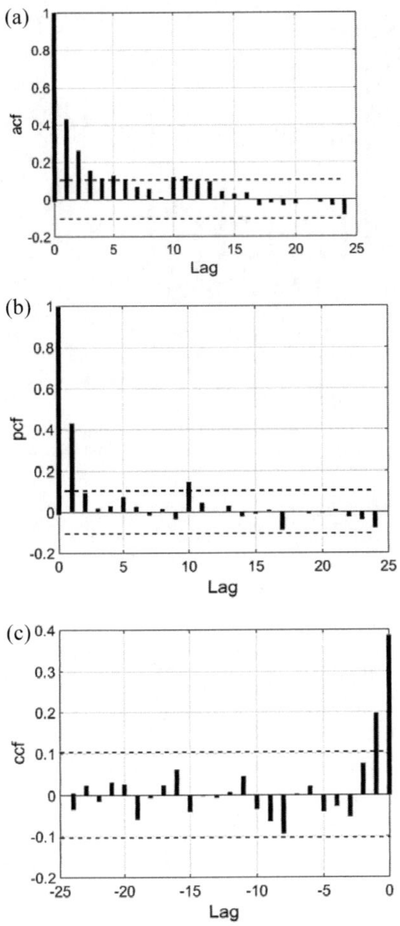

Figure 3. Inflow **(a)** auto-correlation, **(b)** partial autocorrelation functions, and **(c)** inflow-rainfall cross-correlation function for Ubonratana system.

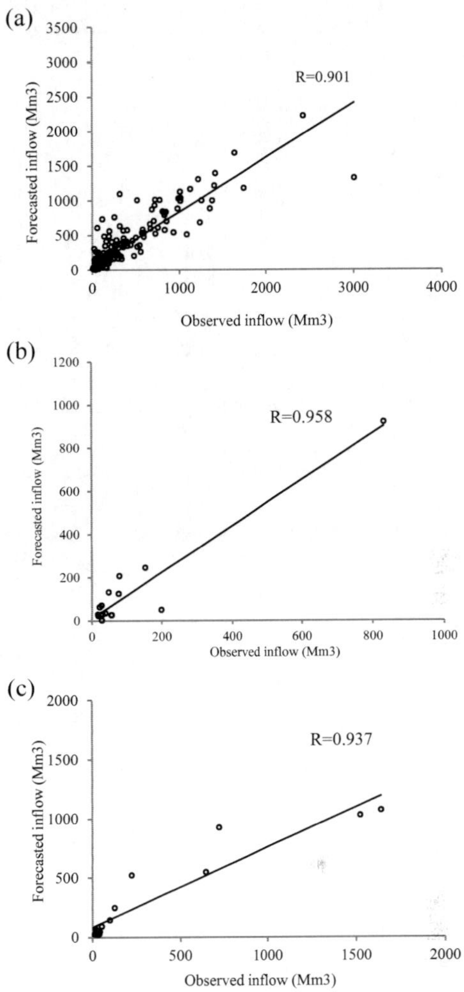

Figure 4. Comparing the 1-month ahead observed and forecast inflow during **(a)** training, **(b)** validation, and **(c)** testing.

the ccf in Fig. 3c indicates that the first two lags of the rainfall are significant. With these, the functional form of the forecast model becomes:

$$Q_t = f(Q_{t-1}, Q_{t-2}, Q_{t-3}, R_{t-1}, R_{t-2}, \overline{Q}_t) \quad (11)$$

where Q_t is the one-month ahead inflow forecast; Q_{t-1}, Q_{t-2} and Q_{t-3} are lagged inflows of one-month, two-month and three-month, respectively; R_{t-1} and R_{t-2} are lagged rainfall of one-month and two-month, respectively; and $\overline{Q}_t$ is historic mean inflow for the current month.

The ESR was used for the ANN training and for this the 360 months of data were split into three (90 : 5 : 5) for training, validation and testing, respectively. The number of hidden neurons was varied between 1 and 35 and based on the R criterion the best architecture had 33 neurons in the hidden layer. Indeed, the final model performed very well with the R exceeding 0.9 in each of the training, validation and testing. Figure 4a, b and c compare the predicted and observed inflow during training, validation and testing, respectively and further confirm the good performance of the forecasting model. The time series of the forecast inflows (April 1982 to March 2012, i.e. 360 months) are also compared in Fig. 5 and this together with the estimated Nash–Sutcliffe efficiency (NSE) of 0.75 is further evidence of the efficacy of the forecasting model. Additionally, the fact that the NSE was higher than zero is an indication that the model has been a better predictor than the mean value of the observed time series.

4.2 Reservoir performance evaluation

The results of the performance evaluation are summarised in Table 1. For convenience, the operating policy with Type A, Type F, Type M and Type N are denoted by P-A, P-F, P-M and P-N, respectively.

As seen in Table 1, in terms of the total amount of water released, P-A, P-F and P-M were significantly better than P-N, which is not surprising given that P-N did not have any

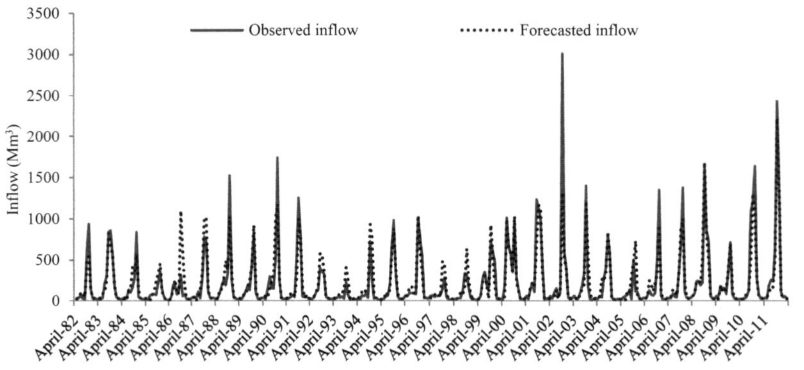

Figure 5. Time series of 1-month ahead of observed and forecast inflows for the complete data record.

Table 1. Summary of evaluated reservoir performance indices for Ubonratana reservoir.

Policy	Water user	Total water shortage (Mm^3)	Excursions of $S_{end,t}$ below LRC	f_d	f_s	Reliability (%)		ϕ	η	λ_{user}	λ_G
						R_t	R_v				
P-A	Domestic	0.0	8.0	0	0	100.00	100.00	–	0.000	1.000	0.557
	Downstream	0.5		1	1	99.72	99.99	1	0.026	0.990	
	Irrigation	309.4		15	3	95.83	98.58	0.200	0.626	0.415	
P-F	Domestic	0.0	14.0	0	0	100.00	100.00	–	0.000	1.000	0.655
	Downstream	0.0		0	0	100.00	100.00	–	0.000	1.000	
	Irrigation	244.5		10	4	97.22	98.88	0.400	0.591	0.542	
P-M	Domestic	0.0	16.0	0	0	100.00	100.00	–	0.000	1.000	0.464
	Downstream	0.0		0	0	100.00	100.00	–	0.000	1.000	
	Irrigation	166.8		6	1	98.33	99.24	0.167	0.853	0.289	
P-N	Domestic	3.2	4.0	6	5	98.33	99.09	0.833	1.000	0.000	0.543
	Downstream	132.7		10	9	97.22	98.04	0.900	0.770	0.586	
	Irrigation	1062.6		28	15	92.22	95.13	0.536	0.684	0.539	

additional water from inflows. In terms of reliability (R_t and R_v), the P-F was marginally better than using P-A and significantly better than P-N; P-F was, however, inferior to P-M. A possible reason for this is that in some of the months, the historic monthly mean and forecast inflows were higher than the actual inflows, implying that more water will be released in those months with P-M and P-F than with the other two inflow situations. However, the net effect of such large releases (based on the upwardly-biased inflow forecasts) is the increased number of excursions of the end-of-period storage ($S_{end,t}$) into the region below the LRC as shown in Table 1 for both the P-F and P-M.

The other performance indices reported in Table 1 all reveal the superiority of P-F relative to the other inflow situations. For example, the group sustainability index for P-F was the highest of all four; indeed, the same better performance of P-F was recorded across all three (public, instream and irrigation) demand sectors supplied by the reservoir. As expected, the conservative nature of P-N resulted in the least number of excursions below the LRC. This is likely to benefit the hydro-power generation potential of the reservoir albeit, as revealed by this study, at the expense of its performance in meeting the consumptive demands.

5 Conclusion

This study has developed MLP-ANN model to forecast one-month-ahead inflow for the Ubonratana reservoir in north-eastern Thailand. Extensive testing of the model showed that it was able to provide inflow forecasts with reasonable accuracy. The performance of the ANN forecasts was tested against those of three other inflow scenarios and the reservoir simulation results showed that the ANN forecasts produced superior reservoir performance. The worst performing inflow situation was when there was complete lack of knowledge about the inflow and release decision was based on the starting storage alone. All this represents an objective demonstration of good inflow forecast knowledge for effective reservoir operation.

Acknowledgements. This study formed part of the PhD research undertaken by the first author with PhD Scholarship provided by the Royal Thai Government. We also thank officials of EGAT for providing the data and other information used for the study.

References

Adeloye, A. J. and De Munari, A.: Artificial neural network based generalized storage-yield-reliability models using Levenberg-Marquardt algorithm, J. Hydrol., 362, 215–230, 2006.

Chiamsathit, C., Adeloye, A. J., and Soundharajan, B.: Assessing competing policies at Ubonratana reservoir, Thailand, Proceedings ICE (Water Management), 167(WM10), 551–560, 2014.

Edossa, D. C. and Babel, M. S.: Forecasting Hydrological Droughts Using Artificial Neural Network Modeling Technique, South Africa: University of Pretoria, Proceedings of 16th SANCIAHS National Hydrology Symposium, 1–3 October 2012, Pretoria, 2012.

EGAT: Improved Rule Curve, Procedure of the Ubonratana reservoir operation: Electricity Generating Authority of Thailand (EGAT) in the Ubonratana dam, EGAT, Khon Kaen, Thailand, 2002.

McMahon, T. A. and Adeloye, A. J.: Water resources yield, Water Resources Publications, LLC, Colorado, USA, 2005.

Mohammadi, K., Eslami, H. R., and Dardashti, S. D.: Comparison of Regression, ARIMA and ANN Models for Reservoir Inflow Forecasting using Snowmelt Equivalent (a Case study of Karaj), J. Agric. Sci. Technol., 7, 17–30, 2005.

Mulia, E. I., Asano, T., and Tkalich, P.: Retrieval of missing values in water temperature series using a data-driven model, Earth Science Informatics, 8, 787–798, 2015.

Sandoval-Soils, S., Mckinney, D. C., and Loucks, D. P.: Sustainability index for water resources planning and management, Water Resources Planning and Management, ASCE, 137, 381–389, 2011.

Shamseldin, A. Y.: Application of Neural Network Technique to Rainfall-Runoff Modelling, Hydrol. J., 199, 272–294, 1997.

Zhang, G., Patuwo, B. E., and Hu, Y. M.: Forecasting with artificial neural networks: The state of the art, Int. J. Forecasting, 14, 35–62, 1998.

Proc. IAHS, 373, 215–219, 2016
proc-iahs.net/373/215/2016/
doi:10.5194/piahs-373-215-2016

Open Access
PIAHS

Towards an optimal integrated reservoir system management for the Awash River Basin, Ethiopia

Ruben Müller, Henok Y. Gebretsadik, and Niels Schütze

Department of Hydrology, Technische Universität Dresden, Dresden, 01069, Germany

Correspondence to: Ruben Müller (ruben.mueller@tu-dresden.de)

Published: 12 May 2016

Abstract. Recently, the Kessem–Tendaho project is completed to bring about socioeconomic development and growth in the Awash River Basin, Ethiopia. To support reservoir Koka, two new reservoirs where built together with extensive infrastructure for new irrigation projects. For best possible socioeconomic benefits under conflicting management goals, like energy production at three hydropower stations and basin wide water supply at various sites, an integrated reservoir system management is required. To satisfy the multi-purpose nature of the reservoir system, multi-objective parameterization-simulation-optimization model is applied. Different Pareto-optimal trade-off solutions between water supply and hydro-power generation are provided for two scenarios (i) recent conditions and (ii) future planned increases for Tendaho and Upper Awash Irrigation projects. Reservoir performance is further assessed under (i) rule curves with a high degree of freedom – this allows for best performance, but may result in rules curves to variable for real word operation and (ii) smooth rule curves, obtained by artificial neuronal networks. The results show no performance penalty for smooth rule curves under future conditions but a notable penalty under recent conditions.

1 Introduction

Recently, the Kessem–Tendaho Project is completed to bring about socioeconomic development and growth in the Awash River Basin, Ethiopia. To support the existing Koka reservoir two new reservoirs where built together with extensive infrastructure for new irrigation projects. Besides the basin wide supply of water of municipal water, irrigation water for various agricultural sites and sugar cane plantations, the reservoirs are also responsible for flood protection. Hydropower production is a critical factor for the local economy. Koka reservoir provides hydropower through hydropower station Awash I and supports the hydro-power stations Awash II and III. Development plans project an increase of 40 000 ha in Tendaho Irrigation project and expansion of upper Awash irrigation site by two fold.

For maximum socioeconomic gains an integrated reservoir system management is crucial. To achieve this, optimal operational policies for all reservoirs are needed. Mathematical optimization models are used widely in water resources management to provide operational policies for optimal integrated reservoir management (Loucks et al., 1981). To account for multi-purpose nature of the Awash River Basin reservoir system a multi-objective parameterization-simulation-optimization (PSO) model is developed in this study. In PSO a reservoir management simulation model is coupled to an optimization algorithm to iteratively search for better operational policies. The advantages of PSO over other common optimization techniques are discussed in Koutsoyiannis and Economou (2003).

Yibetal et al. (2013) analysed the water audit of Awash Basin using WEAP model on the basis of three different scenarios (Expansion of irrigation area, improvement of irrigation and Climate change). Berhe et al. (2013) assessed the water allocation for future development scenarios in a modelling study using MODSIM model (Labadie, 2007). However hydropower production is modelled on purely opportunistic basis, because releases from the reservoir respond only to irrigation demands.

This study is a first step to provide optimal rule curves for an integrated management of the reservoir system for possible compromises between energy production and basin wide water supply for (i) recent conditions and (ii) the planned in-

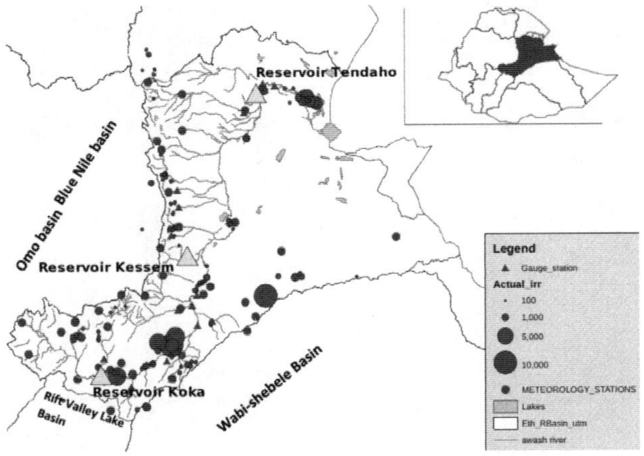

Figure 1. Map of Awash River Basin, highlighting the reservoir system and the local irrigation projects (adopted from Gebretsadik, 2015).

crease of 40 000 ha in Tendaho Irrigation project and expansion of upper Awash irrigation site by two fold. Sustainability is assessed in two steps. The maximum performance of the reservoir system is assessed with curves with a high degree of freedom. This allows for best performance, but results in non-smooth monthly rule curves which are not favoured by reservoir operators. Therefore also smoothness constrained rule curves are optimized by using a surrogate function. Possible losses of reservoir system performance are analysed.

The Awash River originates from a high plateau, which is the central Ethiopian Highland and an elevation up to 3000 m a.s.l. It descends in the Rift Valley after passing Koka Reservoir and flows into Lake Abe near the border of Ethiopia at 250 m a.s.l. (see Fig. 1).

2 Material and methods

2.1 Awash River Basin

Awash River basin is one of the twelve basins in Ethiopia. The basin has a total catchment area of 110 000 km^2 and total length of 1200 km.

Awash River has 15 important tributaries which significantly contribute for the flow of the main course.

The land use is dominated by exposed rock with about 34.9 % followed by cultivated land of about 27 % and open shrub land (20.9 %). The seasonal distribution of rainfall with two distinct rainy periods is caused by a shifting of the Inter Tropical Convergence Zone. The March–May season is the main rainfall season yielding 100–200 mm per month, followed by a lesser rainfall season in October–December with 100 mm per month. More detail about the basin give Berhe et al. (2013).

Table 1. Management zones and total storage in 10^6 m^3, (^{a}) denotes upper constraints and (^{b}) denotes lower constraints.

Reservoir	Dead	Buffer	Conservation	Total
Koka	8	300^b	1000^a	1186
Kessem	200	200^b	500^a	800
Tendaho	364	364.6^b	1610^a	1860

2.2 Setup of the parameterization-simulation-optimization model

For all reservoirs upper rule curves define the top of conservation storage zones and lower rule curves define the top of buffer storage zones. Important reservoir characteristics are summarized in Table 1. These storages mark upper and lower constraints in the optimization as the task is, to find optimal monthly storage values for each zone. Storage above the conservation zone is designated for flood protection.

Awash I hydropower station, located at Koka reservoir houses three units with 14.4 MW capacity. Average and firm production of hydropower are 110 and 80 GWh per annum respectively. Awash II and Awash III run-of-river hydropower stations are located 25 and 28 km of the Koka dam. Both stations feature an installed capacity of 34 MW.

The regulated turbine flow is restricted to 40 $m^3\,s^{-1}$. Reservoir Koka serves water for energy production only from the conservation zone; water in the buffer zone is preserved for supply of irrigation and municipal demands. Several problems arise from an unsatisfactory data basis. Almost no data is available to model the hydrological characteristics of the Gedabbesa swamp, which has high influence in the water allocation (Berhe et al., 2013).

Therefore, Gedabbesa swamp is modelled as follows: average monthly patterns for losses to the swamp and returns from the swamp are calculated from gauging stations upstream and downstream of the swamp. In the model all flow exists at node 702 and the difference between the monthly pattern of losses and the flow to node 702 is returned at node 703. Additional mean monthly returns enter at node 701. Similarly, little data is available for Lake Abe, the terminal lake of Awash River.

Awash I hydropower station, located at Koka reservoir houses three units with 14.4 MW capacity. Average and firm production of hydropower are 110 GWh per annum and 80 GWh per annum respectively. Awash II and Awash III run-of-river hydropower stations are located 25 and 28 km of the Koka dam. Both stations feature an installed capacity of 34 MW. The regulated turbine flow is restricted to 40 $m^3\,s^{-1}$. Reservoir Koka serves water for energy production only from the conservation zone; water in the buffer zone is preserved for supply of irrigation and municipal demands.

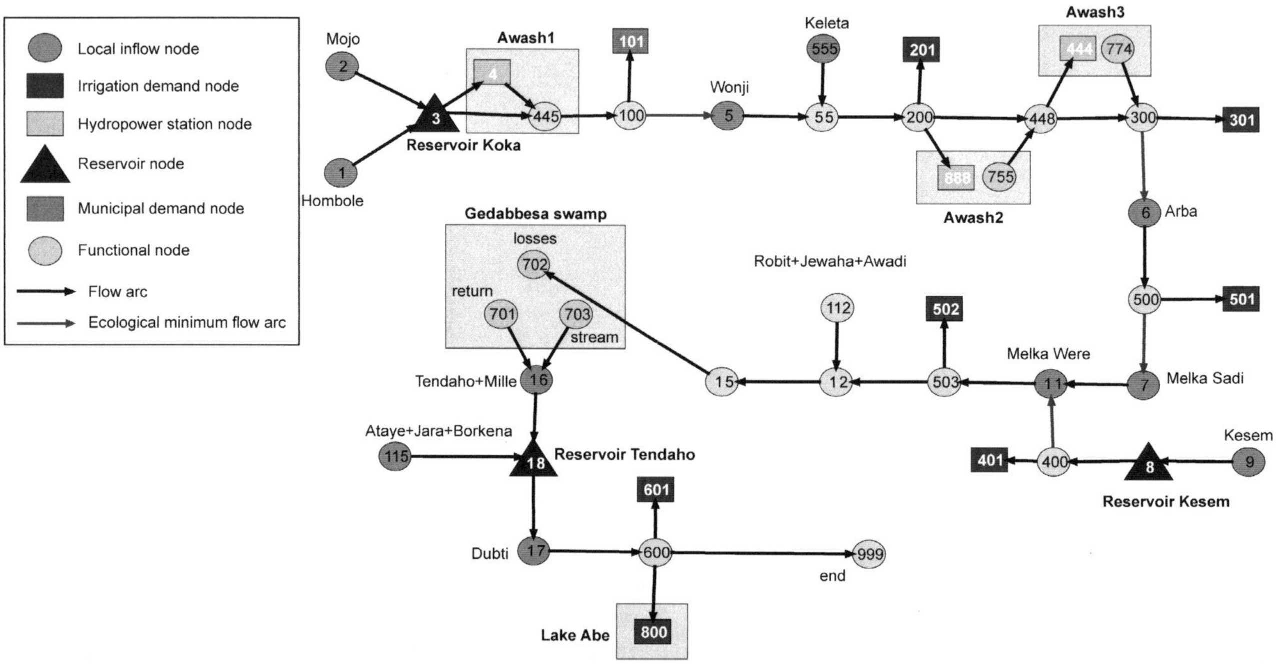

Figure 2. Schematic representation of Awash Basin multi-reservoir system in the modelling software OASIS.

3 Formulation of the optimization problem

The competing management goals of water supply for irrigation projects, municipal, ecology and the production of hydropower in the Awash river basin requires the formulation of two objective functions. Objective function F_1 minimizes the sum of all deficits over all demands

$$\min(F_1) = \min\left\{\sum_{t=1}^{240}\left(D_{\text{Mun},\,t} + D_{\text{Irr},\,t} + D_{\text{Eco},\,t}\right)\right\}, \quad (1)$$

where for each time step t, $D_{\text{Irr},t}$ is the sum of all deficits in the supply for all irrigation projects, $D_{\text{Mun},\,t}$ is the deficit in municipal supply and $D_{\text{Eco},\,t}$ is the sum of all deficits in ecological minimum flow support for all nodes as given in Fig. 2. Objective function F_2 maximizes the energy production of the three hydropower stations in the basin

$$\max(F_2) = -1\cdot\min \cdot\left\{\sum_{t=1}^{240}\left(-E_{\text{Awash1},\,t} - E_{\text{Awash2},\,t} - E_{\text{Awash3},\,t}\right)\right\}, \quad (2)$$

where $E_{\text{Awash1},\,t}$ is the energy production at hydropower unit Awash1 and $E_{\text{Awash2},\,t}$ and $E_{\text{Awash3},\,t}$ at units Awash2 and Awash3, respectively. The produced energy is calculated as $E = \eta g \rho Q_{\text{T}} H$ (MWh), with efficiency of $\eta = 0.9$ (–) for all units, turbine flows Q_{T} ($\text{m}^3\,\text{s}^{-1}$) for all units is given by OASIS model, as is the head H (m) for Awash1. For Awash2 and Awash3 run of river units the head is fixed at 60 m.

Three models are considered, which vary in the formulation of the rule curves. In model MOD1 the lower rule curve of reservoir Koka is kept constant. For model MOD2 this lower rule is seasonally variable. Decision variables for MOD1 and MOD2 are the storage control volumes. A constraint free formulation and bounded formulation from Müller (2014) is used for MOD1 and MOD2. Smooth variable rule curves for MOD3 are obtained by formulating the rule curves as an artificial neuronal network

$$Z_s^k = \alpha_{1,k} + \sum_{n=1}^{2}\cdot\left\{\alpha_{2,k\cdot n}\cdot \text{tansig}\left(\alpha_{3,k\cdot n}\cdot\sin(T) + \alpha_{4,k\cdot n}\cdot\cos(T) + \alpha_{5,k\cdot n}\right)\right\}, \quad \forall(s=1,\ldots,12;\,k=1,\ldots,4) \quad (3)$$

subject to $Z^{\text{buffer}} \geq Z_s^{k=1} \geq Z_s^{k=2} \geq Z^{\text{cons}}$ for reservoir Koka and $Z^{\text{buffer}} \leq Z_s^k \leq Z^{\text{cons}}$ for all other reservoirs. Here, Z_s^k is a storage control volume for $k=1$ the upper rule curve or $k=2$ the lower rule curve of reservoir Koka. With $k=3$ and $k=4$ the upper rule curves for denoted for reservoirs Kessem and Tendaho respectively. $s = 1,\ldots,12$ enumerates the months of a year. In Eq. (3) the storage control volumes are coded using hyper-parameters α. Constraints for the buffer zones Z^{buffer} and conservation zones Z^{cons} are given in Table 1. Variable T codes the cyclostationarity and runs from $-\pi$ to $+\pi$ for $s = 1,\ldots,12$. The approach is adopted from Castelletti et al. (2012), who apply artificial neuronal networks for operational rules in a different context. The decision variables sets are evaluated by simulation model OASIS (Hydrologics Inc, 2009), which handles all further constraints like maximum flows and mass conservation. The evolutionary strategy MO-CMA-ES (Igel et al., 2007) was used as multi-objective optimization algorithm in this study.

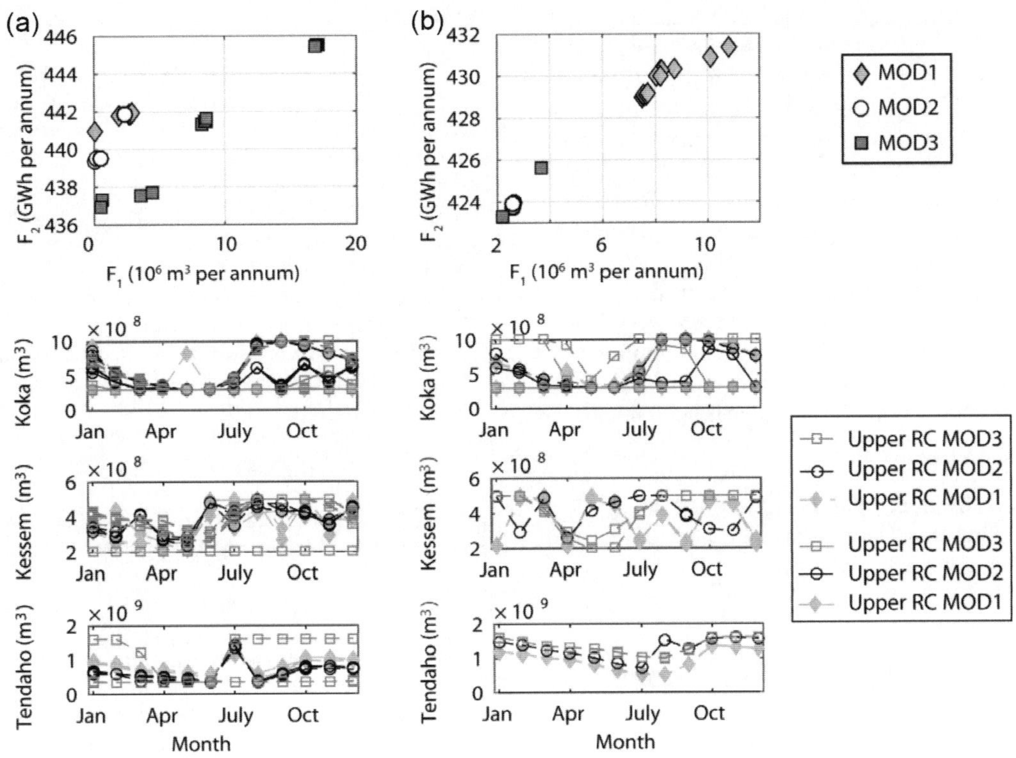

Figure 3. Pareto-Fronts and rule curves (RC) for all solutions from a specific model for **(a)** recent conditions and **(b)** the future development plan. Note for **(a)** and **(b)** that F_1 is to be minimized and F_2 and to be maximized.

4 Results and discussion

Optimization runs for the three models where conducted with 60 000 model evaluations and a population size of 48 each. The resulting Pareto-Fronts are depicted for recent conditions in Fig. 3a and for the future development plan in Fig. 3b.

With nearly no deficits under recent conditions and 440 GWh per annum energy production, MOD1 performs best when preference is set to minimizing deficits (Objective F_1). The rule curve (RC) behind this solution is unique (Fig. 3a, lower part), because of the size of the conservation zone in May. For the high energy production preference up to 442 GWh per annum, the performance of the reservoir under constant lower (MOD1) and varying lower (MOD2) RCs is similar. The size buffer zone of reservoir Koka is 0 m^3 for all solutions of MOD1, MOD2 reserves water for demands especially in January, August, October and December.

For reservoir Tendaho MOD3 proposes a draw down period from March to April and refill in July for low deficits. Surprisingly, RCs for high energy production require an empty reservoir Tendaho.

MOD1 and MOD2 result in lower storages throughout the year and only a major fill in July; this reduces evaporation losses and support from upstream reservoirs. The RCs from all models show the same general course; yet, MOD2 produces the most variable RCs with several draw downs and refills.

Under future development plans no model dominates the others in overall performance, but the models cover different spaces of the Pareto-space. This might be due to the formulation of the RCs or an optimization related problem. In general, deficits under increases demands can be as low as 2×10^6 m^3 when energy production is reduced to 424 GWh per annum. In general, the energy production decreases in average about 3 % under future plans. For reservoir Koka higher storages are proposed from January to April for MOD3. RCs for Kessem reservoir are much smoother for MOD3 in comparison to MOD1. The same shapes of RCs are evident for MOD1 and MOD3, MOD2 proposes an additional refill in August.

5 Conclusions

An integrated reservoir management for the multi-reservoir system in the Awash River Basin is needed to maximize socioeconomic benefits. Multi-objective optimization is carried out for conflicting management goals of energy production and basin wide water supply. Future development plans for expansion in irrigation sites are considered. Reservoir performance is assessed by rule curves with a high degree of freedom and smooth rule curves, obtained by artificial neuronal networks. For recent conditions the smoothness constraint rule curves cause a performance penalty for the reservoir management. The available water resources in the system are sufficient and a high degree of freedom (non-smooth rule curves) allows for a timely precise allocation. Under the development plans with a higher stress on the system, this cannot be observed. The higher degree of freedom cannot provide any additional performance gains. However, the trade-offs under both scenarios and for all considered models are only huge in terms of deficit, while relative gains in energy performance are negligible. A balanced solution will focus on low deficits and the decision maker may choose his preferred management by special consideration of the underlying rule curves.

It is advised to conduct studies to enhance understanding of Gabeddesa Swamp. Additionally shortcomings in the model, like missing translation times for water routing and irrigation efficiencies need to be addressed.

Acknowledgements. The second author would like to thank the German Academic Exchange Service (DAAD, grant no. A/13/90526) for awarding a scholarship.

References

Berhe, F. T., Melesse, A. M., Hailu, D., and Sileshi, Y.: MODSIM-based water allocation modelling of Awash River Basin, Ethiopia, Catena, 109, 118–128, doi:10.1016/j.catena.2013.04.007, 2013.

Castelletti, A., Pianosi, F., Quach, X., and Soncini-Sessa, R.: Assessing water reservoirs management and development in Northern Vietnam, Hydrol. Earth Syst. Sci., 16, 189–199, doi:10.5194/hess-16-189-2012, 2012.

Gebretsadik, H. Y.: Simulation Based Optimization for Operation of Multi-Purpose Reservoir System and potential water allocation: Case Study in Awash River Basin Ethiopia, MS thesis, Fakultät Umweltwissenschaften, Technische Universität Dresden, Dresden, Germany, 76 pp., 2015.

Hydrologics Inc: User manual for OASIS with OCL, available at: http://www.hydrologics.net, last access: 12 December 2015, 2009.

Igel, C., Hansen, N., and Roth, S.: Covariance Matrix Adaptation for Multi-objective Optimization, Evol. Comput., 15, 1–28, 2007.

Koutsoyiannis, A. and Economou, D.: Evaluation of the parameterization-simulation-optimization approach for the control of reservoir systems, Water Resour. Res., 39, 1170–1187, doi:10.1029/2003WR002148, 2003.

Labadie, J. W.: MODSIM 8.1: River Basin Management Decision Support System, User Manual and Documentation, Colorado State University, Fort Collins, CO, USA, 2007.

Loucks, D. P., Stedinger, J. R., and Haith, D. A.: Water Resource Systems Planning and Analysis, Prentice-Hall, Englewood Cliffs, N.J., 1981.

Yibeltal, T., Belte, B., Semu, A., Imeru, T., and Yohannes, T.: Coping with water scarcity, the role of agriculture, developing a water audit for Awash river basin, Synthesis report, GCP/INT/072/ITA, Addis Ababa, Ethiopia, 2013

AUTHOR INDEX

AUTHOR INDEX